AF685824

State and Federal Standards for Mobile-Source Emissions

Committee on State Practices in Setting
Mobile Source Emissions Standards

Board on Environmental Studies and Toxicology

Division on Earth and Life Studies

NATIONAL RESEARCH COUNCIL
OF THE NATIONAL ACADEMIES

THE NATIONAL ACADEMIES PRESS
Washington, D.C.
www.nap.edu

THE NATIONAL ACADEMIES PRESS 500 Fifth Street, NW Washington, DC 20001

NOTICE: The project that is the subject of this report was approved by the Governing Board of the National Research Council, whose members are drawn from the councils of the National Academy of Sciences, the National Academy of Engineering, and the Institute of Medicine. The members of the committee responsible for the report were chosen for their special competences and with regard for appropriate balance.

This project was supported by Contract No. EP-C-04-024 between the National Academy of Sciences and the U.S. Environmental Protection Agency. Any opinions, findings, conclusions, or recommendations expressed in this publication are those of the author(s) and do not necessarily reflect the view of the organizations or agencies that provided support for this project.

International Standard Book Number 0-309-10151-4 (Book)
International Standard Book Number 0-309-65868-3 (PDF)
Library of Congress 2006926248

Additional copies of this report are available from

The National Academies Press
500 Fifth Street, NW
Box 285
Washington, DC 20055

800-624-6242
202-334-3313 (in the Washington metropolitan area)
http://www.nap.edu

Copyright 2006 by the National Academy of Sciences. All rights reserved.

Printed in the United States of America

THE NATIONAL ACADEMIES

Advisers to the Nation on Science, Engineering, and Medicine

The **National Academy of Sciences** is a private, nonprofit, self-perpetuating society of distinguished scholars engaged in scientific and engineering research, dedicated to the furtherance of science and technology and to their use for the general welfare. Upon the authority of the charter granted to it by the Congress in 1863, the Academy has a mandate that requires it to advise the federal government on scientific and technical matters. Dr. Ralph J. Cicerone is president of the National Academy of Sciences.

The **National Academy of Engineering** was established in 1964, under the charter of the National Academy of Sciences, as a parallel organization of outstanding engineers. It is autonomous in its administration and in the selection of its members, sharing with the National Academy of Sciences the responsibility for advising the federal government. The National Academy of Engineering also sponsors engineering programs aimed at meeting national needs, encourages education and research, and recognizes the superior achievements of engineers. Dr. Wm. A. Wulf is president of the National Academy of Engineering.

The **Institute of Medicine** was established in 1970 by the National Academy of Sciences to secure the services of eminent members of appropriate professions in the examination of policy matters pertaining to the health of the public. The Institute acts under the responsibility given to the National Academy of Sciences by its congressional charter to be an adviser to the federal government and, upon its own initiative, to identify issues of medical care, research, and education. Dr. Harvey V. Fineberg is president of the Institute of Medicine.

The **National Research Council** was organized by the National Academy of Sciences in 1916 to associate the broad community of science and technology with the Academy's purposes of furthering knowledge and advising the federal government. Functioning in accordance with general policies determined by the Academy, the Council has become the principal operating agency of both the National Academy of Sciences and the National Academy of Engineering in providing services to the government, the public, and the scientific and engineering communities. The Council is administered jointly by both Academies and the Institute of Medicine. Dr. Ralph J. Cicerone and Dr. Wm. A. Wulf are chair and vice chair, respectively, of the National Research Council.

www.national-academies.org

COMMITTEE ON STATE PRACTICES IN SETTING MOBILE SOURCE EMISSIONS STANDARDS

Members

DAVID ALLEN (*Chair*), University of Texas, Austin
JOHN BAILAR III, University of Chicago (Retired), Washington, DC
HUGH ELLIS, Johns Hopkins University, Baltimore, MD
ALISON GEYH, Johns Hopkins School of Public Health, Baltimore, MD
DAVID GREENE, Oak Ridge National Laboratory, Knoxville, TN
JAMES LENTS, University of California, Riverside, Riverside
GARY MARCHANT, Arizona State University, Tempe
VIRGINIA MCCONNELL, Resources for the Future, Inc., Washington, DC
ALISON POLLACK, ENVIRON International Corporation, Novato, CA
HAROLD SCHOCK, Michigan State University, East Lansing
KARL SPRINGER, Southwest Research Institute (retired), San Antonio, TX

Project Staff

K. JOHN HOLMES, Project Director
RAYMOND A. WASSEL, Senior Program Officer
MATTHEW RUSSELL, Associate Staff Officer
RUTH E. CROSSGROVE, Senior Editor
MIRSADA KARALIC-LONCAREVIC, Research Associate
BRYAN SHIPLEY, Research Associate
RADIAH A. ROSE, Senior Program Assistant
ALEXANDRA STUPPLE, Senior Editorial Assistant
RAHEL MENGHESTAB, Anderson Intern
SAMMY BARDLEY, Librarian

Sponsor

U.S. ENVIRONMENTAL PROTECTION AGENCY

Board on Environmental Studies and Toxicology[1]

Members

Jonathan M. Samet (*Chair*), Johns Hopkins University, Baltimore, MD
Ramón Alvarez, Environmental Defense, Austin, TX
John M. Balbus, Environmental Defense, Washington, DC
Thomas Burke, Johns Hopkins University, Baltimore, MD
Dallas Burtraw, Resources for the Future, Washington, DC
James S. Bus, Dow Chemical Company, Midland, MI
Costel D. Denson, University of Delaware, Newark
E. Donald Elliott, Willkie Farr & Gallagher LLP, Washington, DC
J. Paul Gilman, Oak Ridge National Laboratory, Oak Ridge, TN
Sherri W. Goodman, Center for Naval Analyses, Alexandria, VA
Judith A. Graham, American Chemistry Council, Arlington, VA
Daniel S. Greenbaum, Health Effects Institute, Cambridge, MA
William P. Horn, Birch, Horton, Bittner and Cherot, Washington, DC
Robert Huggett, Michigan State University (emeritus), East Lansing
James H. Johnson Jr., Howard University, Washington, DC
Judith L. Meyer, University of Georgia, Athens
Patrick Y. O'Brien, ChevronTexaco Energy Technology Company, Richmond, CA
Dorothy E. Patton, International Life Sciences Institute, Washington, DC
Steward T.A. Pickett, Institute of Ecosystem Studies, Millbrook, NY
Danny D. Reible, University of Texas, Austin
Joseph V. Rodricks, ENVIRON International Corporation, Arlington, VA
Armistead G. Russell, Georgia Institute of Technology, Atlanta
Lisa Speer, Natural Resources Defense Council, New York, NY
Kimberly M. Thompson, Massachusetts Institute of Technology, Cambridge
Monica G. Turner, University of Wisconsin, Madison
Mark J. Utell, University of Rochester Medical Center, Rochester, NY
Chris G. Whipple, ENVIRON International Corporation, Emeryville, CA
Lauren Zeise, California Environmental Protection Agency, Oakland

Senior Staff

James J. Reisa, Director
David J. Policansky, Scholar
Raymond A. Wassel, Senior Program Officer for Environmental Sciences and Engineering
Kulbir Bakshi, Senior Program Officer for Toxicology
Eileen N. Abt, Senior Program Officer for Risk Analysis
K. John Holmes, Senior Program Officer
Susan N.J. Martel, Senior Program Officer
Suzanne van Drunick, Senior Program Officer
Ellen K. Mantus, Senior Program Officer
Karl E. Gustavson, Senior Program Officer
Ruth E. Crossgrove, Senior Editor

---

[1]This study was planned, overseen, and supported by the Board on Environmental Studies and Toxicology.

OTHER REPORTS OF THE BOARD ON ENVIRONMENTAL STUDIES AND TOXICOLOGY

Fluoride in Drinking Water: A Scientific Review of EPA's Standards (2006)
Superfund and Mining Megasites—Lessons from the Coeur d'Alene River Basin (2005)
Health Implications of Perchlorate Ingestion (2005)
Air Quality Management in the United States (2004)
Endangered and Threatened Species of the Platte River (2004)
Atlantic Salmon in Maine (2004)
Endangered and Threatened Fishes in the Klamath River Basin (2004)
Cumulative Environmental Effects of Alaska North Slope Oil and Gas Development (2003)
Estimating the Public Health Benefits of Proposed Air Pollution Regulations (2002)
Biosolids Applied to Land: Advancing Standards and Practices (2002)
The Airliner Cabin Environment and Health of Passengers and Crew (2002)
Arsenic in Drinking Water: 2001 Update (2001)
Evaluating Vehicle Emissions Inspection and Maintenance Programs (2001)
Compensating for Wetland Losses Under the Clean Water Act (2001)
A Risk-Management Strategy for PCB-Contaminated Sediments (2001)
Acute Exposure Guideline Levels for Selected Airborne Chemicals (4 volumes, 2000-2004)
Toxicological Effects of Methylmercury (2000)
Strengthening Science at the U.S. Environmental Protection Agency (2000)
Scientific Frontiers in Developmental Toxicology and Risk Assessment (2000)
Ecological Indicators for the Nation (2000)
Waste Incineration and Public Health (1999)
Hormonally Active Agents in the Environment (1999)
Research Priorities for Airborne Particulate Matter (4 volumes, 1998-2004)
The National Research Council's Committee on Toxicology: The First 50 Years (1997)
Carcinogens and Anticarcinogens in the Human Diet (1996)
Upstream: Salmon and Society in the Pacific Northwest (1996)
Science and the Endangered Species Act (1995)
Wetlands: Characteristics and Boundaries (1995)
Biologic Markers (5 volumes, 1989-1995)
Review of EPA's Environmental Monitoring and Assessment Program (3 volumes, 1994-1995)
Science and Judgment in Risk Assessment (1994)
Pesticides in the Diets of Infants and Children (1993)
Dolphins and the Tuna Industry (1992)
Science and the National Parks (1992)
Human Exposure Assessment for Airborne Pollutants (1991)
Rethinking the Ozone Problem in Urban and Regional Air Pollution (1991)
Decline of the Sea Turtles (1990)

Copies of these reports may be ordered from the National Academies Press
(800) 624-6242 or (202) 334-3313
www.nap.edu

Preface

The task undertaken by this committee for the National Academies was to review and evaluate the scientific and technical practices used by states in setting emission standards for mobile sources, including those for non-road engines and vehicles. The study assessed the scientific and technical procedures used by states to develop or adopt emissions standards separate from those set by the Environmental Protection Agency under the Clean Air Act as well as the factors that cause states to move to more stringent emissions standards. The committee considered the scientific, technical, and economic rationale and methodologies used by the states in setting standards and how they compare to those used by the EPA. In addition, the committee assessed the direct and indirect impacts that state emissions standards have had on various factors, including compliance costs, energy consumption, air quality, and human health.

The committee received oral and written presentations from the following individuals: Steve Albu, California Air Resources Board; Thomas Austin, Sierra Research, Inc.; Robert Babik, General Motors Company; William Becker, State and Territorial Air Pollution Program Administrators and the Association of Local Air Pollution Control Officials; Thomas Cackette, Air Resources Board; Coralie Cooper, Northeast States Coordinated Air Use Management; Greg Dana, Alliance of Automobile Manufacturers; David Dickinson, U.S. Environmental Protection Agency; Karl Simon, U.S. Environmental Protection Agency; Chet France, U.S. Environmental Protection Agency; Timothy French, Engine Manufacturers Association; Dawn Gallagher, Maine Department of Environmental Protection; John German, Honda Motor Company; Robert Golledge, Massachusetts Department of Environmental Protection;

William Guerry, Outdoor Power Equipment Institute; Patricia Hanz, Briggs & Stratton Corporation; Peter Hotz, Briggs & Stratton Corporation; Roland Hwang, Natural Resources Defense Council; Peter Iwanowicz, American Lung Association of New York State, Inc.; Carl Johnson, New York Department of Environmental Conservation; Robert Jorgensen, Cummins Engine Company; Therese Langer, American Council for an Energy-Efficiency Economy; Peter Lidiak, American Petroleum Institute; Arthur Marin, Northeast States Coordinated Air Use Management; Gina McCarthy, Connecticut Department of Environmental Protection; George Miller, International Consortium for Fire Safety, Health and the Environment; Frederick Postel, International Consortium of Fire Safety, Health and the Environment; Thomas Snyder, Maryland Department of Environmental Protection; Richard Valentinetti, Vermont Department of Environmental Conservation; Barry Wallerstein, South Coast Air Quality Management District; Michael Walsh, Independent Consultant; Catherine Witherspoon, California Air Resources Board; Merrylin Zaw-Mon, U.S. Environmental Protection Agency. The committee thanks all of these individuals for their contributions. A complete list of dates, titles and presenter names can be found in Appendix E.

The committee is also grateful for the assistance of the National Research Council (NRC) staff in the preparation of this report. K. John Holmes played a key role in preparing this report in his role as project director. The committee also thanks Raymond Wassel, senior program director of environmental sciences and engineering in the Board on Environmental Studies and Toxicology (BEST), and the other staff members contributing to this report: James Reisa, director of BEST; Ruth Crossgrove, senior editor; Matthew Russell, associate staff officer; Mirsada Karalic-Loncarevic and Bryan Shipley, research associates; Radiah Rose, senior program assistant; Alexandra Stupple, senior editorial assistant; and Rahel Menghasteb, Anderson intern.

As chair, I thank all the members of the committee for their expertise and dedicated effort throughout the study.

Finally, given the topic of this report, it is appropriate to acknowledge a major supporter of the NRC, Arnold Beckman. Few know of the involvement of Arnold Beckman in the early efforts to reduce air pollution in California. In 1953, Arnold Beckman chaired a five-member committee formed by then Governor Knight to address a comprehensive program needed to eliminate smog. One of the several short- and long-term recommendations was to control automobile exhaust. Perhaps the following best sums up the Beckman Committee's concerns.

> The fact must be faced that cleaning the air in this area is a tremendous task, one that many will find appallingly expensive. This committee would be gravely remiss if it failed to point out that the community must either assume the real hardships imposed by abatement or accept those imposed by the continuing smog nuisance.[1]

NRC committees regularly use the Beckman Center, located on the campus of the University of California, Irvine. This acknowledgment is included in the Preface to our report to express the appreciation on behalf of those who have made use of the Beckman Center and to recognize Arnold Beckman's pioneering leadership to reduce air pollution.

David Allen, Ph.D.
Chair, Committee on State
Practices in Setting Mobile
Source Emissions Standards

[1] Aplet, J.H., Meade, G. Mobile Source Emissions Regulations in California: 1960 to 1995, from Advances in Economics of Environmental Resources Vol. 2, 1997, Edited by Hall, J.V.

Acknowledgment of Review Participants

This report has been reviewed in draft form by individuals chosen for their diverse perspectives and technical expertise, in accordance with procedures approved by the NRC's Report Review Committee. The purpose of this independent review is to provide candid and critical comments that will assist the institution in making its published report as sound as possible and to ensure that the report meets institutional standards for objectivity, evidence, and responsiveness to the study charge. The review comments and draft manuscript remain confidential to protect the integrity of the deliberative process. We wish to thank the following individuals for their review of this report:

Charles Amann, KAB Engineering; Thomas Austin, Sierra Research, Incorporated; Frank Barnes, University of Colorado at Boulder; Michael Bradley, M.J. Bradley & Associates; David Calkins, Sierra Nevada Air Quality Group; Laurence Caretto, California State University, Northridge; Joseph Colucci, Automotive Fuels Consulting, Incorporated; John DeCicco, Environmental Defense; Charles Freed, Independent Consultant; Daniel Greenbaum, Health Effects Institute; John Johnson, Michigan Technological University; Robert Jorgensen, Cummins Engine Company; Joseph Kubsh, Manufacturers of Emission Controls Association; Douglas Lawson, National Renewable Energy Laboratory; James McNew, Outdoor Power Equipment Institute; John Mooney, Environmental & Energy Technology & Policy Institute; and Arthur Winer, University of California, Los Angeles.

Although the reviewers listed above have provided many constructive comments and suggestions, they were not asked to endorse the conclusions or recommendations, nor did they see the final draft of the report before its release. The review of this report was overseen by William Agnew, General Motors Corporation (Retired) and Thomas Graedel, Yale University. Appointed by the NRC, they were responsible for making certain that an independent examination of this report was carried out in accordance with institutional procedures and that all review comments were carefully considered. Responsibility for the final content of this report rests entirely with the authoring committee and the institution.

Contents

TABLES

State and Federal Standards for Mobile-Source Emissions

Summary

Air pollutant emissions from mobile sources have been regulated for almost half a century. During this time, the focus has largely been on tightening emissions standards for on-road vehicles and engines, particularly passenger cars and small trucks (light-duty vehicles). Light-duty-vehicle emissions control grew out of research that implicated the increasing use of vehicles in the deterioration of air quality conditions in the 1950s in Southern California. Control of motor vehicle emissions began in the early 1960s with the introduction of positive crankcase ventilation, a simple approach consisting of a hose and valve that reduced the venting of uncombusted gases to the atmosphere. From that simple beginning, light-duty-vehicle emissions control evolved to today's complex regulation of fuel properties, exhaust emissions, and evaporative emissions, which require the use of sophisticated engine and emissions-control technologies. These strategies enabled per-mile-exhaust emissions of new, properly operating light-duty vehicles to decrease by 95-99% in 2004 compared with emissions of 1967 model-year vehicles.

The focus of mobile-source emissions control expanded to include on-road heavy-duty engines and later nonroad engines. The broadening in regulatory attention arose from the increasing fraction of mobile-source emissions that come from sources other than light-duty vehicles and the relative lack of emissions controls on these sources. On-road heavy-duty-vehicle engines were first regulated for air pollutants in the 1970s, and engines used in off-road applications were first regulated in the mid-1990s. Over the next decade, regulations already approved for new sources will substantially reduce emissions from on-road diesel vehicles, nonroad diesel engines, and gasoline-powered nonroad engines.

The federal Clean Air Act (CAA) establishes the framework for controlling mobile-source emissions in the United States. During the development of the CAA in 1967, Congress recognized that the imposition of many different state standards could result in inefficiencies in vehicle markets. Therefore, state-established emissions standards were preempted by federal emissions standards in what is now section 209 of the CAA. A special exemption to this federal preemption was made in section 209 for California because of the state's special air quality problems and pioneering efforts in the control of air pollutants. This exemption, still in existence, gives the state of California the authority to set on-road vehicle standards that differ from the federal standards as long as they are as protective in the aggregate as federal standards. Later amendments to section 209 granted California the authority to set emissions standards and regulations for some nonroad engines, and section 177 was added to allow other states to adopt California standards.

The National Research Council (NRC) convened the Committee on State Practices in Setting Mobile Source Emissions Standards in response to a request from Congress in its fiscal 2003 appropriations report for the U.S. Environmental Protection Agency (EPA) to arrange for an independent study of the practices and procedures by which states develop separate emission standards. For this report, the committee was asked to assess the scientific and technical procedures used by states to develop or adopt different emissions standards and to compare those policies and practices with those used by EPA. The committee was also asked to consider the factors that caused states to move toward more stringent emissions standards and to consider the impacts of state emissions standards on various factors, including emissions, compliance costs, energy consumption, air quality, and human health. As part of its work plan, the committee was directed to consider the effects of California's experience in setting separate emissions standards and the effects of California's standards on federal emissions standards. The full Statement of Task for the committee is provided in Chapter 1.

CONCLUSIONS AND RECOMMENDATIONS

Role of New Mobile-Source Emissions Standards

Despite the substantial progress made over the past few decades in reducing air pollutant emissions from many sources, including mobile sources, some locations continue to exceed National Ambient Air Quality Standards (NAAQS).[1] Further improvements in air quality will be needed, particularly to attain the new ambient standards for fine particulate matter and ozone at concentrations averaged over 8 hours. Although many emissionscontrol programs have been developed and regions with air quality problems have implemented a variety of programs, stricter new mobile-source emissions standards are an important component of overall emissions-control plans for locations that need air quality improvements. Federal mobile-source emissions standards set by EPA ensure that all regions of the country have some emissions reductions and that the mobility of these emissions sources does not undermine other air quality initiatives. California emissions standards, which are set by the California Air Resources Board (CARB), provide additional emissions reductions for the state's most populated and worst polluted regions, including the Los Angeles area and San Joaquin Valley. In many cases, CARB has tightened mobile-source emissions standards earlier and to a greater extent than the federal government. Other states that seek mobile-source emissions reductions from new-vehicle standards beyond those provided by federal standards have adopted California standards to supply the additional benefits.

While this study was in progress, CARB adopted light-duty-vehicle emissions standards for greenhouse gases. These standards have been challenged in the courts. The committee did not develop findings and recommendations specific to these standards because of their timing, the uncertainty surrounding their standing, and the lack of comparable federal standards.

California's Role In Mobile-Source Emissions Regulation

The CAA gives California the authority to set its own mobile-source emissions standards. Over the history of mobile-source regulation to date, California has usually led EPA in establishing emissions standards on light-duty vehicles and small nonroad gasoline engines, and

---

[1] NAAQS set maximum allowable ambient air concentrations for six so-called "criteria" pollutants; the standards are to be protective of public health (primary standards) and welfare (secondary standards). The six criteria pollutants are carbon monoxide, lead, nitrogen dioxide, ozone, particulate matter, and sulfur dioxide.

EPA has usually led California in establishing standards for on-road heavy-duty diesel vehicles and off-road diesel engines. This shared leadership promotes improvements in the efficiency of EPA's and CARB's regulatory efforts and allows sharing of expertise.

The mobile-source emissions standards developed by CARB, like those developed by EPA, have typically been "technology forcing."[2] In forcing technology development, California has been a laboratory for emissions-control innovations. An advantage of having a state laboratory for innovation is that the risk of failure to develop the required technologies is restricted to a limited geographic area. CARB's regulatory process is supportive of this laboratory role in that California's standards can be amended rapidly in the face of changing market and technological conditions in contrast to EPA's regulatory process.

The original reasons for which Congress authorized California to have a separate set of standards remain valid. California still has some of the worst air quality conditions in the country, and certain emission-reduction needs are greater in California than in the rest of the country. California has used its authority as Congress envisioned: to implement more aggressive measures than the rest of the country and to serve as a laboratory for technological innovation. These have resulted in successes, such as CARB's early recognition of the need to couple fuel composition with emissions control, and failures, such as the promotion of widespread use of electric vehicles under the original zero-emissions vehicle mandate.

California's authority to set its own mobile-source emissions standards inevitably imposes additional risks and costs, such as design, production, and distribution costs, although the costs and benefits are difficult to quantify. However, experience to date indicates that the California program has been beneficial overall for air quality by improving mobile-source emissions control.

Recommendation

California should continue its pioneering role in setting mobile-source emissions standards. The role will aid the state's efforts to achieve

---

[2] "Technology forcing" refers to the establishment by a regulatory agency of a requirement to achieve an emissions limit, within a specified time frame, that can be reached through use of unspecified technology or technologies that have not yet been developed for widespread commercial applications and have been shown to be feasible on an experimental or pilot-demonstration basis.

air quality goals and will allow it to continue to be a proving ground for new emissions-control technologies that benefit California and the rest of the nation.

EPA and CARB Technical and Scientific Practices in Setting Standards

CARB and EPA have essentially the same starting point and motivation for setting new or stricter standards: attainment of the NAAQS. Each agency follows a series of procedural steps leading to a finalized regulation. These steps include identification of the need for new emissions standards, evaluation of potential control strategies, publication of proposed regulations, and solicitation of public comments on proposals before promulgating the regulations. Some differences exist in the scope of CARB and EPA regulatory assessments as a result of the different procedures that the agencies must follow.

Some important similarities in the practices of setting standards followed by the two agencies are the following:

- CARB and EPA establish emissions standards based on assessments of technological feasibility and estimated engineering costs.
- CARB and EPA periodically update their practices as emissions estimation models and procedures continue to evolve and improve over time.
- CARB and EPA study technical practices of industry and perform engineering and market cost analyses.
- CARB and EPA test new technologies in the laboratory, using their own staffs, as well as outside contractors.

Some important differences in the practices of setting standards followed by the two agencies are the following:

- EPA's rule-making practices are subject to federal requirements defined in multiple acts and executive orders. Because emissions-standard regulations are typically deemed "significant,"[3] the rule-making process is overseen by the Office of Management and Budget. CARB's emission-standard rule-making is subject to state laws and to oversight

[3] "Significant rules" are defined as those that have an annual impact of $100 million or greater, raise novel regulatory issues, or have other significant impacts.

by the California Office of Administrative Law. As a result of the particular requirements, each agency's rule making includes some special components. For example, EPA is required to perform a cost-benefit analysis in which it estimates the monetary benefits of improved air quality to public health. CARB in turn is required to perform various California-specific economic impact assessments, such as employment impacts on in-state businesses.

- California is required to submit state implementation plans (SIPs) to EPA. In 1988, California passed its own California CAA, which includes additional standards with which the state must comply. The SIPs must describe the emissions reductions required to reach NAAQS attainment. Within these individual SIPs, the air quality impacts of California mobile-source emissions standards are assessed, although not in isolation from other emissions-control strategies. In recent years, EPA assessed the air quality impacts of its major mobile-source emissions standards in regulatory impact assessments (RIAs), which accompany all major federal regulations. The assessments estimate the air quality impacts in some or all of the country for each proposed set of standards in isolation from other emissions-control policies.
- EPA's RIAs have evolved to require assessments of the public health effects and estimates of monetary benefits. CARB does not directly consider public health benefits in its regulatory analysis of emissions standards because it uses its proposed standards to attain health-based NAAQS, which EPA has already assessed for public health benefits. California estimates health impacts of air pollutants in its reviews of California ambient air quality standards.[4]
- CARB routinely considers only the costs or impacts of its standards in its jurisdiction (California) and not in other states that might later adopt California standards, whereas EPA accounts for the costs and benefits for the entire nation in its assessments.
- CARB adopts emission-standard regulations in a public meeting with a public vote by board members. Public comments during this hearing can result in modifications to final standards. CARB may also include requirements for periodic review of standards during which standards can be modified. EPA's emissions standards, although subject to lengthy public-comment and technical-review periods, are issued through a finalized notice in the *Federal Register*. (A 2004 EPA regulation in-

---

[4] California has its own ambient air quality standards that are lower than the NAAQS for some pollutants. However, California ambient air quality standards do not have any deadlines for attainment.

cludes a requirement to conduct a periodic review of technical progress in attaining an emission standard.) In contrast to the flexibility CARB has in revising standards based on new scientific and technical information, EPA has historically developed new or revised mobile-source emissions standards only when directed or authorized to do so by Congress.

Recommendations

Consistent with a 2000 NRC report on modeling mobile-source emissions,[5] CARB and EPA should work in tandem to improve mobile-source emissions models. In particular, consistent with the NRC report, CARB and EPA should complete long-range plans that address improvements or new approaches to mobile-source emissions models. Such plans will improve estimations of emissions reductions. The estimations are a major part of assessing the impacts of emissions standards. The committee also recommends that CARB and EPA include, to the extent possible, air quality impact assessments as part of each rule-making, because the effect of reducing mobile-source emissions on ambient pollutant concentrations will vary from region to region.

Although the committee did not have sufficient information to evaluate the safety issues associated with past regulations, it recommends that safety issues continue to be given careful consideration by EPA and CARB when setting mobile-source emissions standards.

Given that CARB and EPA emissions standards tend to require new technological developments, the committee also recommends that periodic assessments of technological feasibility be continued by the agencies for some of the more important standards. Examples of such assessment include CARB's biennial review of the zero-emission-vehicle mandate and EPA's biennial review of on-road diesel-engine standards. Periodic assessments will allow the standards to be based on the most current understanding of the science and technology.

The Waiver Process

Each time CARB sets or substantially revises a California mobile-source emission standard, it must seek a waiver from EPA. The waiver

---

[5] NRC (National Research Council). 2000. Modeling Mobile Source Emissions. Washington, DC: National Acadmies Press.

review process usually takes several years to complete, and waivers are often granted shortly before the vehicles and engines that meet the standards are in the market. In some cases, waivers have been approved after vehicles and engines that meet the standards are already in the market.

EPA's consideration of a California waiver request requires substantial EPA resources in terms of personnel and time. EPA is required to provide an opportunity for interested parties to provide comment and to participate in public hearings, if hearings are requested. Each of these steps is time-consuming and perhaps duplicative. EPA must also conduct technical analyses of all comments provided by California, manufacturers, and other interested parties, further extending the time needed to issue a waiver decision. Although many California waiver requests are relatively straightforward and uncontroversial, EPA must nevertheless provide the opportunity for full public participation and subsequent technical analyses. This time-consuming process creates uncertainty for California, other states considering adopting those California standards, and manufacturers.

Recommendations

California, other states, and manufacturers all have a strong interest in obtaining EPA waiver decisions well before the applicable standards take effect. The committee recommends establishment of a two-track system for waiver requests. Many California waiver requests have not been controversial, and EPA has not received any significant comments. EPA could expedite waiver requests that it considers noncontroversial, approving the waiver with a minimal analysis in a direct final decision without a full notice-and-comment process. The final decision would be published in the *Federal Register*, and if any interested party raised a substantive objection to the decision, it would be withdrawn and subjected to the full waiver process. This expedited process would allow EPA to process quickly and efficiently those waiver requests that are noncontroversial, freeing up resources to focus on those that require more time and discussion.

The committee also recommends consideration of a mandatory time limit for EPA to review and issue a waiver decision for controversial waiver requests. The time limit could be based on existing timetables for the EPA waiver process. California is required to provide adequate lead time between adoption of state regulations and their implementation:

usually at least 2 years for on-road sources and at least 2 years for non-road sources. A time limit of 2 years or less for EPA review would place the review process between the adoption of the standards by California and the time that the standards take effect. Given the importance of the EPA waiver review and the need to conclude such reviews more quickly, EPA should ensure that sufficient resources are devoted to the waiver review process so that the quality of the review is not sacrificed to comply with new time limits.

Adoption of California Emissions Standards by Other States

The primary reason that other states adopt California emissions standards is to obtain additional emissions reductions to help attain and maintain the NAAQS. States first began using their authority under section 177 of the CAA in the early 1990s when New York and Massachusetts adopted California emissions standards for new light-duty vehicles. To date, section 177 authority has mostly been used to adopt light-duty-vehicle standards by various northeastern states, although a growing number of other states have adopted or expressed an interest in using this authority to adopt California standards for both light-duty and heavy-duty vehicles.

Some states have cited additional rationales for adopting California standards. When considering emissions standards for on-road heavy-duty diesel vehicles, some states have indicated that they consider the adoption of California standards to be a safety net in case EPA delays similar federal standards. When considering emissions standards for light-duty vehicles, even when current federal standards provide emissions reductions similar to those in California, some states expect that California will continue to reduce standards earlier than the federal program. In addition, some states have adopted or expressed interest in adopting the California greenhouse gas emissions standards.

Manufacturers of mobile sources have raised objections to the adoption of California standards by other states. Manufacturers contend that states overestimate the emissions benefits of adopting California standards and that California standards often provide no significant air quality benefits over the applicable federal standards. Other objections include the claims of incremental costs of producing additional California-certified engines, the risks of expanding technology-forcing experiments to a greater share of the national market, and the additional complexity of having to distribute products that attain different standards in different states. Disputes have also arisen between the states and the manufacturers over the ability of California-certified vehicles to meet emissions standards and function properly under conditions in their states.

Up to this point, adopting states and manufacturers have resorted to the courts to resolve their technical and legal disputes when direct negotiations have failed. Among the issues that have been litigated are whether adopting states also had to adopt California fuel regulations, whether electric vehicles designed for California (under the zero-emission-vehicle [ZEV] mandate) could be mandated in northeastern states where their batteries might not function properly in wintertime, and whether the California ZEV mandate met the definition of a standard that could be separately adopted by other states. Although EPA is an appropriate entity to comment on some of these disputes, it has no authority over states' adoption decisions.

Recommendations

The process by which a state adopts California emissions standards should be improved to aid in the resolution of the legal and technical disputes that often arise. As the agency that has the overall authority for implementing the CAA, including the mobile-source provisions, EPA should consistently participate in the process of the adoption of California standards by another state. EPA's current role in the state adoption process includes the authority to approve or disapprove the state SIP claims for emissions benefits from California emissions standards. The committee discussed additional roles for EPA to improve the state adoption process and considered two possible alternatives.

1. Each time a state intends to adopt a California emission standard, EPA would provide formal guidance to aid the state's adoption decision. EPA would determine whether any new issues have arisen that were not considered in the California waiver for the same standard (for example, issues related to technological feasibility, lead time, identicality, and cost) and whether these issues provide cause for states to reject the standard. EPA would further determine whether the state action is consistent with the requirements specified in section 177 of the CAA. EPA's determinations would be developed with the aim of deterring litigation over potential disputes. However, EPA's determinations would not be bind-

ing, and states would retain their ability to adopt California standards at their discretion.

2. EPA is given the authority to review and, under limited circumstances, deny a state adoption decision using a truncated waiver determination process. In its review, EPA would consider whether the state's adoption of the California standard raises any issues not considered in the original California waiver and whether the state action is consistent with section 177 of the CAA. In this scenario, it is important that EPA's waiver determination not delay or otherwise impede adoption of a California standard. EPA would be required to approve automatically any state's adoption request that had not been denied after 18 months of submittal. It is also important that EPA give the same deference to section 177 state findings as it does to California's findings when making a waiver determination. Under this alternative, EPA's determinations would be binding in the same manner as EPA's determination of a California waiver application.

The committee also discussed whether EPA's review under alternative 2 should include an assessment of the necessity or usefulness of the adoption for states to attain their air quality goals. Such an assessment would have to balance the benefits of additional emissions reductions, increased flexibility for states to develop air quality management plans, and wider distribution of new technologies against the costs to industry and consumers.

What role EPA is to have in the state adoption process is a policy decision that goes beyond scientific and technical considerations. The committee disagreed as to which of the two approaches described above would be most effective. However, even if there is no change in the adoption process, non-California states should continue their efforts to work with manufacturers to minimize compliance burdens. As an example, the committee encourages northeastern states that have adopted California light-duty-vehicle emissions standards to implement a regionwide fleet-average emission standard rather than having each state meet a separate fleet-average standard.

Technical and Scientific Practices of States That Adopt California Standards

States that adopt California light-duty-vehicle emissions standards have supported the adoption by estimating the emissions reductions and in some cases the in-state economic impact of the regulations. The methods used to estimate emissions impacts in general rely on the same basic emissions models used by EPA and CARB.

Engineering-cost estimates of California emissions standards are typically adopted from CARB. States with larger populations, such as New York and Massachusetts, tend to perform their own analyses. Other states have relied on outside analyses, such as those conducted by the Northeast States for Coordinated Air Use Management.

Small-Engine Emissions Standards

An area of active interest for emissions control is in small gasoline-powered engines[6] used in an array of equipment and applications. Compared with light-duty-vehicle emissions control, small-engine emissions control poses special design, production, and distribution challenges. Small-engine manufacturers sell engines as well as equipment that use their engines, such as lawn mowers and chain saws. Small-engine equipment is often sold through a multistep distribution chain—from manufacturer to retail distributor to retail dealer. In addition, there is no state registration process for most small-engine products that can be used to ensure compliance with emissions standards. CARB has demonstrated some flexibility in setting emissions standards for small engines to deal with some of the difficulties inherent in the non-integrated industry. CARB has also worked with industry to reduce some of the burden in the compliance testing and certification process.

Recent federal legislation prohibits other states from adopting the California standards for small gasoline-powered engines and mandates that EPA issue new standards for small engines for the rest of the country. Thus, small engines form a new regulatory category different from other mobile sources. All other mobile sources fall into one of two categories: sources such as light-duty and heavy-duty vehicles for which California can set and other states can adopt emissions standards and sources such as airplanes and locomotives for which standards are entirely federally preempted.

---

[6] Here the committee refers to engines smaller than 25 horsepower that are used mainly in lawn and garden equipment.

Recommendations

California should continue its pioneering role when setting emissions standards for small engines to aid its efforts to improve air quality and be a proving ground for new emissions-control technologies. The committee encourages CARB to use the flexibility it has shown in revising standards based on new scientific and technical information for regulating small engines. The committee also recommends that the suggested alternatives for improving the state adoption process be used if a decision is made in the future to allow states to adopt California small-engine standards.

Cost Analyses

CARB and EPA estimate the costs to meet emissions standards. Both agencies look at variable parts costs; fixed costs, such as research and development costs; and testing and certification costs. States that adopt California's standards typically rely on California's cost estimates. One element of relying on technology-forcing regulations and, in California's case, of serving as a laboratory for mobile-source emissions-control technologies is the considerable uncertainty in estimating the cost of complying with emissions standards. Future technologies assumed during standards development and thus included in the regulatory estimates are not always the ones used for compliance. Even when technological assumptions turn out to be correct, estimated cost might not be correct. Some costs, such as the costs to other states to implement and maintain a program, are excluded from California's economic impact analyses of their own standards. CARB and EPA have used cost estimates to calculate cost-effectiveness in units of dollars per mass of pollutant reduced to weigh mobile-source emissions control against other emissions-control strategies. In the past decade, EPA estimated the monetized benefits of its major rules to compare with costs.

The committee finds that it is difficult to determine what parties bear what fraction of the costs of emissions standards. Manufacturers closely guard cost and pricing data to avoid placing themselves at a competitive disadvantage. The majority of available estimates of the cost of emissions standards are for light-duty vehicles, but these estimates vary substantially and are uncertain. Vehicles are not priced to recoup directly the costs of meeting emissions standards because costs are difficult to allocate to a single model, especially research and design costs. Additional costs are also difficult to incorporate directly into the sticker price when two models that differ only in emissions equipment are being sold side by side or when a competitor does not include such a surcharge for its vehicles. Under such conditions, the cost of emissions controls must be absorbed in reduced profits, in reduced costs, or in distribution of costs throughout the whole product line.

Recommendations

To address the uncertainty inherent in prospectively estimating costs to comply with mobile-source emissions standards, the committee recommends that agencies and stakeholders attempt to improve communication about the uncertainty by providing a range of costs rather than a single point estimate, especially for new technologies. In addition, because costs are such an important element for understanding the impacts of state emissions standards, the committee finds a need for a comprehensive study of the costs of state standards. This study should include the difference in costs for the states that adopt California standards compared with costs for California, the distribution of those costs, and their cost-effectiveness. Costs should be viewed broadly and include the costs to manufacturers and distributors to develop and distribute products certified under two emissions standards and the costs to states to implement, enforce, and maintain the program.

Harmonization of Standards and Procedures

Recognizing the needs of some states to adopt more stringent mobile-source emissions standards to help meet air quality goals, a desirable objective is harmonization of CARB's and EPA's certification procedures. Although meaningful differences in standards can be important in achieving clean air, superficial differences in areas such as certification procedures can be wasteful.

Harmonization of standards and testing procedures also has a global context. Since the beginning of emissions controls on mobile sources almost 50 years ago, there has been a profound shift in the manufacturing of mobile sources. Increased globalization means that foreign manufacturers are producing and selling their products within the United States, and domestic manufacturers are producing and selling globally. In addition, countries around the world have adopted mobile-source emissions standards, with those of the United States serving as one model.

Although harmonization is a worthy pursuit when the interests of the federal government and the states coincide, there are areas where their priorities diverge. A high-profile example is California's recent development of greenhouse gas emissions standards for light-duty vehicles. The California greenhouse gas standards, which have been promulgated by CARB and have been adopted or considered by several other states, reflect a difference in policy between the federal government and some states in addressing climate change with light-duty-vehicle emissions standards.

Recommendations

Regulators should make a determined effort to harmonize the procedures for testing and certification and look for opportunities to harmonize the emissions standards. Domestically, CARB and EPA should conduct a biennial assessment, either through a written report or public meeting, of where emissions testing and certification procedures can be harmonized and what emissions standards can be harmonized. The committee recognizes that EPA is leading the U.S. participation in international efforts to harmonize emissions standards and testing procedures. EPA should continue these efforts and encourage international participation in the biennial harmonization assessments. The committee recognizes that many countries will lag in the adoption of mobile-source emissions standards; therefore, global efforts to harmonize may need to focus initially on emissions testing and certification procedures.

Overview of Conclusions and Recommendations

Despite the substantial progress made over the past decades in reducing emissions from mobile sources, further progress is needed to attain air quality standards in many parts of the country. Separate California mobile-source emissions standards provide emissions control, air quality benefits, and innovation beyond federal standards. California should continue its pioneering role when setting mobile-source emissions standards. Although a second set of standards imposes additional costs and complexity to manufacturers, the committee concludes that the California program has been beneficial overall.

The committee recommends that CARB and EPA continue to review their scientific and technical practices in tandem to address areas for possible improvements and harmonization. The committee also recommends improvements to EPA's waiver process to provide timelier waiver decisions for California emissions standards.

State decisions to adopt California emissions standards have resulted in several disputes between states and industry, which typically led to extensive litigation. Although EPA is an appropriate entity to resolve or comment on some of these disputes, it has no authority over states' adoption decisions. The committee discussed two alternative roles that EPA could play to improve the state adoption process but did not reach a consensus to recommend either one.

1

Introduction

Emissions from mobile sources ranging in size from small, handheld gardening equipment to on-road passenger vehicles, heavy-duty trucks, and large construction equipment contribute significantly to air pollution in the United States. The federal Clean Air Act (CAA) mandates, with specific exceptions, that the U.S. Environmental Protection Agency (EPA) regulate mobile sources by uniform national emissions standards (CAA section 209(a)). An exception to federal preemption of these standards is granted to the state of California. California's exemption was granted in the early years of mobile-source regulation because air pollution was more severe in California than in the rest of the nation, and the state had a long history of establishing its own emissions standards for on-road vehicles and other mobile sources (CAA sections 209(b) and 209(e)). Other states may choose to adopt California emissions standards as a mitigation strategy but may not establish their own standards.

California and the federal government have set mobile-source emissions standards and have tightened those standards over time. This evolution has led to important technological advances in emission-control systems and substantial reductions in mobile-source emissions. Several states, particularly in the Northeast region, have adopted California emissions standards in the past 2 decades as part of their plans to improve air quality and protect public health. Because mobile-source emissions still represent a substantial fraction of air pollutant emissions, mobile-source emissions reductions can still contribute substantially to improvements in air quality. The role of state versus federal government in establishing mobile-source emissions standards has been and remains an important environmental policy issue.

ORIGIN OF STUDY AND COMMITTEE CHARGE

In January 2003, a Senate omnibus bill included a provision directing EPA to "submit a report … on the practices and procedures by which States develop separate emission standards, including standards for nonroad engines or vehicles, as compared to the development by EPA of national emission standards under the Clean Air Act. This report shall include an assessment of the procedures, practices, standards and requirements used by States as opposed to those used by EPA, including how States and the EPA take into account technological feasibility, economic feasibility, impact on the economy, costs, safety, noise and energy factors associated in the development of these standards." The provision was later modified, directing EPA to contract with the National Academy of Sciences to conduct the review and submit a report of its findings to the Committees on Appropriations. In response to EPA's request, the National Research Council (NRC) established the Committee on State Practices in Setting Mobile Source Emissions Standards. The NRC study began in March 2004. The Statement of Task set forth to the committee is as follows:

> This committee will review and evaluate the scientific and technical practices used by states in setting emissions standards for mobile sources, including those for nonroad engines and vehicles. The committee will assess the scientific and technical procedures used by states to develop or adopt emissions standards separate from those set by EPA under the Clean Air Act and assess the factors that cause states to move to more stringent emissions standards. The committee will consider the scientific, technical, and economic rationale and methods used by the states in setting standards and how they compare with those used by EPA. In addition, the committee will take into consideration the direct and indirect impacts and benefits that state emissions standards have had on various factors, including compliance costs, energy consumption, air quality, and human health. Specifically, the committee will assess the following:

- How states assess their need for more stringent emissions standards.
- How California develops and other states adopt separate emissions standards.
- How states take into consideration technical and economic feasibility of compliance with separate emissions standards.
- How states take into account expected impacts on cost to consumers, employment, safety, noise, and energy as part of the standards-setting process.
- How expected emissions reductions and air quality benefits are accounted for in the state emissions standards-setting process, and how those benefits help states to meet their air quality attainment requirements.
- How the rationale and methods used by states compare with those used by EPA.
- How California emissions standards affect those set by EPA.

The committee was also asked to consider the effects of over 30 years of experience by California in setting separate emissions standards, not only in terms of the impacts and benefits on such factors as compliance costs, energy use, air quality and human health but also in terms of the effects on the timing of federal emissions standards. The committee was also asked to examine the intersection of state emissions standards with fuel standards.

THE COMMITTEE'S APPROACH TO THE CHARGE AND THE REPORT ORGANIZATION

The committee began its work with a historical review of mobile-source regulation by California and the federal government. The review includes the significance of mobile-source emissions to air quality and health (Chapter 2), the history and status of mobile-source regulation in the United States (Chapter 3), and the evolution of technology to meet emissions standards (Chapter 4). The report continues with an analysis of the reasoning behind the current system of mobile-source regulation in the United States (Chapter 5). The committee then considers several case studies of mobile-source regulation (Chapters 6 and 7) to understand and explain the specific aspects of regulating the various sectors of mobile sources. The report concludes with a chapter of findings and recommendations (Chapter 8).

The committee interpreted its charge to include any mobile-source emission standard that has been established as a rule by California. The committee did not consider emissions standards for federally preempted sources, such as aircraft, locomotives, commercial marine vessels, and certain other nonroad equipment. The committee limited its considerations to emissions standards for new mobile sources or engines and did not consider in-use emissions control programs, such as motor vehicle emissions inspection and maintenance programs (I/M). Furthermore, the committee did not consider fuel composition or refueling emissions standards.

Historically, emissions standards set by California and the federal government have aimed to reduce ambient concentrations of criteria pollutants, including nitrogen dioxide (NO_2), ground-level ozone (O_3), carbon monoxide (CO), particulate matter (PM), and sulfur dioxide (SO_2) (see "criteria air pollutants" in Glossary). Emissions standards have also been developed or proposed for hazardous air pollutants (also called air toxics) by EPA and California and for greenhouses gases by California. The committee's assessment focuses on emissions standards for criteria pollutants. However, in Chapter 6, the committee also summarizes standards that have been recently set for greenhouse gas emissions, recognizing that no such standards have been established by the federal government. The committee did not perform an in-depth analysis of the recent greenhouse gas standards or develop findings and recommendations specific to these standards. The committee decided that it would be premature to consider the greenhouse gas emissions standards because these standards were passed into law by California after the committee had begun its analysis and because the legal standing of the standards was in question during this study.

In the course of its work, the committee found that, historically, EPA and CARB have focused on the technical feasibility of proposed standards, the emissions benefits of the standards, and the monetary costs to manufacturers to comply with the standards. EPA has been required to assess and monetize the health benefits of its major regulations in recent years and utilizes air quality analysis as part of this process. With some exceptions, the impacts of emissions standards on employment, safety, noise, and energy are dealt with qualitatively or not at all. Therefore the main focus in this report is on the technical feasibility of proposed standards, the emissions and air quality benefits of those standards, and the monetary costs to manufacturers to comply with the standards.

2

Air Quality, Emissions, and Health Impacts Overview

OVERVIEW OF POLLUTANTS AND STANDARDS

The primary goal of air quality management is the protection of public health and welfare. The governing legislation that exists today is the federal Clean Air Act (CAA), which was established in 1963 (PL 86-493) and substantially amended in 1967, 1970, 1977, and 1990. Several amendments to those CAAs also occurred between full CAA reviews and formal legislative revisions, particularly during the 1960-1980 period. The CAA provides the regulatory framework for air quality management, including mobile-source emissions. The management framework has five goals: mitigate ambient concentrations of criteria pollutants (described below), limit exposure to hazardous air pollutants (HAPs), protect and improve visibility in pristine areas, reduce emissions that cause acid deposition, and curb the use of stratospheric ozone-depleting chemicals (NRC 2004). The reduction of mobile-source emissions plays a key role in attaining those goals and is dealt with explicitly in the CAA, as discussed in Chapter 3 of this report. Regulation of mobile-source emissions is aimed predominantly at mitigating criteria pollutants; however, these sources also emit pollutants that contribute to air toxic exposures (also called hazardous air pollutants or HAPs), acid deposition, visibility degradation, and greenhouse gas concentrations.

The benefits from the CAA and its amendments have been substantial. The economic benefits to public health from improved air quality have outweighed the overall costs required to implement all mitigation strategies (OMB 2004). EPA estimated that the benefits of implementation of the CAA between 1970 and 1990 were $5-50 trillion greater than the costs (EPA 1997). EPA (1997) estimated that these benefits include about 100,000 to 300,000 fewer premature deaths per year and 30,000 to 60,000 fewer children each year with intelligence quotients below 70. In addition, regulations to improve air quality have helped propel the development of the emission-control industry.

Criteria Pollutants

The six criteria pollutants are identified as those reasonably anticipated to endanger public health or welfare and those whose presence results from numerous or diverse mobile and stationary sources (CAA section 108(A) (1)). The federal CAA Amendments of 1970 directed EPA to set National Ambient Air Quality Standards (NAAQS) for criteria pollutants and to review the NAAQS at intervals of not more than 5 years and to update them as needed (CAA section 109). Standard concentrations for each pollutant are set at two levels: a primary standard that protects the public health and a secondary standard that protects public welfare, including effects on visibility and agriculture.[1] NAAQS were first established in 1971 on the basis of the current scientific knowledge of the effects of the pollutants on health and welfare. The NAAQS of 1971 included carbon monoxide (CO), sulfur dioxide (SO_2), nitrogen dioxide (NO_2), total photochemical oxidants, total suspended particles (TSP), and hydrocarbons (HC). Over time, a standard for lead (Pb) was added, TSP was revised to a standard for PM_{10},[2] a standard for ground-level ozone replaced the oxidants standard, and the standard for HC was removed. The most recent NAAQS were promulgated in 1997 when a standard was added for fine PM ($PM_{2.5}$),[3] and the standard for ozone was lowered

[1] The term primary and secondary is also used in some instances in air pollution to differentiate between pollutants emitted directly by sources and pollutants formed in the atmosphere. For example, particulate matter emitted from a factory is called primary particulate matter, and particulate matter formed in the atmosphere from sulfur emissions from the same source is called secondary particulate matter.

[2] PM_{10} refers to a subset of particulate matter collected by a sampling device with a size-selective inlet that has a 50% collection efficiency for particles with an aerodynamic diameter of 10 micrometers (μm).

[3] $PM_{2.5}$ refers to a subset of particulate matter collected by a sampling device with a size-selective inlet that has a 50% collection efficiency for particles with an aerodynamic diameter of 2.5 μm.

and changed from a 1-hr average basis to an 8-hr average basis, effectively making the standard more stringent. Mobile sources contribute to ambient concentrations of all criteria pollutants.

Air Quality Standards

Some of the earliest, most severe, and most persistent air pollution episodes in the United States have been in California, particularly in the Los Angeles basin, and parts of California continue to have the most severe air pollution in the country. State legislation to control air pollution was passed as early as 1947, well before the federal CAA of 1963, and California has often preceded the federal government in establishing air quality and emissions standards. California's Department of Public Health set ambient standards as early as 1959, and later the California Air Resources Board (CARB), formed in 1969, became the agency authorized by California law to set state ambient air quality standards. The federal CAA authorizes states to adopt ambient air quality standards to protect public health that are more protective than the EPA standards. California is among several states that have adopted separate ambient standards. California's standards are more stringent than the NAAQS, and the state has adopted standards for additional criteria pollutants. In contrast to the federal standards, California's ambient standards do not have attainment deadlines. The current federal and California standards are presented in Table 2-1.

Nonattainment Areas

The CAA mandates that ambient concentrations of criteria pollutants be monitored in urban and rural areas throughout the United States. EPA then determines whether the monitored regions attain the NAAQS based on statistical analysis of monitored data. Figure 2-1 shows nonattainment counties as of September 2005, indicating how many criteria pollutants are in nonattainment in each county. Figure 2-2 shows nonattainment counties for $PM_{2.5}$ and ozone (with concentrations averaged over 8 hr) and, when compared with Figure 2-1, shows that the majority of nonattainment counties in the United States violate either the ozone or $PM_{2.5}$ NAAQS. The figures show that nonattainment of criteria pollutants is especially problematic in much of California and the Northeast region

TABLE 2-1 Federal and California Ambient Air Quality Standards

Pollutant	Averaging Time	Federal Standards Primary	Federal Standards Secondary	California Standards[a] Primary
Ozone	1 hr	0.12 parts per million (ppm) (235 μg/m^3)	Same as primary	0.09 ppm (180 μg/m^3)
	8 hr	0.08 ppm (157 μg/m^3)	Same as primary	0.070 ppm (137 μg/m3)
PM_{10}	24 hr	150 μg/m^3	Same as primary	50 μg/m^3
	Annual arithmetic mean	50 μg/m^3	Same as primary	20 μg/m^3
$PM_{2.5}$	24 hr	65 μg/m^3	Same as primary	No separate state standard
	Annual arithmetic mean	15 μg/m^3		12 μg/m^3
CO	8 hr	9 ppm (10 mg/m^3)	None	9.0 ppm (10 mg/m^3), Lake Tahoe 6 ppm (7 mg/m^3)
	1 hr	35 ppm (40 mg/m^3)		20 ppm (23 mg/m^3)
NO_2	Annual arithmetic mean	0.053 ppm (100 μg/m^3)	Same as primary	
	1 hr			0.25 ppm (470 μg/m^3)
SO_2	Annual arithmetic mean	0.030 ppm (80 μg/m^3)		
	24 hr	0.14 ppm (365 μg/m^3)		0.04 ppm (105 μg/m^3)
	3 hr		0.5 ppm (1,300 μg/m^3)	
	1 hr			0.25 ppm (655 μg/m^3)
Pb[a]	30-day average			1.5 μg/m^3
	Calendar quarter	1.5 μg/m^3	Same as primary	
Visibility	8 hr	No federal standards		Extinction coefficient of 0.23 per kilometer; visibility of 10 miles or more

Sulfates	24 hr	No federal standards	25 μg/m^3
Hydrogen sulfide	1 hr	No federal standards	0.03 ppm (42 μg/m^3)
Vinyl chloride[a]	24 hr	No federal standards	0.01 ppm (26 μg/m^3)

[a]CARB has identified lead and vinyl chloride as "toxic air contaminants" with no threshold level of exposure for adverse health effects determined. These actions allow for the implementation of control measures at levels below the ambient concentrations specified for these pollutants.

Source: Adapted from CARB 2003a.

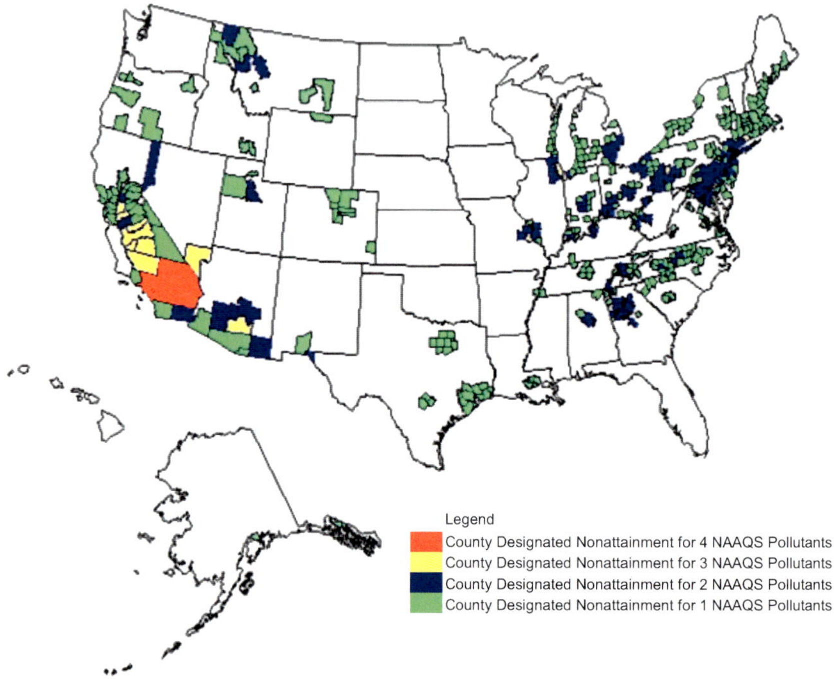

FIGURE 2-1 Counties designated "nonattainment" for NAAQS as of September 2005. Guam—Piti and Tanguisson Counties are designated nonattainment for the SO2 NAAQS. Partial counties, those with part of the county designated nonattainment and part attainment, are shown as full counties on the map. Source: EPA 2005a.

of the United States. Many counties in the United States are nonattainment designated for both ozone and $PM_{2.5}$ concentrations.

Trends

On average, monitored concentrations for all criteria pollutants have decreased throughout the nation since 1970. A National Research Council (NRC 2004) assessment of air quality management credits the CAA with these substantial emissions reductions despite growth in population, energy use, and vehicle activity. The exceedances of the NAAQS, however, differ in relative severity from location to location and over time, depending on emissions sources, prevailing meteorology, and ef-

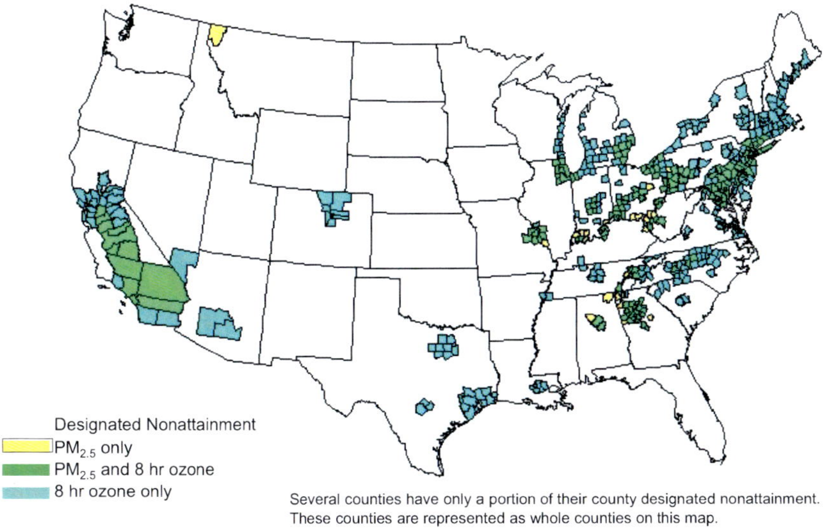

FIGURE 2-2 Counties designated nonattainment for $PM_{2.5}$ and/or ozone with concentrations averaged over 8 hrs. Designations for ozone were made in June and September 2004. Designations for the $PM_{2.5}$ were made in April 2005. Source: EPA 2005b.

fectiveness of regulation. Ambient Pb concentrations, for example, have been dramatically reduced throughout the nation since the 1980s as a result of a national policy to remove Pb from gasoline. Because of effective regulation of CO emissions, especially from mobile sources, the number of CO nonattainment areas has been reduced from many to only a few, representing another area of success due to the CAA (NRC 2003; Holmes and Russell 2004).

In contrast, ground-level ozone remains above the NAAQS in many areas despite decades of precursor reductions. Figure 2-3 shows the number of days per year that ozone concentrations exceeded the NAAQS from 2001 to 2003. The figure shows that areas in California exceed the NAAQS most frequently and that exceedances are common in Texas, the Midwest, and the entire Northeast region. Figures 2-4 and 2-5 presents historical air quality trends in exceedances of the 1-hr maximum ozone concentration and the 8-hr maximum ozone concentration in Los Angeles and New York. Both New York and Los Angeles are approaching the 1-hr NAAQS as a result of air quality improvement programs. The new standard for ozone, however, places both locations in a much more diffi-

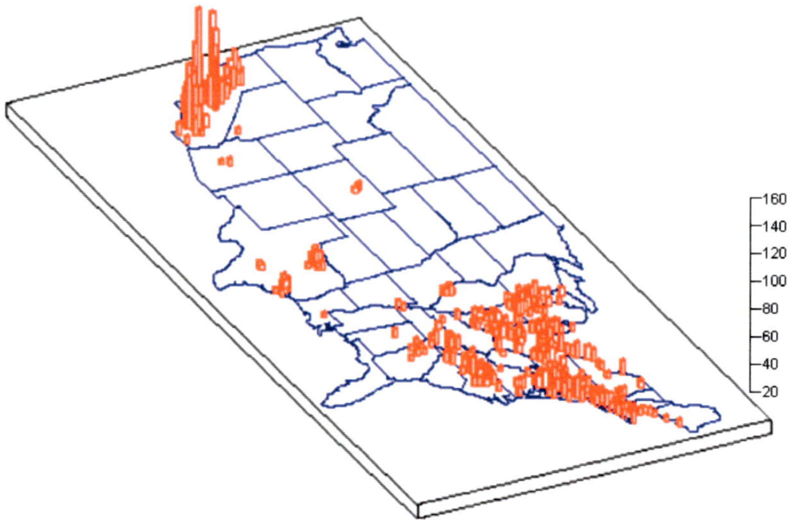

FIGURE 2-3 Frequency with which the NAAQS for ozone (with concentrations averaged over 8 hr) was exceeded 2001-2003. Source: Witherspoon 2004.

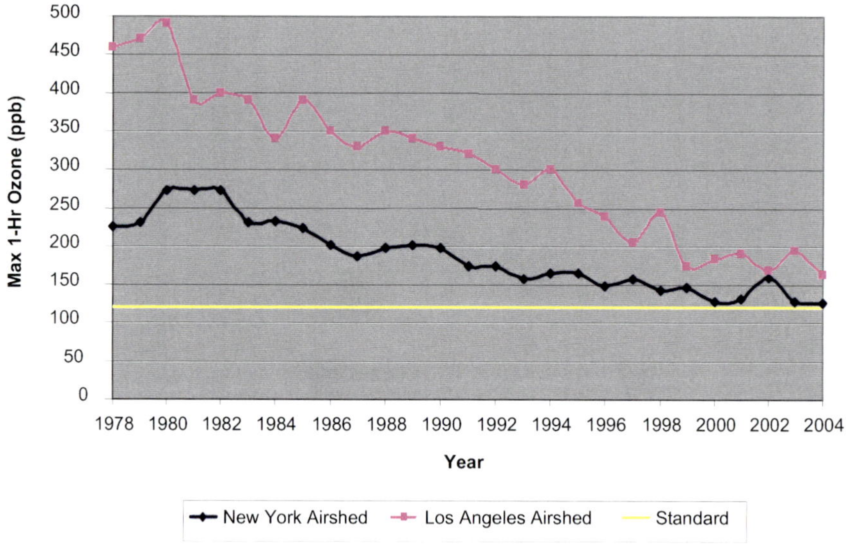

FIGURE 2-4 Trends in maximum ozone concentrations averaged over 1 hr for New York and Los Angeles airsheds from 1978 to 2004. Source: EPA 2004a; CARB 2005a.

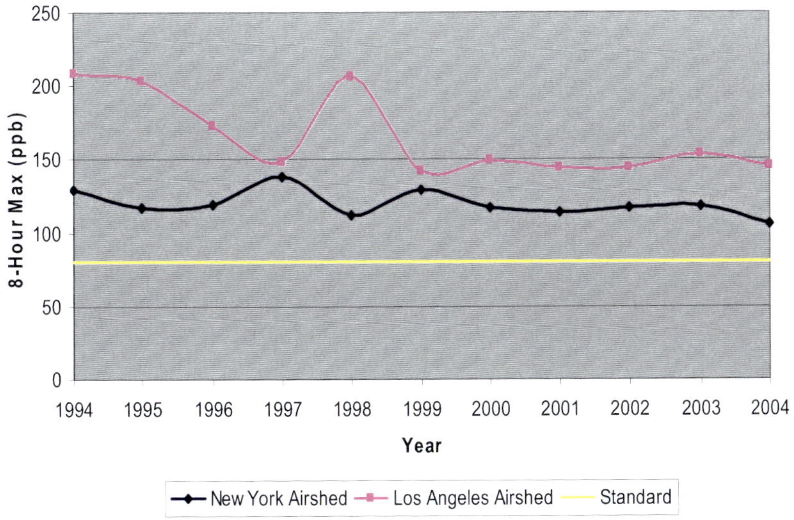

FIGURE 2-5 Trends in maximum ozone concentrations averaged over 8 hr for New York and Los Angeles airsheds from 1994 to 2004. Source: EPA 2004a; CARB 2005a.

cult position. Meeting this standard will require more aggressive ozone precursor control strategies for each region. That position is typical of other nonattainment areas on both the east and the west coasts of the United States.

More generally, further progress in improving air quality in the United States is likely to be challenging, especially in meeting the new NAAQS for $PM_{2.5}$ and ozone and in addressing issues such as regional haze, HAPs, and greenhouse gas emissions (NRC 2004; Chameides et al. 2005). Mobile-source emissions standards, the focus of this report, are promulgated primarily to meet NAAQS, but such standards will also affect other air quality issues, such as mobile-source HAPs and greenhouse gas emissions. For example, EPA analyses show that mobile-source emissions standards programs already in place will yield significant reductions of mobile-source HAPs (EPA 2000a).

GROUND-LEVEL OZONE AND FINE PARTICULATE MATTER

Ground-level ozone and fine PM currently account for the majority of nonattainment areas in the United States and will be the focus of many

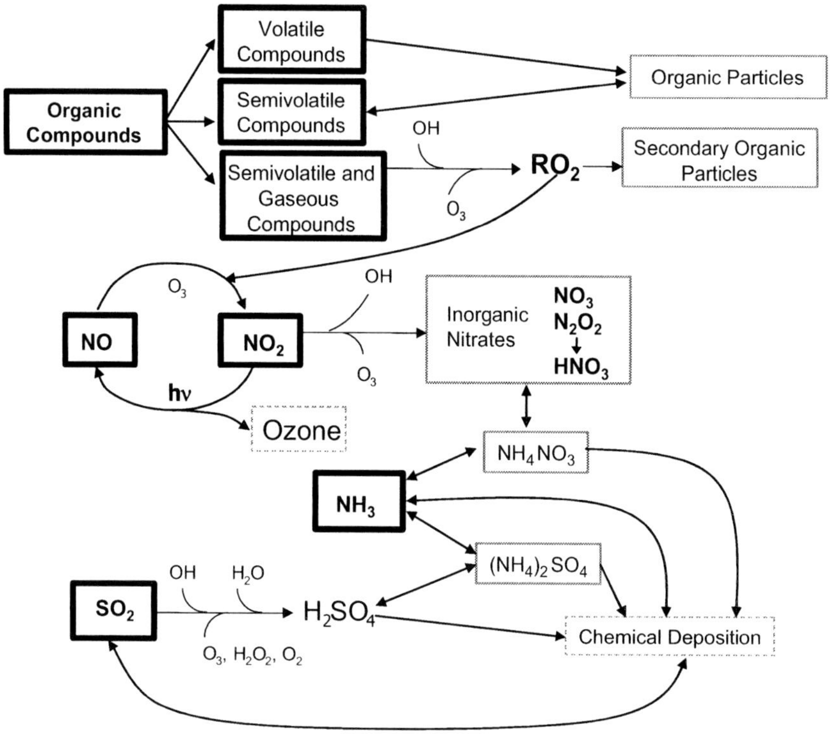

FIGURE 2-6 Schematic of the atmospheric processes involved in the formation of O_3 and secondary PM. Major precursors are shown in the boxes with thick sides. Secondary particle components are shown in the boxes with thin, solid sides. Source: NARSTO 2004. Reprinted with permission; copyright 2004, Cambridge University Press.

future mitigation efforts. General characteristics of these two pollutants relevant to mobile-source management are presented here. Figure 2-6 shows a schematic representation of ozone and secondary PM formation and the relationships among the atmospheric processes involved in their formation. Later sections expand on the contribution of mobile sources to the concentrations of ozone and fine PM.

Ozone

Some criteria pollutants accumulate in the atmosphere due to direct emissions and are characterized as primary pollutants. Ozone, in contrast, is characterized as a secondary pollutant because it is formed almost entirely in the atmosphere. Ozone formation is a highly nonlinear process that depends on sunlight intensity, meteorology, and the emissions and transport of its two major precursors, NO_x and HCs. Understanding tropospheric ozone chemistry is a key to understanding how ozone formation depends on concentrations and emissions of its major precursors.[4] A detailed description of this chemistry can be found elsewhere (Seinfeld and Pandis 1998; Finlayson-Pitts and Pitts 1999), and the following is meant to give a brief overview of the relationship of ozone to NO_x and HC concentrations.

Ozone is formed when NO_2 disassociates in the presence of sunlight to form NO and a single reactive oxygen atom (O), which can then combine with molecular oxygen (O_2) to produce ozone. However, NO can remove ozone by reacting with it to recreate NO_2 in a cycle that by itself would not necessarily result in ozone accumulation. Ozone accumulates when NO is converted to NO_2 by alternate pathways, thereby eliminating an ozone sink (reaction with NO) and creating a new ozone source (more NO_2). The alternate NO to NO_2 conversion pathways are driven by HCs and reactive, short-lived species called radicals. The hydroxyl radical (OH) can react with HCs to form new organic (carbon-containing) radicals and inorganic (noncarbon-containing) radicals through multistep reactions with such species as oxygen. These new radicals convert NO to NO_2 while regenerating more organic and inorganic radicals in a self-propagating process, including regeneration of OH that can then oxidize a new HC. The NO_2 is then available to form ozone as described above. Ozone accumulates when conditions favor this recycling of radicals and NO_x. Organic radicals and NO_2 can be removed from the system by termination reactions that result in formation of less-reactive or stable compounds, thereby reducing their ability to promote ozone formation. Ozone formation slows or reverses when conditions favor these termination processes. NO_x and thousands of different HC species may participate in this process.

Because ozone formation is driven by sunlight, as are other important radical reactions, ozone formation (and concentrations) is typically at a maximum in the afternoon and at a minimum before sunrise; however, ozone concentrations also depend on the temporal patterns of emissions and concentrations of available NO_x and HCs.

---

[4] The discussion in this report focuses on air pollution of the troposphere or lower (ground-level) atmosphere.

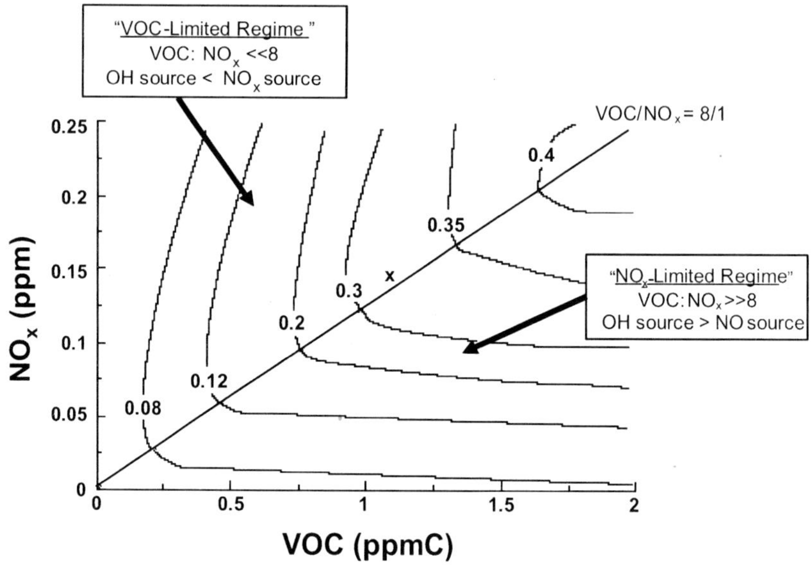

FIGURE 2-7 Ozone isopleths (lines of constant ozone concentration). Source: NARSTO 2000. Reprinted with permission; copyright 2000, Cambridge University Press

A common way of understanding the relationship between ozone, NO_x, and HC is through the use of ozone isopleths diagrams (lines of constant ozone concentration), which relate relative amounts of HC or volatile organic carbon (VOC)[5] and NO_x to ozone concentrations in a particular location. Ozone isopleths typical of an urban area in the United States are presented in Figure 2-7. The x- and y-axes represent initial VOC and NO_x concentrations, respectively, in a mixture of precursors. The maximum ozone concentration (averaged over 1 hr) that results when the mixture reacts is plotted as a function of different initial precursor concentrations, and isopleths are drawn through these resultant concentrations.

A characteristic feature of Figure 2-7 is the ridgeline (in this example, at VOC/NO_x ratio of 8:1), which passes through the local maxima in the ozone isopleths and separates conditions into two regimes: VOC-

[5] Hydrocarbons are classified in a number of ways. EPA typically reports total VOC emissions. The California Air Resources Board uses reactive organic gas (ROG) emissions. Appendix A in this report lists various ways of reporting HC emissions.

limited conditions above the ridgeline and NO_x-limited conditions below the ridgeline. At VOC-limited conditions reducing NO_x either will not affect ozone concentrations or will even increase ozone concentrations. At NO_x-limited conditions, reducing VOC concentrations have little to no effect on ozone concentrations. The isopleths provide some insight into the complexities in designing appropriate control strategies. VOC emissions reductions might be appropriate if conditions in an urban area are in the VOC-limited regime, whereas NO_x emissions reductions might be appropriate if conditions are NO_x-limited. In some cases, both NO_x and VOC control might be appropriate. Because VOC and NO_x concentrations and their ratios can vary within a single airshed and throughout the day and week, appropriate emissions-reduction strategies are complex. The effectiveness of NO_x versus VOC emissions reductions has been an important science-policy question for ozone mitigation.

Particulate Matter

Atmospheric PM refers to all suspended solid and solid-liquid particles in the atmosphere. Particles vary in size and chemical composition and originate from natural and anthropogenic sources. Some particles are emitted directly (primary PM) and some are formed in the atmosphere as a result of chemical and physical processes involving gaseous species (secondary PM). Figure 2-8 shows a typical urban size distribution of PM in the United States and shows both particle number concentrations and mass concentrations as a function of particle diameter. Particles are often divided by size into coarse (>2.5 μm), fine (≤2.5 μm), and ultrafine (<0.1 μm) particles. Figure 2-8 shows that the majority of particles (as indicated by number concentration $n°_N$) are ultrafine in size; however, the majority of particle mass (as indicated by volume concentration $n°_V$) is coarse or fine in size.

The NAAQS for $PM_{2.5}$ is expressed in terms of total mass concentration; however, regulatory agencies measure both total mass and composition of fine PM to provide information about its sources and potential health effects. Figure 2-9 shows the bulk composition of $PM_{2.5}$ for different U.S. regions. The size of the pie charts also gives an indication of average total $PM_{2.5}$ mass concentrations in the different regions. Major components of ambient $PM_{2.5}$ are ammonium sulfates, ammonium nitrates, and elemental and organic carbon compounds. Elemental carbon refers to nonvolatile materials that are almost exclusively carbon, where-

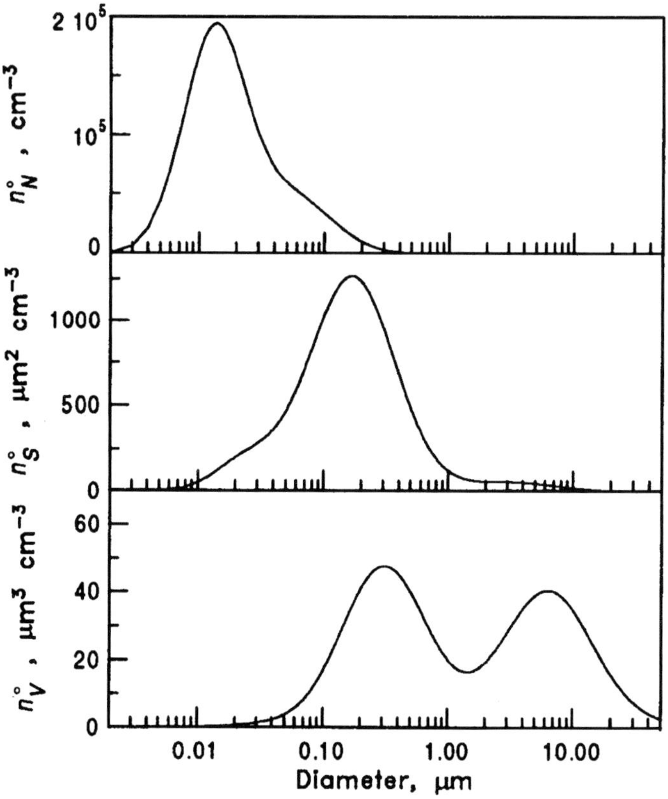

FIGURE 2-8 Typical particle volume ($n°_V$), surface area ($n°_S$), and number ($n°_N$) concentrations of urban aerosol as a function of particle diameter. Source: Seinfeld and Pandis 1998. Reprinted with permission; copyright 1998, Wiley.

as organic carbon refers to organic compounds in the particle phase. Ammonium sulfates and ammonium nitrates are formed as a result of gas-to-particle conversion of ammonia (NH_3), SO_2, and NO_x, and are considered secondary PM. Elemental carbon is exclusively primary; however, a fraction of organic carbon in $PM_{2.5}$ was found to be secondary. The compositions presented in Figure 2-9 apply to total $PM_{2.5}$. The chemical composition of the ultrafine fraction of $PM_{2.5}$ was found to be different from that of total $PM_{2.5}$ (EPA 2004f), suggesting that different size particles have different compositions and origins.

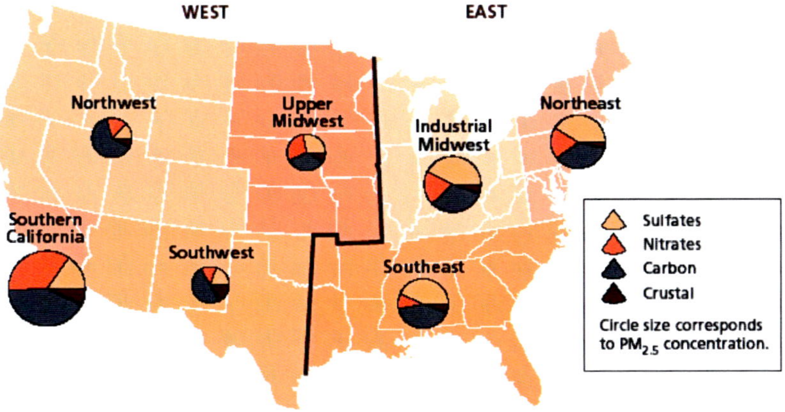

FIGURE 2-9 Mean $PM_{2.5}$ compositions in the United States. Source: EPA 2004b.

MOBILE-SOURCE EMISSIONS CHARACTERIZATION

Fraction of Overall Inventory

Emissions of air pollutants are generally divided into the source categories shown in Table 2-2. Mobile sources compose a large fraction of the overall inventory. The National Air Quality and Emission Trends Report 2003 (EPA 2004b), Special Studies Edition reports that the percentage of national emissions due to on-road and nonroad mobile sources are 82% for CO, 56% for NO_x, and 45% for VOCs. Although mobile sources emit SO_2, point sources tend to dominate these emissions (EPA 2003a). Since phasing out leaded gasoline, Pb nonattainment has been virtually eliminated in the United States, and point sources are now more important than mobile (EPA 2003a). During 1993-2002, across the continental United States, transportation sources were estimated to have contributed on average about 20% of PM_{10} and 30% of $PM_{2.5}$ (EPA 2003a). Data in this report are mostly limited to CO, NO_x, VOCs, and PM. CO, NO_x, and VOCs have been the primary focus of emissions control from light-duty vehicles. NO_x and PM have been the primary focus of emissions control from diesel-fueled mobile sources, such as heavy-duty on-road vehicles.

Figure 2-10a-c shows the national trends in total anthropogenic emissions by major sector from 1970 through 2002 for CO, NO_x and VOC, respectively. Biogenic emissions are not included in these summa-

TABLE 2-2 Air Pollutant Emissions Sources

- *Point sources* are major, localized, and stationary sources that typically have their emissions regulated through permits, such as power plants, refineries, and large manufacturing facilities.
- *Area sources* do not individually produce sufficient emissions to qualify or to be reported as a single point source, but collectively the emissions from all the small sources of the same type in an area may be significant and are reported as a category. These sources are numerous, including dry-cleaning facilities, architectural coatings, consumer products, commercial and residential heating, fugitive dust, solvent usage, and gasoline-dispensing facilities.
- *On-road mobile sources* are emissions from vehicles certified for highway use, including passenger cars, buses, trucks, and motorcycles.
- *Nonroad mobile sources* encompass a wide variety of equipment types that either move under their own power or are capable of being moved from site to site. Nonroad mobile equipment sources are not licensed or certified as highway vehicles but are defined as those that move or are moved within a 12-month period and are covered under EPA's emissions regulations as nonroad mobile sources. Major nonroad sources include aircraft, marine vessels, locomotives, farm and agricultural equipment, lawn and garden equipment, construction and mining equipment, and recreational vehicles, such as boats, all-terrain vehicles and snowmobiles. Regulation of aircraft, commercial marine vessels, locomotives, and certain farm and agricultural equipment is entirely federally preempted. These sources are not considered in this study because no state, including California, has authority to regulate them.
- *Biogenic emissions* are dominated by hydrocarbons released by natural and cultivated vegetation. These emissions are not typically subject to emission-control strategies, and their spatial distribution (highest in heavily forested areas) differs from that of other emissions sources, which are often highest in populated areas.

ries because they are typically not subject to controls. The figure illustrates the substantial contribution made by on-road and nonroad mobile sources for all three pollutants. The figure shows that total emissions and on-road emissions, especially of CO and VOC, have decreased substantially since 1970. As shown later in this chapter, this decrease has occurred despite increased on-road vehicle activity. Total nonroad emissions, in contrast, have remained fairly constant, and as a result, the fraction of total emissions that come from nonroad sources has increased. Nonroad CO increased from 6% to 22% of the total CO inventory from 1970 to 2002. Likewise, nonroad NO_x increased from 10% to 19%, and nonroad VOC increased from 5% to 16% of total inventories from 1970 to 2002.

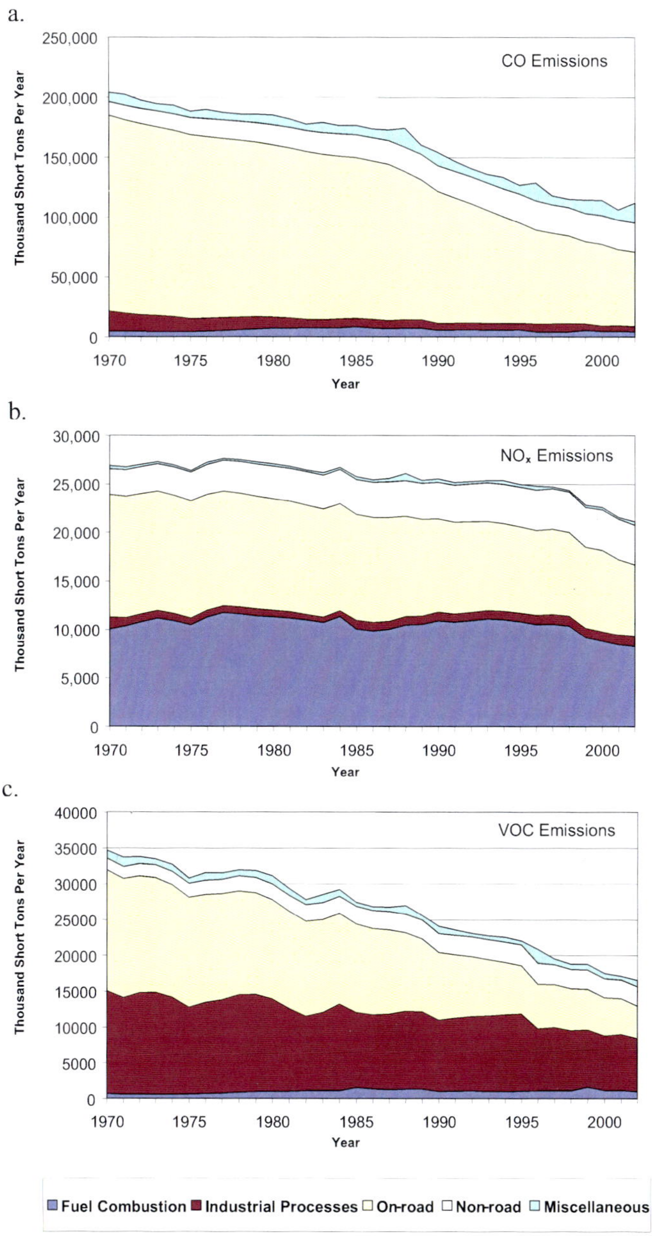

FIGURE 2-10 Trends in (a) CO, (b) NO_x, and (c) VOC emissions in the United States by major sector, 1970-2002. Source: EPA 2005c.

Composition of Mobile-Source Emissions

Mobile-source emissions are a complex mixture of gases and particles. In general, the gas-phase component includes CO, NO_x, sulfur compounds, ammonia, and a vast array of volatile and semivolatile HCs, although the latter are often associated with particle emissions. There are some characteristic differences in gas-phase emissions composition between normal operating gasoline and diesel vehicles. Diesel engines, which burn a mixture rich with air, emit more NO_x per unit activity than do normal gasoline vehicles. Sulfur is found in the United States in higher levels in diesel fuel than in gasoline, resulting in higher SO_2 emissions from these engines. Gasoline-powered engines operating under nonstoichiometric[6] conditions, such as at cold start and during idling, as well as older high-emitting engines are important contributors to mobile-source CO and organic compounds. Organic compounds typically found in gas-phase emissions include *n*, branched, cyclic alkanes and alkenes; other compounds found are 1,3-butadiene, aldehydes, ethers, ketones, steranes, hopanes, and aromatic compounds ranging from benzene to polycyclic aromatic hydrocarbons (PAHs) (Schauer et al. 2002).

Particles in mobile-source exhaust are generally quite small, from 0.001 to 1 μm, and are a complex mixture of a high number of compounds. Diesel exhaust is a significant contributor of particles in the ultrafine size range. Studies have shown diesel exhaust to include particles ranging from 0.01 to 0.3 μm, with the most found in the lower end of the size distribution (Park et al. 2003; Sakurai et al. 2003). The relative contribution of gasoline light-duty vehicles versus diesel heavy-duty vehicles to PM concentrations is an area of active research. A study of $PM_{2.5}$ in Denver, Colorado, found that light-duty vehicles contributed a much larger fraction of $PM_{2.5}$ emissions than did diesel vehicles (Fujita et al. 1998; Norton et al. 1998); however, the Southern California Air Quality Study found diesel heavy-duty vehicles to be the dominant contributor of mobile-source-emitted $PM_{2.5}$ (Schauer et al. 1996).

The particle exhaust mixture includes elemental carbon resulting from incomplete combustion and organic and other compounds resulting from chemical reactions between combustion products. A large number of organic compounds are generated by the combustion of both gasoline and diesel fuel (see sample compositions in Zielinska et al. 2004; Shah et al. 2004). The semivolatile fraction of the emissions mixture for both

---

[6] Chapter 3 discusses stoichiometric engine operations.

fuels includes a large number of PAHs containing two to seven rings, as well as PAHs that include nitrogen, oxygen, and sulfur. Additional semi-volatile HCs identified included hopanes and steranes, which are considered signature compounds for lubricating oil used in mobile sources. Both gasoline and diesel particles may also contain inorganic compounds and a range of trace metals. Twelve elements—Mg, Al, Si, P, S, Cl, Ca, Fe, Cu, Zn, Br, and Pb—have been regularly identified in the PM produced by light-duty vehicles (Gillies and Gertler 2000).

ON-ROAD MOBILE-SOURCE EMISSIONS

On-Road Vehicle Types

Approximately 236 million vehicles were registered in the United States in 2003 (FHWA 2005). As shown in Figure 2-11, roughly 136 million were passenger cars, 95 million were trucks, and 5 million were motorcycles. In 2002, roughly 86% of the truck fleet is composed of light-duty pickups, vans, and sport utility vehicles (SUVs) at a gross vehicle weight (GVW)[7] of less than 8,500 lb; 7.5% comprised heavy-duty pickups, vans, and SUVs at a GVW of 8,500-10,000 lb; and 6.5% at a GVW of greater than 10,000 lb (Davis and Diegel 2004).[8] Approximately 1.8 million truck tractors (heavy-duty trucks) and 800,000 buses were registered in 2003 (FHWA 2005). California registers nearly 13% of the nation's vehicles; another 13% are registered in the Northeast (Maine, Vermont, New Hampshire, Connecticut, Massachusetts, Rhode Island, New York, and New Jersey) (FHWA 2005).

Passenger cars, as well as light- and heavy-duty pickups, vans, and SUVs are dominated by spark-ignition gasoline-powered engines. As described in later sections, passenger cars have historically been certified to more stringent emissions standards than light-duty pickups, vans, and SUVs. Because different vehicle classes have different emissions stan-

---

[7] GVW rated by manufacturers is actual weight of the vehicle plus fluids, passengers, and payload.

[8] For reporting on-road mobile-source emissions, vehicles are categorized by GVW and are broadly divided by EPA into light-duty and heavy-duty vehicles, weighing less than 8,500 lb GVW and more than 8,500 lb GVW, respectively. CARB and EPA also use medium-duty vehicle (MDV) (MDV 8,500–14,000 lb GVW) and medium-duty passenger vehicle (MDPV) (MDPV 8,500–10,000 lb GVW), respectively to refer to some large pickups and sport utility vehicles.

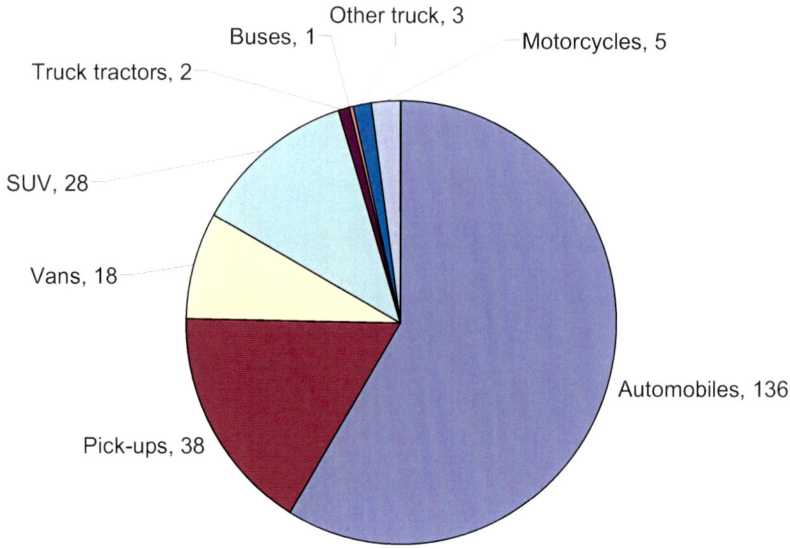

FIGURE 2-11 Number of registered vehicles in various on-road vehicle classes in 2003. A fraction of pickups, vans, and SUVs in the above figure are heavy-duty vehicles. Source: FHWA 2005.

dards, the mix of vehicle classes in the fleet plays an important role in fleet-wide emissions. Figure 2-12 shows how the mix of light-duty vehicles has changed with time. Pickups, vans, and SUVs accounted for only 20% of sales in 1976; sales of those vehicles have steadily increased over time and accounted for 50% of sales in 2003 (ORNL 2004).

The heaviest classes of trucks are powered predominantly by compression-ignition (diesel) engines. Although those vehicles are a small fraction of the total fleet, they contribute proportionally more to total vehicle miles traveled (VMT) and consume more fuel per mile, as shown in Table 2-3. Trucks greater than 26,000 lb GVW make up approximately 3% of the truck fleet, but contribute over 10% of all truck VMT and over 23% of all truck fuel use (U.S. Census Bureau 1999). A substantial number of heavy-duty vehicles on U.S. roads are also registered in other countries, especially Canada and Mexico, and those are not included in the reported totals.

Roughly 520,000 vehicles, or 0.2 % of the fleet, in 2002 used alternative fuels, such as liquid petroleum gas (LPG), compressed natural gas (CNG), alcohol-gasoline blends, or electricity (battery) (Davis and

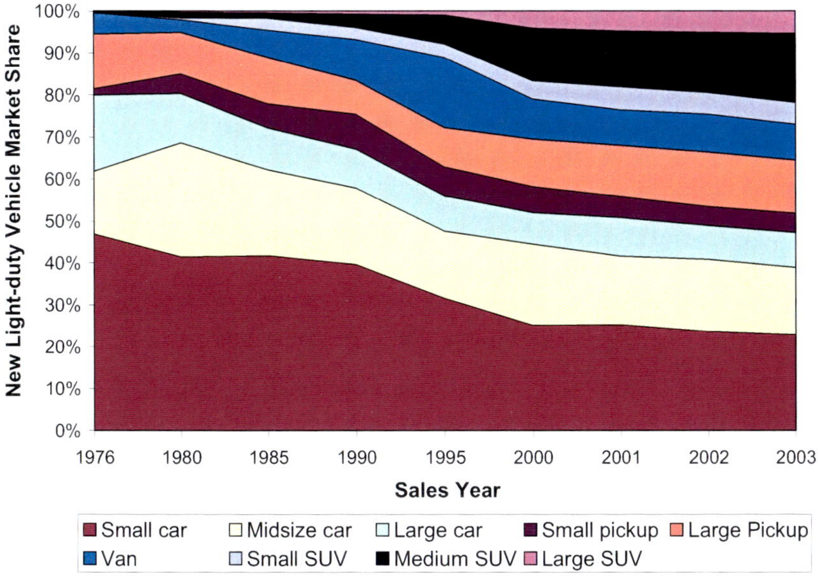

FIGURE 2-12 New light-duty vehicle market share by vehicle class from 1976 to 2002. The small-car category includes minicompact, subcompact, compact, and two-seater vehicles. Source: ORNL 2004.

Diegel 2004). These vehicles contribute little to the fleetwide emissions, although the number of hybrid vehicles, combined gasoline and battery powered, is increasing rapidly. New hybrid vehicle registrations totaled 83,153 in 2004, an increase of 81% over 2003, although this is less than 1% of the 17 million total new vehicles sold in 2004 (Durbin 2005).

On-Road Emissions Estimation

Emissions Modes

Mobile-source emissions are generally characterized as exhaust or evaporative emissions. Other categories of mobile-source emissions include tire and brake wear; however, these categories of emissions are not yet regulated. Exhaust emissions are the products of combustion or incomplete combustion and include HC, NO_x, PM, SO_2, CO, and Pb, as well as greenhouse gases and HAPs. The volume and composition of

TABLE 2-3 Comparison of 2003 Activity and Fuel Consumption by Vehicle Type

	Passenger Cars and Other 2-Axle, 4-Tire Vehicles	Single-Unit 2-Axle, 6-Tire or More, and Combination Trucks	Motorcycles	Buses
Average yearly miles traveled per vehicle	11,939	27,286	1,776	8,548
Average miles traveled per gallon fuel consumed	20.3	5.7	50.0	6.9

Source: FHWA 2005.

exhaust emissions depend on such factors as engine type and age, engine load, engine temperature, and quality of maintenance. Exhaust emissions are further divided into cold-start and running emissions. The former occur when a vehicle is initially started and the emission-control equipment has not yet reached a temperature of optimal pollutant removal, as discussed in Chapter 4.

Evaporative emissions include HCs and some HAPs and are divided into five types:

1. Resting loss: Vapor permeation and leaks from the evaporative emissions control system.
2. Diurnal: Evaporative emissions from idle equipment due to daily ambient temperature and pressure variations.
3. Running loss: Evaporative emissions during vehicle operation.
4. Hot soak: Evaporative emissions due to engine heat that persists after the vehicle has stopped operating.
5. Refueling loss: Includes vapors displaced in fuel tanks during refueling as well as volatilized spilled fuel.

Estimates from CARB's current statewide inventory indicate that, on a typical day, the breakdown of light- and medium-duty reactive organic gas (ROG)[9] emissions are 58% from exhaust; 30% from evaporative running losses; 5% from diurnal losses; 5% from hot-soak losses; and 2% from evaporative resting losses (refueling evaporative emissions not included in the total) (CARB 2005b). These distributions of emission types are broadly representative of actual conditions, but as discussed below, the models used to estimate exhaust and evaporative emissions include important uncertainties. For example, discrepancies have been found between data used in emissions models and data used in ambient measurements of the ratio of evaporative to exhaust emissions (Pierson et al. 1990, 1999).

Emissions Rates

The standard practice for estimating emissions rates for a source is to combine an emissions factor (EF) with an activity rate:

---

[9] Refer to Appendix A on hydrocarbon classifications

Emissions (mass/time) =
EF (mass/unit activity) × activity rate (unit activity/time).

Emissions factors for light-duty on-road mobile sources are typically reported as mass pollutant per distance traveled (for example, grams of NO_x/mile) and emissions standards are developed on such a basis. Evaporative emissions factors that do not depend on use, such as hot soak and diurnal evaporative emissions, are reported as mass/trip or mass/time. For heavy-duty on-road vehicles and nonroad vehicles and equipment, emissions factors are typically reported as mass pollutant per unit energy output (kilowatt-hour or brake horsepower-hour[10]) and emissions standards are developed on such a basis. Emissions factors are developed by the MOBILE and EMFAC models described in a later section.

Uncertainty and Variability in Emissions

Emissions vary considerably by vehicle and engine type, driving speeds, engine loads, fuel formulation, and ambient driving conditions. There is a great deal of uncertainty in how emissions change as a vehicle ages (emissions deterioration) and in how the change depends on vehicle make and model, where the vehicle is located, and maintenance factors. In-use measurement programs, such as exhaust remote sensing, state-run inspection and maintenance programs, and tunnel studies, are often used to characterize real-world emissions from vehicles. These efforts have confirmed that a small percentage of the fleet from each model year, called high emitters, contributes disproportional amounts to total emissions (Wayne and Horie 1983; Ashbaugh and Lawson 1991; Stedman et al. 1994; Bishop et al. 1999, 2000; Popp et al. 1999; Pokharel et al. 2000). (See NRC [2001] for more details on in-use measurements.) High-emitter vehicles have either been tampered with or have emission-control systems that have rapidly deteriorated or malfunctioned. Although emissions tend to increase with vehicle age, remote-sensing data have shown in the past that high emitters from newer model years may emit as much or more than the majority of older model year vehicles (Ashbaugh and Lawson 1991; Stedman et al. 1994; Bishop et al. 1999). A typical number reported in the literature is that approximately 10% of

---

[10] Emissions testing for heavy-duty on-road engines is reported in grams per brake horsepower-hour (g/bhp-hr).

the light-duty fleet emits 50-60% of the exhaust emissions. Pokharel et al. (2003) speculate that the combination of more durable components in newer vehicles and advanced technologies, such as on-board diagnostics, might be reducing the number of high emitters in newer generations of vehicles. Variability and uncertainty of vehicle emissions factors, especially in high-emitting, poorly maintained vehicles, remain important uncertainties in the estimation of vehicle fleet emissions.

On-Road Vehicle Activity

On-road vehicle activity is typically estimated in units of VMT. A widely used approach for calculating on-road VMT in urban areas is travel-demand modeling, which involves estimating the trips on each road in a transportation network by modeling travel requirements and preferences of the population. The travel-demand modeling results are provided by local metropolitan planning agencies. Travel-demand models can provide VMT for individual road segments by hour or multihour time periods (for example, a morning peak). Travel-demand models typically estimate total VMT across all vehicle types; VMT by vehicle type can then be determined from vehicle registration data in combination with assumptions about average mileage accumulation by vehicle type and age. The VMT from travel-demand models can be compared with estimates derived by traffic counters, thus improving model accuracy. Still, there is uncertainty in estimates which, in combination with uncertainty in emissions factors, leads to uncertainty in the estimated emissions.

Emissions Modeling

EPA has developed the MOBILE model for estimating on-road mobile-source emissions in the United States. The model was established in the 1970s to estimate the variability in vehicle emissions factors across the fleet. The estimates were then used to estimate changes in emissions resulting from planned changes in regional transportation.[11] MOBILE has since evolved into a tool to develop emissions inventories ranging from yearly, nationwide, fleetwide estimates to hourly, roadway level,

---

[11] Emissions that result from planned transportation projects may not counter any emissions reductions that the region requires to improve air quality (also referred to as conformity requirements).

vehicle-class-specific estimates. The most recent version, MOBILE6.2, released by EPA in 2004 (EPA 2003b), models the variation in emissions factors across roadway type, vehicle type, vehicle age, driving speeds, ambient conditions, fuel types, and different types of emissions-control programs.

Vehicle emissions rates are typically estimated as a function of a zero-mileage, new-car emissions factor and the deterioration of the emissions factor over time. The deterioration rates are determined from statistical analyses of extensive test programs on in-use vehicles and data from vehicle inspection and maintenance (I/M) programs. Adjustments are then made for in-use conditions, such as speed (by roadway type), ambient temperature and humidity, local fuel characteristics and regulations, and other control programs. These adjustments are typically based on test data under a broad range of conditions. (See NRC [2000] for a more complete description of how emissions rates are estimated.)

MOBILE is used to estimate historical emissions factors and future factors that are expected from possible changes in regulations, such as changes in vehicle emissions standards, changes in fuel composition, and I/M programs. California developed a similar model, EMFAC, to estimate on-road mobile-source emissions; the latest version of this model is EMFAC 2002. Although the EMFAC 2002 and MOBILE6.2 models are conceptually similar, there are a number of important differences:

- EMFAC provides emissions factors (for example, g/mi) as well as mass emissions (for example, tons/day). Activity estimates for all mobile emissions sources by county are incorporated into the model. MOBILE6.2 estimates emissions factors only; mass emissions are estimated outside the model with activity estimates derived from local data.
- EMFAC models California emissions standards in the California fleet and uses a larger number of vehicle classes, especially for later model years for which California's low-emitting vehicle (LEV) standards are in place.
- MOBILE6.2 estimates emissions factors by vehicle class and roadway type; EMFAC 2002 estimates emissions factors and emissions by vehicle class but not by roadway type.
- MOBILE6 estimates ammonia emissions factors and mobile-source HAPs emissions factors; EMFAC does not.
- MOBILE6 allows the user to estimate the benefits of equipping fleets with natural gas; EMFAC does not.

Uncertainties in Emissions Modeling

MOBILE is widely used in air quality management, although MOBILE and other mobile-source emissions models have important limitations. Underestimation of mobile-source emissions inventories, especially for CO and HCs, by a factor of two or more have been described (Ingalls 1989; Fujita et al. 1992; see also references in NRC [1991] and NRC [2000]). NRC (2000) identified several important fleet features for which test data were either insufficient or not treated adequately in mobile-source emission-estimation models. These features include the following:

- Real-world changes in emissions with vehicle age and maintenance history compared with accelerated aging typically used in testing for both older and newer vehicles.
- Emissions factors, activity levels, and population of high-emitting vehicles and how these vary by vehicle age, model, and geographic location.
- Emissions factors for heavy-duty diesel vehicles and nonroad vehicles.
- Emissions factors for HAPs and PM from all mobile sources.

The review also found that the lack of data has resulted in poor evaluation of the MOBILE model's performance, and that the uncertainty in the model is not quantified or understood adequately (NRC 2000). EPA's sixth-generation model of MOBILE (MOBILE6) is based on analyses of more data from EPA, CARB, and manufacturer exhaust and evaporative test programs, as well as data from in-use emissions measurements. EPA has recently released a draft version of their next generation of mobile-source model called "motor vehicle emission simulator" (MOVES).[12]

On-Road Mobile Emissions Inventory

Figures 2-13 and 2-14 show historical trends in on-road mobile CO, NO_x, and VOC (or ROG) emissions by category for the United States and California, respectively. The figures show substantial reductions in

[12] Accessed July 18, 2005, EPA 2005d.

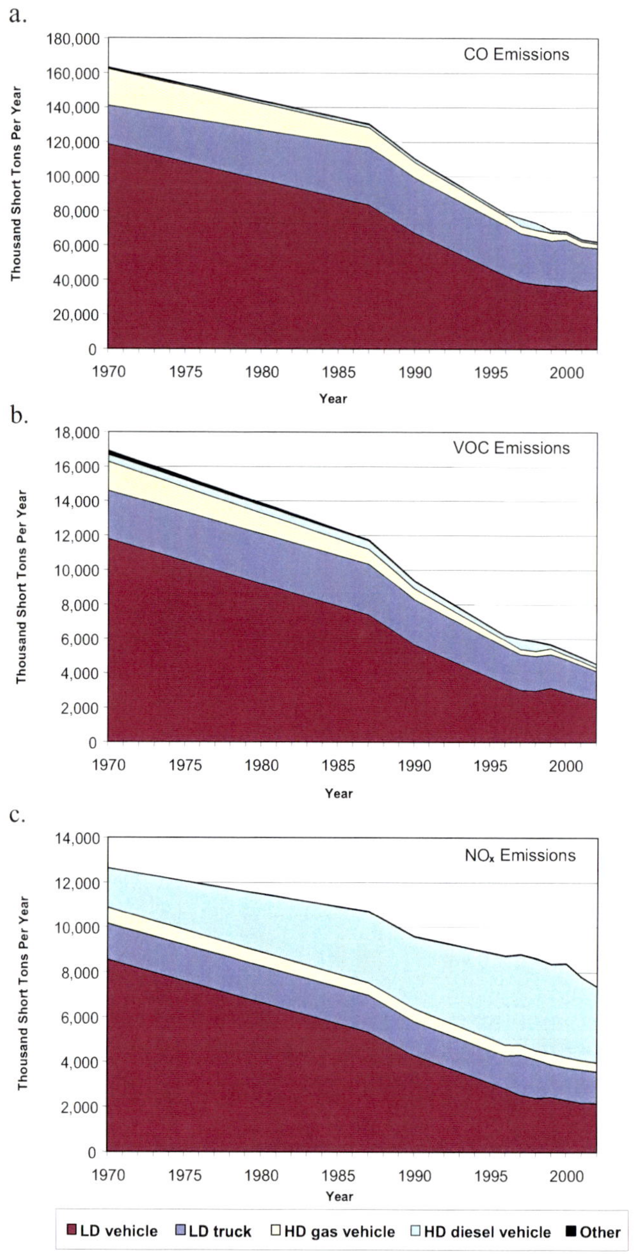

FIGURE 2-13 Historical trends in mobile-source (a) CO, (b) VOC, and (c) NO_x emissions in the United States by vehicle class, 1970-2002. Source: EPA 2005c.

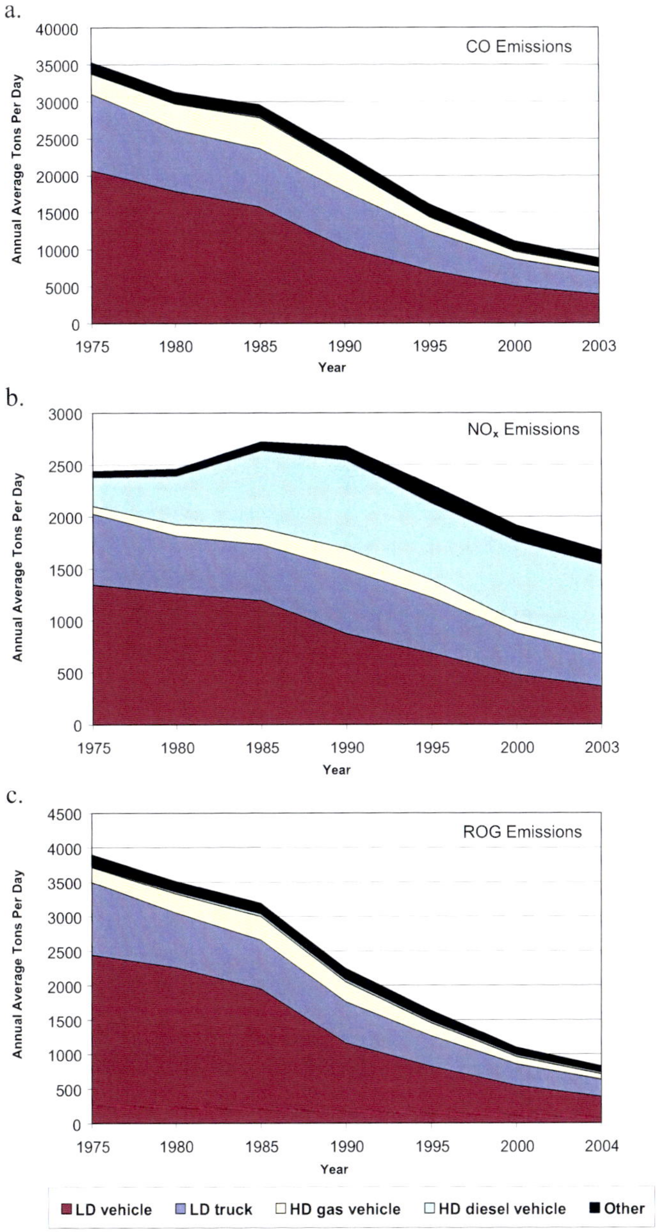

FIGURE 2-14 Historical trends in mobile-source (a) CO, (b) NO_x, and (c) ROG emissions in California by vehicle class, 1975-2003 (2004 for ROG). Source: CARB 2005b,c.

TABLE 2-4 Percent Change in Annual Emissions from 1980 to 2000 for NO_x, VOC, and CO, and Percent Change in Annual Vehicle Miles Traveled from 1980 to 2000

	Emissions	National	California
	NO_x	-27 %	-22 %
	VOC	-62 %	-69 %
	CO	-53 %	-65 %
VMT		+ 80 %	+ 105 %

Sources: Davis and Diegel 2004 (vehicle miles traveled data); CARB 2005b, EPA 2005c (emissions data).

on-road light-duty vehicle emissions throughout the nation for all three pollutants since 1970, reductions that can be attributed to the regulation and control of emissions from these sources. Total NO_x emissions have declined less than CO and VOC or ROG emissions because of the increase in diesel heavy-duty vehicle emissions resulting from the rapid increase in heavy-duty vehicle VMT that occurred in this period and from less stringent regulations compared with light-duty vehicles for most of this period. The figures highlight how emissions from light-duty trucks and diesel heavy-duty trucks became a more significant fraction of the on-road emissions inventories as light-duty vehicle emissions have been reduced. Table 2-4 shows the percent change in annual on-road emissions and VMT between 1980 and 2000 in the United States and in California. The table highlights the significant decreases in on-road emissions despite increases in vehicle activity. The uncertainties related to emissions modeling and VMT estimates discussed above apply to these emissions estimates.

NONROAD MOBILE SOURCES AND EMISSIONS

Nonroad Vehicle and Equipment Types and Emissions Estimation

Nonroad sources comprise a wide variety of vehicles and equipment with many different uses as listed in Table 2-5. Nonroad vehicle and equipment population and activity are not known as accurately as they are for on-road vehicles because only some sources, such as recreational boats and snowmobiles, are registered, and fewer data have been collected on nonroad source activity. Figure 2-15 shows estimated U.S.

TABLE 2-5 Nonroad Mobile-Sources Categories

- Agricultural equipment, such as tractors, combines, and balers.
- Aircraft, including both jet and internal combustion engines.
- Airport ground support equipment, such as terminal tractors.
- Commercial and industrial equipment, such as forklifts and sweepers.
- Construction and mining equipment, such as graders and backhoes.
- Lawn and garden equipment, such as leaf blowers and lawn mowers.
- Locomotives and switching and line-haul trains.
- Logging equipment, such as shredders and large chain saws.
- Pleasure craft, such as powerboats and personal watercraft.
- Commercial marine vessels.
- Railway maintenance equipment, such as rail straighteners.
- Recreational equipment, such as all-terrain vehicles, off-road motorcycles, and snowmobiles.
- Oil-field and underground-mining equipment, such as mechanical drilling engines.

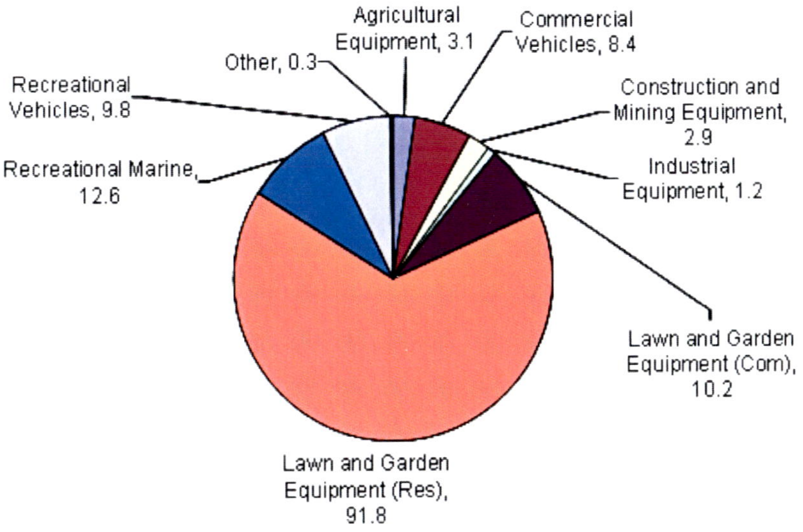

FIGURE 2-15 National 2002 estimates for nonroad equipment population by source category (in millions). The total population for all source categories is approximately 140 million. The basic equation for estimating emissions is the NONROAD model is the following: Emissions = Pop × Power × LF × A × EF, where Pop = engine population, Power = average power (hp), LF = load factor (fraction of available power), A = activity (hr/yr), and EF = emission factor (g/hp-hr). Source: Data from EPA 2006.

nonroad populations by equipment type for 2002. The estimates in Figure 2-15 are from the NONROAD2004 model that EPA used to inventory and estimate emissions from nonroad sources other than aircraft, locomotive, and commercial marine sources (EPA 2006).[13] Equipment populations and activity data in the NONROAD model are largely based on proprietary data from Power Systems Research (PSR), a market research company (EPA 2004c). PSR provides national population estimates, and the NONROAD model estimates county populations for each equipment type by using socioeconomic and other spatial allocation surrogates. Because of the uncertainty in the county population and activity estimates, EPA encourages state and local agencies to perform surveys, or use other measures to derive local estimates. Such surveys are costly, however, and only a few have been performed.

Of the 140 million estimated nonroad engines, 66% are associated with residential lawn and garden equipment, which are dominated by small spark-ignition engines. Ninety-five percent of all engines in the NONROAD model use gasoline, 4% use diesel, and less than 1% use compressed natural gas (CNG) or liquefied petroleum gas (LPG).

Nonroad Emissions Inventories

As shown in Figure 2-10, nonroad source emissions are an increasing fraction of total emissions of CO, NO_x, and VOCs in the United States. Figures 2-16 and 2-17 show historical trends in nonroad mobile-source emissions by major category for the United States and California, respectively. The figures show the diversity of sources that contribute to the inventories of all three pollutants. Lawn and garden equipment, recreational land, and recreational marine equipment are the three largest sources of CO and VOC or ROG nonroad inventories nationwide and in California. In contrast, nonroad NO_x emissions are dominated by locomotive, commercial marine, construction, and mining equipment. The figures also show that all three pollutants' emissions rose until the early to mid-1990s, owing to limited regulation before 1990, and that reductions were achieved in California earlier than in the entire nation. The uncertainties related to emissions and activity estimates discussed in relation to on-road sources also apply to these emissions estimates.

[13] In December 2005, EPA released the final NONROAD2005 model. CARB uses their OFFROAD model to estimate nonroad emissions in California.

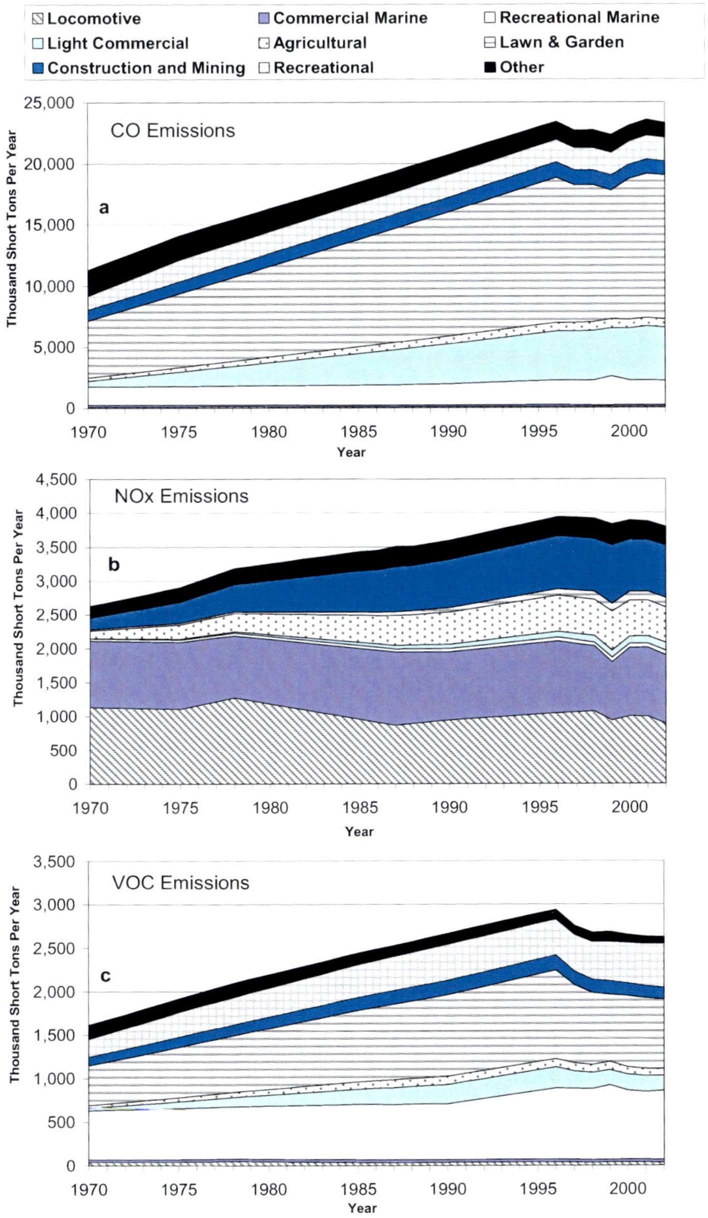

FIGURE 2-16 Historical trends in nonroad sources of (a) CO, (b) NO_x, and (c) VOC emissions in the United States by equipment class, 1970-2002. Source: EPA 2005c.

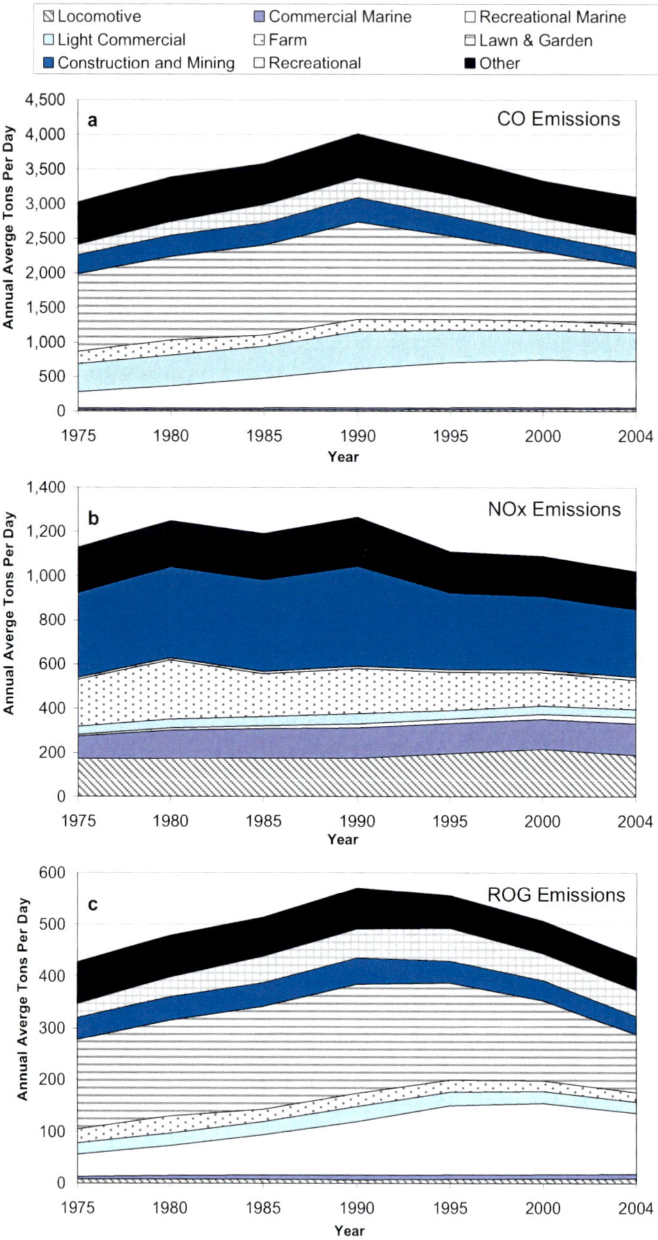

FIGURE 2-17 Historical trends in nonroad sources of (a) CO, (b) NO_x, and (c) ROG emissions in California by equipment class, 1975-2004. Source: CARB 2005b.

LINK BETWEEN EMISSIONS AND AIR QUALITY

Factors Influencing Variability in Air Quality

Mixing, transport, chemical reaction, and physical removal of emitted pollutants directly influence the ambient concentrations to which populations are ultimately exposed. These processes are complex, given the dynamic nature of the atmosphere and the numerous gas and particle air pollutants with different physical and chemical properties. They can also vary significantly from region to region, leading to different ambient pollution levels in regions that have similar emissions.

Meteorology and topography influence air pollutant concentrations through effects on vertical mixing, wind speeds, temperature, humidity, and emissions. Atmospheric inversions occur when the temperature of the atmosphere increases with altitude, greatly reducing vertical mixing in the atmosphere. Combined with low wind speeds, inversions prevent air circulation because colder air is trapped near the ground by warmer air above. The California South Coast Air Basin, Fairbanks (Alaska), and Denver (Colorado) are examples of urban areas where vertical and horizontal transport is limited by topography and meteorology (NRC 2003). Such conditions allow pollutants to accumulate and enhance chemical and photochemical transformations due to longer residence times of precursors. In regions with limited ventilation, exceedances of the NAAQS can occur at emissions concentrations that generally do not lead to NAAQS exceedances in other regions.

Another phenomenon that can cause regional differences in ambient air pollution concentrations is the long-range transport of pollutants. The northeastern United States experiences increased air pollutant concentrations in part because of long-range precursor and pollutant transport from industrialized regions in the Midwest. Regions affected by long-range transport of pollutants might need to reduce local emissions to a greater extent to meet the NAAQS than areas not affected by regional transport.

Spatial and Temporal Variability in Mobile-Source Emissions

In addition to the factors discussed above, variability in emissions inventories may affect variability in air quality between and within regions. On-road and nonroad mobile sources are significant contributors to emissions inventories in all major urban areas. Figure 2-18 shows that

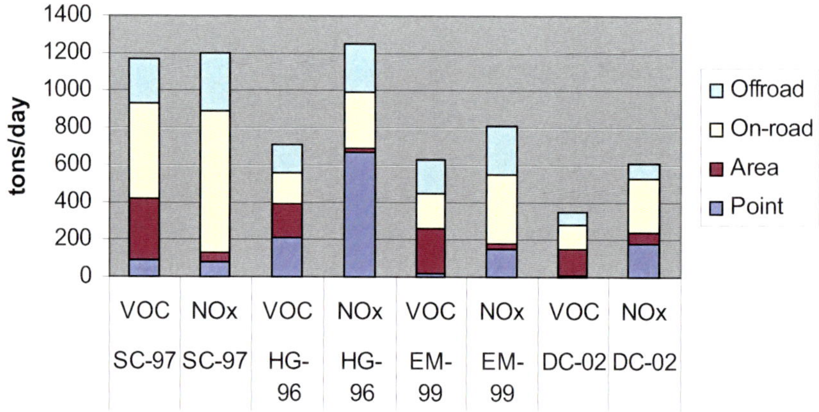

FIGURE 2-18 Representative emissions inventories (tons/day) from areas that are in nonattainment of the ozone NAAQS, based on 1-hr averaged concentrations. The California South Coast (SC) and Houston (HG) have average daily inventories. Eastern Massachusetts (EM) and District of Columbia (DC) areas have "ozone season" daily inventories. See Table 2-2 for definitions of emission-source categories in legend. Sources: MADEP 2002; TNRCC 2002; MWCG 2003; SCAQMP 2003.

the fraction of total emissions due to mobile sources varies significantly from region to region. For the four regions reported in the Figure, the fraction of total emissions due to mobile sources ranges from 45% to 65% for VOCs and 45% to 89% for NO_x. Factors influencing regional differences in mobile emissions include population, activity patterns, regional characteristics of the fleet (the distribution of vehicle types and ages), and control programs in place.

The spatial variation of some primary pollutants within a single region can follow the spatial pattern of emissions, including mobile-source emissions. Figure 2-19 shows how CO, black carbon (similar to elemental carbon), and particle number vary in close proximity to a major highway (Zhu et al. 2002). Such primary pollutants emitted from motor vehicles will not have uniform concentrations across a region, and spots with higher concentrations of ambient pollutants can occur at discrete locations. These locations often are in places with high vehicle traffic, although topographical and meteorological conditions also play a role.

Mobile-source emissions also vary by day of week and throughout the day, which influences ambient concentrations of primary pollutants,

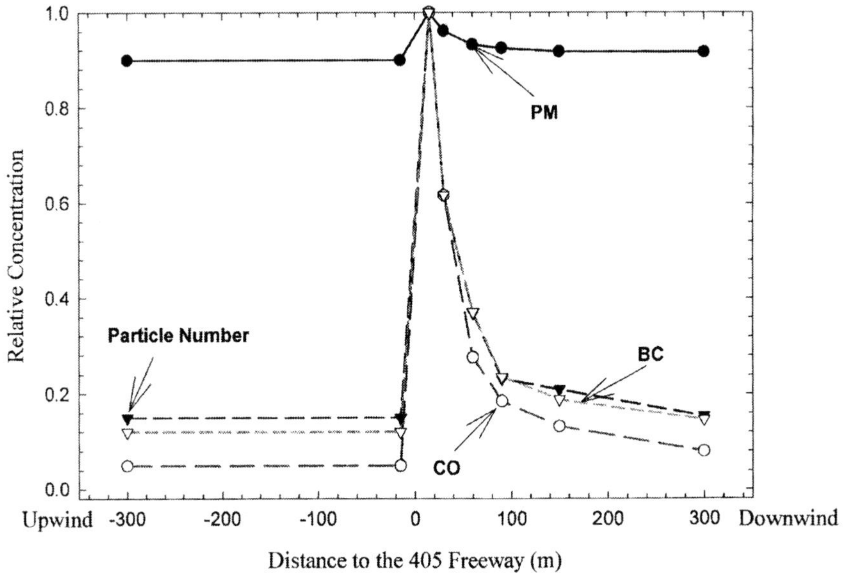

FIGURE 2-19 Total particle number, black carbon (similar to elemental carbon), and CO concentrations versus downwind distance from a freeway. Source: Zhu et al. 2002. Reprinted with permission; copyright 2002, Air & Waste Management Association.

such as CO, and secondary pollutants, such as ground-level ozone. In areas influenced by commuter traffic, emissions from light-duty vehicles typically peak during the morning and afternoon rush hours. Minimal ozone concentrations in urban areas on weekday mornings are common and are attributed to commuter traffic NO emissions that chemically remove ozone, as described previously. Ozone has also been observed to be as high or higher on weekend days than on weekdays in some urban areas, which is known as the weekend ozone effect. (See recent review by Heuss et al. [2003], and analyses by Pun et al. 2003, and Fujita et al. [2003], Blanchard and Tanenbaum [2003, 2005]). These analyses have also found that NO_x and, in some cases, CO concentrations are significantly lower on weekends than weekdays. One explanation for reduced precursor concentrations on weekends is reduced emissions on weekends, particularly mobile-source NO_x (Chinkin et al. 2003). Because NO_x can result in both ozone formation and removal as well as secondary PM formation, there is some debate as to the effectiveness of NO_x emissions reduction for ozone and PM control (Croes et al. 2003; Lawson 2003).

HAZARDOUS AIR POLLUTANTS

HAPs (also called air toxics) are compounds or compound groups that may cause serious health effects, even at low concentrations. Unlike criteria pollutants, HAPs do not have NAAQS. In the 1990 CAA, Congress mandated that EPA regulate a list of approximately 190 HAPs. The CAA amendments further direct EPA to assess the need and feasibility for emissions standards of HAPs from mobile sources and to regulate these emissions as necessary (CAA section 202(l)). In 2001, EPA issued a rule that identifies 21 HAPs associated with mobile sources, including benzene, 1,3-butadiene, formaldehyde, acetaldehyde, and total PM from diesel exhaust. The EPA rule projects that current mobile emissions standards and fuel formulations that address criteria pollutants are sufficient mitigation strategies for HAPs. Thus, there are no separate federal standards for mobile-source HAPs; however, reducing HAP benefits regulation of criteria pollutants and their precursor emissions from mobile sources. Ambient concentrations of HAPs have not been monitored for as long as criteria pollutants. Ambient concentrations of benzene, an important mobile-source HAP, from 95 urban locations decreased by 47% on average from 1994 to 2000 (EPA 2003a). It should be noted that indoor sources also contribute to exposure of some HAPs.

AIR QUALITY HEALTH EFFECTS FROM EXPOSURE TO MOBILE SOURCES

Effects on health resulting from exposure to air pollutants depend on a large number of variables, including the contaminants present during the exposure, the toxicity of the contaminants, their concentrations and durations of exposure, the dose, and the health status of the person exposed. Exposure to mobile-source air pollutants typically occurs within the context of exposure to a host of air pollutants from a host of sources, and health effects are probably due to exposure to the mixture and not to any one contaminant or source. However, health effects related to exposure to several key constituents of the mobile-source emissions mixture are well understood. There also is research correlating air pollution health impacts to proximity to major roadways, although an assessment of the research is beyond the scope of this committee.

Carbon Monoxide

The toxic properties of CO have been extensively characterized and are directly related to the ability of CO to competitively bind to the heme group of hemoglobin. The binding mechanism of CO is identical to that of O_2, but the affinity of CO for heme is 234 times as great as that of O_2. As a result, when CO is present, it is more likely to bind and remain attached to hemoglobin, creating the potential for hypoxia if the concentrations are high enough (Townsend and Maynard 2002). The acute toxic effects of CO are well known and range from headache and shortness of breath at low percentage of hemoglobin bound to CO (%COHb) to death when the %COHb reaches 50% to 90%. Typical ambient concentrations of CO generally do not produce acute toxic effects; however, in the NAAQS nonattainment area of Southern California, a relationship between chronic exposure to CO and birth outcomes was reported (Ritz et al. 2002). CO at ambient concentrations is also hypothesized to affect the health of persons suffering from cardiovascular disease. For persons with preexisting heart disease, exposure to CO at low concentrations has been shown to have an impact on cardiac function (Allred et al. 1989). Although the evidence is inconsistent, studies of the relationship between exposure to ambient air pollutants and cardiovascular disease have suggested an association between some outcomes and CO (Morris et al. 1995; Schwartz and Morris 1995).

CO is unique among the criteria air pollutants because of its significance for both ambient air quality management and public safety. Outdoor exposures to CO at very high and lethal concentrations were reported from motor boat exhaust (CDC 2004), and from farm equipment (CDC 1997). Control of CO through new-vehicle emissions standards has had a significant collateral public-safety benefit through the reduction of accidental CO poisoning (Cobb and Etzel 1991; Shelef 1994; Marr et al. 1998). For example, Mott et al. (2002) discussed how exposure to motor vehicle CO emissions results in a substantial number of accidental deaths and estimates the number of accidental CO poisonings that have been prevented as a result of more stringent CO emissions standards.

Oxides of Nitrogen

Evidence for health effects associated with exposure to NO_2 remains inconclusive. The health effects of exposure to NO_x are largely related to exacerbation of symptoms of respiratory disease. Short-term exposures (for example, less than 3 hr) to low concentrations of NO_2 might result in changes in airway responsiveness and lung function in persons with preexisting respiratory illnesses. Such exposures might also increase respiratory illnesses in children. Long-term exposures to NO_2 might result in increased susceptibility to respiratory infection and might cause irreversible alterations in lung structure (EPA 2004d).

Hydrocarbons

A broad range of VOCs have been identified in both diesel and gasoline emissions. Individual VOCs found in the emissions mixture are themselves toxic. Although health effects of exposure to the entire mixture have not been characterized, the toxicities of many of the components of the mixture are well understood. The health effects of a few mobile-source organic compounds that are considered HAPs are described below.

1,3 Butadiene

1,3-Butadiene is generated by incomplete combustion of gasoline and diesel fuel. Butadiene is a mildly irritating gas that can cause neurological symptoms at high exposure concentrations. Epidemiological studies of workers exposed to butadiene in rubber plants found an increased risk for cardiovascular disease (ATSDR 1993). Animal studies found developmental and reproductive effects related to inhalation exposure (EPA 2002a). Butadiene exposures were also associated with the development of leukemia and other lymphomas in both epidemiological and animal studies, and EPA has classified butadiene as a human carcinogen (EPA 2002a).

Benzene

Both on-road and nonroad gasoline-powered mobile sources contribute benzene to the ambient air. Benzene is a known human carcinogen and is associated with the development of leukemia. Neurological symptoms of inhalation exposure to benzene include drowsiness, dizziness, headaches, and unconsciousness in humans (EPA 2005e).

Formaldehyde

Formaldehyde is a nearly colorless gas with a pungent, irritating odor even at concentrations below 1 ppm. Formaldehyde is an eye, skin, and respiratory tract irritant. Inhalation of vapors can produce narrowing of the bronchi and an accumulation of fluid in the lungs. The systemic effects of formaldehyde may include metabolic acidosis, circulatory shock, respiratory insufficiency, and acute renal failure. Formaldehyde is a potent sensitizer at high concentrations and a probable human carcinogen (ATSDR 2004).

Benzo(*a*)pyrene

Acute effects of benzo(*a*)pyrene at increased concentrations potentially include red-blood cell damage, resulting in anemia and a suppressed immune system. Long-term exposure to benzo(*a*)pyrene might result in developmental and reproductive effects. It was classified as a carcinogen by the International Association of Cancer Research (IARC 1973).

Particulate Matter

Epidemiological studies over the last several years consistently demonstrated a statistical relationship between exposure to particles and cardiopulmonary morbidity and mortality (Pope et al. 1991; Dockery et al. 1993; Samet et al. 2000; Pope et al. 2002; Metzger et al. 2004). Toxicological studies exploring specific mechanisms of injury from exposure to PM supported the epidemiological findings (Ghio et al. 2000; Frampton et al. 2002; Utell et al. 2002; Devlin et al. 2003). For persons suffering from chronic obstructive pulmonary disease (COPD), which includes chronic bronchitis, emphysema, and some cases of chronic asthma, exposure to PM can induce inflammatory responses resulting in exacerbation of symptoms. Adverse physiological effects in individuals diagnosed with cardiovascular disease have been associated with increased PM exposure. These effects include increased blood pressure, cardiac arrhythmias, increased oxidative stress and inflammation, and progression of atherosclerosis. Pope et al. (2002) also found a statistically significant increase in the risk of developing lung cancer associated with exposure to particulate air pollution.

Studies examining health effects related to exposure to PM from mobile sources have primarily focused on diesel exhaust. Diesel exhaust, which includes hundreds of organic compounds and includes particles in the ultrafine size range, was designated as probable human carcinogen by IARC (1989) and EPA (2003c). Some studies suggested that ultrafine particles, which do not contribute substantially to the total particle mass, might carry a greater risk per weight than particles of other sizes (HEI 2002). Several specific components of mobile-source particles were shown to be toxic. A large number of polycyclic aromatic hydrocarbons (PAHs) have been identified in the exhaust of both diesel- and gasoline-powered vehicles, including benzo(*a*)pyrene. Additional carcinogenic compounds, such as dioxins, have also been identified in the exhaust stream from both diesel and gasoline vehicles. Measurements of TCDD (2,3,7,8-tetrachlorodibenzo-*p*-dioxin) ranged from about 2 picograms (pg) of toxicity equivalent quotient per kilometer (TEQ/km) to 5,100 pg TEQ/km for a diesel light-duty vehicle (HEI 2002). A recent study of exhaust particles plus semi-VOCs showed no difference in the toxicity of gasoline and diesel exhaust from light-duty vehicles, and more potent on an equivalent mass basis from high-emitter vehicles than normal emitters (Seagrave et al. 2002).

Ozone

Although ozone is not a component of mobile-source emissions, components of the emission mixture result in the formation of ozone. Ozone is the product of photochemical processes involving VOCs and NO_x, and is formed in greatest abundance during summer months when sunlight is strongest. Ozone is a known respiratory irritant and can cause inflammation of the airways resulting in breathing restrictions. For persons suffering from respiratory illnesses, such as COPD and asthma, exposure to ozone can be especially problematic. Significant association between daily variations in ambient ozone concentrations and adverse health outcomes, such as lung function decrements, aggravation of pre-existing disease, increases in hospital admissions and emergency room visits for respiratory symptoms, and increases in mortality have been found by a large number of epidemiological studies (Thurston and Ito 1999). A recent study has also found an association between short-term changes in ozone and short-term mortality for 95 large urban areas in the United States, although the potential for statistical confounders is a concern (Bell et al. 2004).

CONCLUSIONS

Substantial progress has been made over the past few decades in reducing air pollutant emissions from many sources, including mobile sources. However, some locations in the Unites States continue to experience ambient concentrations of criteria pollutants above the NAAQS. Further improvements in air quality will be needed, particularly to attain the recently adopted standards for fine particulate matter and ozone. Evidence presented in the chapter suggests that mobile-source emissions contribute to poor air quality and have important health effects. The following conclusions are drawn based on the evidence:

- Mobile sources, both on-road and nonroad, are major sources of precursors for ground-level ozone and $PM_{2.5}$. Estimates of emissions suggest that mobile-source emissions are composed of approximately 50% anthropogenic NO_x and VOC emissions inventories and approximately 75% anthropogenic CO inventories. On-road light-duty vehicles are still the largest contributor to total mobile emissions, even though their emissions have been decreasing despite increases in vehicle activity. Nonroad emissions have remained relatively constant since 1970. As a result, the relative amount of nonroad to on-road emissions has been increasing over this period.
- There are many uncertainties in the methods and data used to estimate mobile- source emissions and thus in the estimates themselves. The technical, research, and regulatory community continue to fill data gaps, thereby improving the methods.

- The contribution of mobile-source emissions to air pollution varies from area to area. Ambient concentrations of pollutants depend on several factors, such as the relative mix of sources, the extent of pollutant transport, the meteorology, and the topography of the area. The result is that different levels of controls on mobile and nonmobile sources are typically required for different areas in the Unites States.

3

Regulation of Emissions from New Mobile Sources

The federal Clean Air Act (CAA) makes state and the federal governments partners in regulating air pollution. The CAA explicitly requires that control of air pollution be primarily the responsibility of states and local governments (CAA § 101(a)(3) [1990], 42 USC § 7401(a)(3) [2005]). To that end, section 116 of the act affirms the general authority of states to adopt or enforce (1) any standard or limitation of air pollutant emissions or (2) any requirement related to the control or abatement of air pollution. An exception to the presumption in favor of state authority to meet federal standards in section 116 is for state regulation of emissions from new[1] mobile sources. Thus, in contrast to federally mandated state control over stationary sources, regulation of new mobile-source emissions has been principally a federal project (Engine Mfrs. Ass'n v. EPA, 88 F.3d 1075, 1079 [D.C. Cir. 1996]). There is one important exception—the ability of California to set separate standards. This chapter describes how the federal government and California set emissions standards for new mobile sources and how the emissions reductions from those standards are incorporated into air quality management.

---

[1] The term "new mobile sources" is used throughout the report to refer to newly manufactured sources not yet in use by the consumer.

THE EVOLUTION OF THE EXISTING STATUTORY FRAMEWORK

Motor Vehicle Air Pollution Control Act of 1965

Congress first addressed the need for emissions controls for motor vehicles in the Motor Vehicle Air Pollution Control Act of 1965 (Pub. L. No. 89-272, § 202(a), 79 Stat. 992, [1965]). This act includes section 202 that authorized the federal government to set "standards, applicable to the emission of any kind of substance, from any class or classes of new motor vehicles or new motor vehicle engines." Congress delegated this new standard-setting authority to the U.S. Department of Health, Education, and Welfare (HEW). (The U.S. Environmental Protection Agency [EPA] would not be established until 1970, 5 years later). Congress's provision for national emissions standards was based primarily on testimony by the assistant secretary of HEW about the potential problems that would be created for vehicle manufacturers by divergent state standards:

> The problem of automobile exhaust cannot be solved on a local or State basis. I would think, if I were in the automobile manufacturing business, that is absolutely the last thing I would hope to see happen where Delaware would pass this kind of legislation, and the District of Columbia this, and Maryland that, and Pennsylvania something else. You would go out of your mind, if you were trying to design devices or engines to meet these varying standards (Hearings before the Subcommittee on Air and Water Pollution of the Senate Comm. on Public Works, 89th Cong., 1st Sess. 33 [1965]; testimony of James M. Quigley, Assistant Secretary of HEW.)

Clean Air Act of 1967

Congress revisited the issue 2 years later in 1967. In congressional hearings, evidence was presented that California had adopted its own state vehicle emissions standards and that several other states were in the process of following California's lead and preparing to adopt their own state standards (Air Pollution—1967 [Automotive Air Pollution]: Hearings before the Subcommittee on Air and Water Pollution of the Senate Comm. on Public Works, 90th Cong., 1st Sess. 395-399 [1967]; NCAPC

1967; Currie 1970). Motor vehicle manufacturers argued that the nature of their manufacturing required a single national standard to eliminate undue economic strain on the automobile industry (S. Rep. No. 403, 90th Cong., 1st Sess. 33 [1967]). HEW testified that divergent state standards would ultimately result in confusion and called upon Congress to explicitly preempt state vehicle emissions standards (Air Pollution—1967 [Automotive Air Pollution]: Hearings before the Subcommittee on Air and Water Pollution of the Senate Comm. on Public Works, 90th Cong., 1st Sess. 107 [1967]; testimony of Dean Coston, Deputy Undersecretary of HEW). Both chambers of Congress concurred. The Senate found that divergent state standards would result in economic disruption and increased costs to consumers (S. Rep. No. 403, 90th Cong., 1st Sess. 33 [1967]). The House elaborated that the nature of motor vehicle manufacturing required the consistency and certainty that could only be provided by uniform federal standards:

> The manufacture of automobiles is a complex matter, requiring decisions to be made far in advance of their actual execution. The ability of those engaged in the manufacture of automobiles to obtain clear and consistent answers concerning emission controls and standards is of considerable importance so as to permit economies in production (H.R. Rep. No. 728, 90th Cong., 1st Sess. 21 [1967]).

On the basis of these findings, Congress explicitly preempted states from adopting or enforcing new motor vehicle emission standards in the CAA of 1967 (Pub. L. No. 90-148, § 208, 81 Stat. 485, 501 [1967]). This preemption provision remains in effect today as section 209(a) of the CAA, which provides the following:

> No State or any political subdivision thereof shall adopt or attempt to enforce any standard relating to the control of emissions from new motor vehicles or new motor vehicle engines subject to this part (42 USC § 7543(a)[2003]).

Appendix D provides the complete text of section 209.

Although the initial congressional discussions in 1967 favored preempting new motor vehicle emissions standards in all 50 states, the California congressional delegation successfully persuaded their House and Senate colleagues to make a special exception for California because of its pollution problems and pioneering efforts in regulating vehicle emissions. Congress concluded that "although the situation may change, in the 15 years that auto emission standards have been debated and discussed, only the State of California has demonstrated compelling and extraordinary circumstances sufficiently different from the Nation as a whole to justify standards on automobile emissions which may, from time to time, need to be more stringent than national standards" (S. Rep. No. 403, 90th Cong., 1st Sess. 33 [1967]). Moreover, California had begun regulating emissions from motor vehicles in 1957, almost a decade before the federal government began developing a national program. House of Representative John E. Moss stated that continuation of those "pioneering" efforts "offer a unique laboratory, with all the resources necessary, to develop effective control devices which can become a part of the resources of this Nation …" (113 Cong. Rec. 30975 [1967]).

The legislation eventually adopted by Congress in 1967 included a compromise provision that directed the secretary of HEW to waive the preemption of state standards, provided the conditions specified in section 209(b) of the statute are met (discussed below), for "any State which has adopted standards (other than crankcase emission standards) for the control of emissions from new motor vehicles or new motor vehicle engines prior to March 30, 1966" (Pub. L. No. 90-148, § 208(b), 81 Stat. 501). California was the only state to have adopted in final form such standards by March 30, 1966, thus being the only state to have the preemption exclusion purposely applied. Congress noted that the conditional exception for California was an acceptable compromise that maintains "that State's right to set more stringent standards to meet peculiar local conditions," while "the industry, confronted with only one potential variation, will be able to minimize economic disruption and therefore provide emission control systems at lower costs to the people of the Nation" (S. Rep. No. 403 at 33 [1969]). Moreover, according to Senator George Murphy, this compromise would permit California to "act as a testing agent for various types of control and the country as a whole will be a beneficiary of this research" (113 Cong. Rec. 32478 [1967]). Thus, as described by the nation's second highest court, Congress intended California to "act as a kind of laboratory for innovation" for mobile-source emissions control (Motor & Eqpt. Mfrs. Assn v. EPA, 627 F.2d 1095 [D.C. Cir. 1979]).

Clean Air Act Amendments of 1977

In 1970, Congress enacted the basic framework of today's CAA and kept in place the preemption of state motor vehicle emissions standards adopted in 1967 and the conditional exemption of California. Although the 1970 amendments enacted some tough new emissions standards for light-duty vehicles, which will be discussed in other sections in the report, it did not alter the preemption requirements. In 1977, when Congress next revisited the CAA it again retained the general structure of preempting state standards and reaffirmed the special exemption for California. As Congress explained, "California was afforded special status due to that State's pioneering role in regulating automobile-related emissions, which pre-dated the Federal effort. In addition, California's air pollution problem was then, and still appears to be, among the most pervasive and acute in the Nation" (H.R. Rep. No. 294, 95th Cong., 1st Sess. 301 [1977]).

Protectiveness Requirements

Congress made two important changes in 1977 in the mobile-source preemption scheme. First, Congress revised the criteria for a federal waiver of vehicle emissions standards for California, which were now administered by EPA, to require that the California standards be at least as protective as the applicable federal standards in the aggregate. In the original version of the waiver provisions adopted in 1967, Congress had required every California standard to be at least as protective as the equivalent federal standard. By 1977, however, Congress recognized that there were trade-offs in regulating emissions of different pollutants and that more stringent standards for one pollutant could necessitate less stringent standards for another pollutant (H.R. Rep. No. 294, 95th Cong., 1st Sess. 302 [1977]). Requiring the protectiveness of the California standards to be evaluated as a package permitted California "to weigh the degree of health hazards from various pollutants and the degree of emission reduction achievable for various pollutants with various emission control technologies and standards" (H.R. Rep. No. 294, 23 [1977]).

The revised waiver criteria, as amended in 1977, are now codified in section 209(b) of the CAA and provide the following:

> The Administrator shall, after notice and opportunity for public hearing, waive application of this section to any State which has adopted standards (other than crankcase emission standards) for the control of emissions from new motor vehicles or new motor vehicle engines prior to March 30, 1966, if the State determines that the State standards will be, in the aggregate, at least as protective of public health and welfare as applicable Federal standards. No such waiver shall be granted if the Administrator finds that—
>
> (A) the determination of the State is arbitrary and capricious,
> (B) such State does not need such State standards to meet compelling and extraordinary conditions, or
> (C) such state standards and accompanying enforcement procedures are not consistent with section 202(a) of this part.

The House Committee on Interstate and Foreign Commerce that drafted this amendment relaxing the California waiver criteria stated that the amendment was "intended to ratify and strengthen the California waiver provision and to affirm the underlying intent of that provision, i.e., to afford California the broadest possible discretion in selecting the best means to protect the health of its citizens and the public welfare" (H.R. Rep. No. 294, 301-302 [1977]). The House committee also made clear that EPA was to be highly deferential in reviewing California's waiver requests:

> The Administrator ... is not to overturn California's judgment lightly. Nor is he to substitute his judgment for that of the State. There must be clear and compelling evidence that the State acted unreasonably in evaluating the relative risks of various pollutants in light of the air quality, topography, photochemistry, and climate in that State, before EPA may deny a waiver (H.R. Rep. No. 294, 302 [1977]).

Adoption of California Standards by Other States

An even more fundamental change adopted by Congress in 1977 was a decision to allow other states to adopt the California standards. This amendment was one of several adopted by Congress in 1977 to provide a greater role and greater assistance for state and local governments in the administration of the CAA. Many states were having difficulty coming into compliance with the National Ambient Air Quality Standards (NAAQS) by the statutory deadlines and called upon Congress to give them greater flexibility to reduce emissions to meet the ambient standards, including the right to adopt the cleaner California vehicle emissions standards (H.R. Rep. No. 294, 309-310 [1977]). Congress expressed concern that the preemption provisions of section 209(a) were unduly restricting the capability of non-California states to obtain emissions reductions from new motor vehicles, which contributed to their inability to meet the NAAQS. As the House committee stated in 1977,

> The Committee is concerned that this preemption (section 209(a) of the Act) now interferes with legitimate policy powers of States, prevents effective protection of public health, limits economic growth and employment opportunities in nonattainment areas for automotive pollutants, and unduly stifles enforcement of present federal emission standards (H.R. Rep. No. 294, 309 [1977]).

Congress responded by enacting section 177 of the CAA 1977 amendments, which provided the following:

> Notwithstanding section 209(a), any State which has plan provisions approved under this part may adopt and enforce for any model year standards relating to control of emissions from new motor vehicles or new motor vehicle engines and take such other actions as are referred to in section 209(a) respecting such vehicles if—
>
> (1) such standards are identical to the California standards for which a waiver has been granted for such model year, and
> (2) California and such state adopt such standards at least two years before commencement of such model year (as determined by regulations of the Administrator).

Appendix D provides the complete text of section 177.

Although Congress gave states new authority to adopt and enforce the California standards, it emphasized that the concerns that motivated the general preemption of state vehicle emissions standards remained valid. Thus, the House Committee on Interstate and Foreign Commerce, which originally drafted and proposed the language that became section 177, stated,

> In 1967 amendments to the Clean Air Act, Congress preempted States other than California from establishing or enforcing new motor vehicle emission standards or test procedures. Congress' concern at that time was that vehicle manufacturers not be subject to 50 different sets of requirements relating to emission controls which would unduly burden interstate commerce. In the Committee's view, that concern remains a valid one today (H.R. Rep. No. 294 309 [1977]).

Similarly, in the Senate, the chief sponsor of the 1977 amendments to the CAA, Senator Edmund Muskie, stated,

> Congress recognized in 1965 that, as a national industry, automobiles required national emission regulation. Except for California, which is unique both from a product distribution and an air pollution point of view, the argument in 1967 for preemptive national standards was defensible. The underlying principle of national emission standards was, and should continue to be, that those national standards would be adequate to achieve health-related air quality standards in the areas with the most difficult problems. Statutory standards established in 1970 reflected that policy. This legislation continues that policy (Senate Debate on S. 252 [June 8, 1977] reprinted in 1977 Legis. Hist., 741).

To avoid imposing undue burdens on vehicle manufacturers, Congress imposed "strict limits" on the authority of states to adopt the California new motor vehicle emissions standards under section 177 (H.R. Rep. No. 294, 310 [1977]). The most important of the restrictions was that the California standards adopted by states must be "identical" to the California standards, a point repeatedly emphasized in the legislative history (Joint Explanatory Statement of the Committee of Conference, reprinted in 1977 Legis. Hist., 536; H.R. Rep. No. 294, 310 [1977]). By ensuring that other states adopted and enforced only standards identical to California's, Congress concluded that the new state authority under section 177 "should not place an undue burden on vehicle manufacturers who will be required, in any event, to produce vehicles meeting the California standards for sale in California" (H.R. Rep. No. 294, 310 [1977]).

Congress also specified that the use of the new state authority under section 177 was discretionary and that the states were not obligated to use this authority, nor could EPA mandate that a state exercise its section 177 authority to adopt the California standards. Thus, the House committee specified that it "intends these provisions as grants of authority to the States. They are not intended as requirements" (H.R. Rep. No. 294, 311 [1977]). The committee also prohibited EPA from requiring states to adopt California standards under section 177 (H.R. Rep. No. 294, 311 [1977]).

Clean Air Act Amendments of 1990

Third-Vehicle Prohibition

The most recent comprehensive amendments to the CAA were in 1990. Although the basic preemption framework of section 177 was retained, some conditions for adoption of California emission standards were added. Section 177 was amended by adding the following provision to its existing language:

> Nothing in this section or in title II of this Act shall be construed as authorizing any such State to prohibit or limit, directly or indirectly, the manufacture or sale of a new motor vehicle or motor vehicle engine that is certified in California as meeting California standards, or to take any action of any kind to create, or have the effect of creating, a motor vehicle or motor vehicle engine different than a motor vehicle or engine certified in California under California standards (a 'third vehicle') or otherwise create such a 'third vehicle' (Pub. L. No. 101-549, § 232, 104 Stat. 2399, 2529 [1990]).

This provision was added by the Conference Committee late in the process and was not included in the original bills passed by the House or Senate, nor was it considered or explained by either the relevant House or Senate committees. The conference committee explained that it was adopting additional restrictions in section 177 "to make plain that States exercising this section 177 option may not, in such adoption and enforcement, create a 'third vehicle' that is not a California vehicle or a 49-state Federal vehicle, because of the burden it would place on the motor vehicle manufacturers" (H.R. Conf. Rep. No. 952, 101st Sess., 2d Sess. 337 [1990]).

Nonroad Preemptions

In the 1990 amendments, Congress also enacted a revised section 209(e), which addresses state preemption of emissions regulations for nonroad engines. Before the 1990 amendments, nonroad engines were largely unregulated by EPA or the states (42 USC § 7543(e) [2003]). The 1990 amendments require EPA to study emissions from nonroad vehicles and, if certain statutory criteria are met, to proceed with promulgating emissions standards for such sources (42 USC § 7547 [2003]). Congress also enacted a new preemption provision in section 209(e) for nonroad engines that differs in important ways from the preemption of on-road mobile sources in section 209(a). For example, in section 209(e)(1), Congress explicitly preempts states, including California, from regulating emissions from two specific categories of nonroad sources: engines smaller than 175 horsepower (hp) used in construction or agriculture and new engines used in locomotives. For other nonroad sources, section 209(e)(2) implicitly preempts state emissions standards by requiring EPA to provide a preemption waiver for California standards and other emissions requirements similar to the waiver provided for motor vehicle emissions standards under section 209(b). Legislation was passed in 2003 that explicitly preempts states except California from adopting separate emissions standards for spark-ignition engines smaller than 50 hp, such as those used in lawn and garden equipment. California maintains its ability to set standards for these engines, but other states may not adopt California's standards. The complete text of this legislation is provided in Appendix D.

Another difference is that section 209(a) preempts "standards relating to the control of emissions from new motor vehicles," section 209(e) preempts any state or local "standard *or other requirement* relating to the control of emissions from non-road sources" (emphasis added) (42 USC § 7543(e)(2) [2003]). Just as section 177 permits non-California states to adopt new motor vehicle emissions standards "identical" to California's, section 209(e)(2)(B) allows other states to adopt and enforce California regulations for nonroad emissions but with two differences. First, although the "identicality" (identical-standard) requirement of section 177 applies only to standards, implementation, and enforcement are explicitly required to be identical to California's for nonroad emissions (42 USC § 7543(e)(2)(B)(i) [2003]). Second, unlike section 177, a state choosing to adopt California's standards for nonroad sources must notify the EPA administrator of its decision, although EPA still has no approval role in the adoption of the state standards (42 USC § 7543(e)(2)(B) [2003]).

ADMINISTRATIVE AND JUDICIAL INTERPRETATIONS

Congress is the ultimate decision maker on the preemption of state standards and, as discussed above, has enacted and fine-tuned the federal framework for preemption of state mobile-source emissions standards. Over the years, Congress has not directly addressed several issues that have arisen about the scope and limits of the federal preemption. EPA, the federal agency charged by Congress to administer the CAA, and the federal courts have had to resolve those issues based on their interpretation of congressional intent. Some of the more important of these administrative and judicial rulings are summarized below.

The Scope of Preemption under Section 209(a) of the 1990 CAA Amendment

The preemption of state mobile-source emissions in section 209(a) has been described by the courts as the cornerstone of the CAA's regulation of mobile-source emissions (Motor Vehicle Mfrs. Ass'n v. NY Dept. Envtl. Conservation, 17 F.3d 521, 526 [2d Cir. 1994]; Engine Mfrs. Ass'n v. EPA, 88 F.3d 1075, 1079 [D.C. Cir. 1996]). Consistent with that understanding, the courts have generally construed the scope of the preemption in section 209(a) broadly. However, the courts have also recognized that the federal preemption provisions detailed in sections 177 and 209 represent an attempt by Congress to balance the interests of states to attain healthful air and the interests of vehicle manufacturers and their customers from being unduly burdened by a patchwork of divergent state and local standards. According to the District Court Circuit, "Rather than being faced with 51 different standards, as they had feared, or with only one as they had sought, manufacturers must cope with two regulatory standards under the legislative compromise embodied in section 209(a)" (Engine Mfrs. Ass'n v. EPA, 88 F.3d 1075, 1080 [D.C. Cir. 1996]). Finally, the First Circuit Court of Appeals likewise noted "Both the statutory language and the legislative history suggest that section 177 was the result of a compromise between the competing interests of the states and the automakers, giving states greater flexibility in dealing with the control of emissions without overburdening automakers with too many separate emissions-control standards" (American Auto. Mfrs. Ass'n v. Mass. Dept. Envtl. Protection [1st Cir. 1998]).

Is the Zero Emissions Vehicle Mandate an Emissions Standard?

One major dispute in applying the preemption provisions is whether state regulations that require a percentage of a given manufacturer's sales fleet to meet specified emissions levels are "standards relating to the control of emissions" that are preempted by section 209(a). Examples include the California zero-emission vehicle (ZEV) mandate, which requires a specified percentage of a manufacturer's fleet to be certified to the ZEV standard, and the California nonmethane organic gas (NMOG) fleet average, which requires a manufacturer's total volume of vehicles sold in the state in a given year to meet an overall and declining average hydrocarbon emission concentration. EPA initially took the position in 1993 that the ZEV mandate was not a standard under sections 209 and 177, because it did not directly limit emissions but rather limited the "flexibility otherwise accorded manufacturers to choose the mix of vehicles produced to meet the NMOG fleet average requirement" (Reilly 1993). One year later, EPA changed its position on the basis of its findings that the ZEV mandate directly resulted in some reductions of evaporative emissions and nitrogen oxides (NO_x) that were not accounted for by the NMOG fleet average (59 Fed. Reg. 48664, 48691 [1994]). In 1999, EPA advanced a firm position that the ZEV mandate was a standard under sections 209 and 177. In a letter explaining its views at the request of the First Circuit Court of Appeals, EPA stated that a standard under section 209(a) is a "requirement to produce a certain number or percentage of vehicles ('production requirement') to meet a numerical emissions limitation," and that this "requirement to produce vehicles is in fact part of the emission standard" (Guzy and Perciasepe 1999).

The courts have also concluded that a ZEV mandate is a standard. As the Second Circuit Court of Appeals explained,

> We view 'standards relating to the control of emissions' as describing regulatory measures intended to lower the level for auto emissions, while 'enforcement mechanisms' describe regulatory devices intended to ensure that the 'standards' are effective. For example, the LEV [low-emission vehicle] program is clearly a 'standard,' whereas periodic testing and maintenance requirements are 'enforcement mechanisms.' Although this distinction, like the distinction between substance and procedure, can be less than a bright line in some cases, we find it of relatively easy application in [the case of the ZEV sales mandate].... The ZEV sales requirement is, therefore, in the nature of a command having a different effect on the level of emissions, rather than in the nature of a means of enforcing, or testing the effectiveness of, a command (American Auto Mfrs. Ass'n v. Cahill, 152 F.3d 196, 200 [2d Cir. 1998]).

The First Circuit has likewise concurred that the ZEV mandate is a "standard" (Ass'n of Int'l Auto. Mfrrs. V. Commissioner, Mass, DEP, 208 F.3d 1 [2000]).

Are Other Mandates Considered Emissions Standards?

In 2003, the United States Solicitor General, in a brief filed with the U.S. Supreme Court on behalf of EPA, suggested perhaps an even broader definition of "standard" under CAA section 209(a) by arguing that "section 209(a)'s reference to 'any standard' is constrained primarily by context and embraces both quantitative and non-quantitative emission criteria that new vehicles and engines are required to meet" (Amicus Curiae Brief of the United States at 15 [2003]).

In 2004, the U.S. Supreme Court addressed the definition of a "standard" preempted under section 209 in the *Engine Manufacturers Association v. South Coast Air Quality Management District* case (EMA 2004). The case involved regulations adopted by the South Coast Air Quality Management District, a local government agency responsible for air pollution control in the Los Angeles metropolitan area. The regulations required local public and private fleet operators to purchase or lease alternative-fueled vehicles or other categories of vehicles on the basis of their emissions performance. In finding that these regulations were standards preempted by section 209, the U.S. Supreme Court rejected a narrow definition of a standard that applied only to regulations that compel manufacturers to meet specified emissions limits (541 U.S. 1761 [2004]). The court also rejected the argument that section 209 standards included only mandates imposed on manufacturers and not those imposed on purchasers of vehicles. The court held that "a command, accompanied by sanctions, that certain purchasers may buy only vehicles with particular emission characteristics is as much an 'attempt to enforce' a 'standard' as a command, accompanied by sanctions, that a certain percentage of a manufacturer's sales volume must consist of such vehicles" (541 U.S. 1763 [2004]).

Several disputes have occurred over whether other types of mobile-source emissions-related regulations are preempted by CAA section 209(a). Consistent with the broad preemptive meaning noted above, EPA and the courts have generally held that such regulations are preempted. EPA's regulation defining the scope of preemption under section 209(a) is written broadly as follows: "These standards or other requirements which are preempted include, but are not limited to, the following: emission standards, mandatory fleet average standards, certification requirements, after market equipment requirements, and non federal in-use testing requirements" (40 CFR § 85.1603 [2003]). The First Circuit Court held that section 209(a) preempts all state regulations, "the purpose and effect of [which] is to affect a quantitative reduction in emissions" (Ass'n Int'l Auto. Mfrs. v. Commissioner, 208 F.3d 1, 7 [1st Cir. 2000]). Likewise, the Second Circuit ruled that any state regulatory "command having a direct effect on the level of emissions" is preempted by section 209(a) (American Auto. Mfrs. Ass'n v. Cahill, 152 F.3d 196, 200 [2d Cir. 1998]).

California Waiver Proceedings under Section 209(b) of the 1990 CAA Amendments

As discussed above, Congress required EPA to review California's emissions standards and, provided they meet certain statutory criteria, to waive federal preemption of those standards. Examples of waivers for on-road vehicles include the waiver for the California LEV standards (58 Fed. Reg. 4166 [1993]). This waiver covered the LEV program's 1996 to 1998 model-year passenger cars, light-duty trucks, medium-duty vehicles, and heavy-duty vehicles and engines. The waiver for the LEV program considered various issues, including the burden of proof needed to grant the waiver, the protectiveness for public health and welfare, the consistency with sections 202 and 209 of the CAA, and the technological feasibility. The waivers for the California LEV II program considered some of the same issues in a more abbreviated manner (68 Fed. Reg. 19811 [2003]). An example of a waiver for California nonroad emissions standards is EPA's authorization for California to enforce regulations for exhaust emissions standards and test procedures for small engines (less than 25 hp) used in lawn and garden equipment (60 Fed. Reg. 37440 [1995]). As with the waivers for on-road sources, the nonroad-source waiver considered various issues, including the burden of proof needed to grant the waiver, the protectiveness for public health and welfare, the consistency with sections 202 and 209 of the CAA amendments, and the technological feasibility.

Waiver Criteria for On-Road Standards

Pursuant to section 209(b), California must first find that its proposed standards for motor vehicles are, in the aggregate, at least as protective of public health and welfare as applicable federal standards. EPA then reviews California's "protectiveness" finding and must deny a waiver if it determines that California's finding was "arbitrary and capricious." Section 209(b) also requires EPA to deny the waiver if it finds that California does not need its standards to meet "compelling and extraordinary" air quality conditions or if it finds that the California standards are not consistent with the requirements of section 202(a) of the CAA, which requires that the standards be feasible. As applied by EPA, the "feasibility" requirement means that a waiver may be issued only if EPA finds that "adequate technology exists with which to meet the California standards" and that "adequate lead time is available in which to implement that technology" (40 Fed. Reg. 23, 102 [1975]; 42 Fed. Reg. 2337, 2340 [1977]). Alternatively, EPA may grant the waiver if it finds that "technology does not exist but there is, or appears to be, adequate lead time to permit the development and application of the requisite technology, giving appropriate consideration to the cost of compliance within that time frame" (41 Fed. Reg. 44209 [1976]). EPA must also deny a waiver if it finds that the costs of the California standards will be "excessive" (43 Fed. Reg. 15490, 15492 [1978]). Even though EPA considers technological feasibility, lead time, and cost in deciding whether

California emissions standards are consistent with section 202(a), it generally only considers test-procedure consistency in determining whether a California enforcement procedure is consistent with section 202(a) (Motor & Eqpt. Mfrs. Ass'n v. EPA, 627 F.2d 1095 [D.C. Cir. 1979]).

Waiver Criteria for Nonroad Standards

A similar waiver procedure is provided for California standards for nonroad engines and vehicles in section 209(e)(2)(A). EPA must make the same three findings relating to the protectiveness, need, and feasibility of the California standards as it does under section 209(b) for on-road vehicle standards. In addition, EPA's regulations governing section 209(e) waiver decisions require that EPA also make findings that the California nonroad standards are consistent with section 209(e)(1), which preempts all states including California from regulating certain categories of nonroad sources, and section 209(a), the motor vehicle preemption provisions (40 CFR §§ 85.1601–85.1606 [2004]).

Waiver Practices

Over the more than 30 years of EPA waiver practice, EPA has granted federal preemption waivers for most of the requests submitted by California. The courts have made clear that California's standards "are presumed to satisfy the waiver requirement and that the burden of proving otherwise is on whoever attacks them" (Motor & Eqpt. Mfrs. Ass'n v. EPA, 627 F.2d 1095 [D.C. Cir. 1979]). In a few cases, particularly in the first few years in which waivers were considered in the early 1970s, EPA denied California's waiver requests in whole or in part for failure to meet the statutory criteria (36 Fed. Reg. 8172 [1971]; 37 Fed. Reg. 8128 [1972]; 38 Fed. Reg. 10317, 10319 [1973]; 40 Fed. Reg. 30311 [1975]). In a few other cases, EPA issued a conditional preemption waiver in which the waiver would apply only to the extent California limited its program as specified by EPA (36 Fed. Reg. 17458, 17459 [1971]); 42 Fed. Reg. 31639, 31641 [1977]). In several other cases, California modified its regulations while the waiver request was pending in response to criticisms by commentators or EPA during the waiver proceeding (42 Fed. Reg. 1503, 1504, 3192, 3193 [1977]; 43 Fed. Reg. 9344, 9346 [1978]). These examples represent the exception rather than the rule, however, as most waiver requests are routinely approved by EPA as submitted by California. EPA has stated that "Congress intended that the standard of EPA review of the [California] decision be a narrow one" (49 Fed. Reg. 18887 [1984]).

Although not explicitly stated in the statute, EPA has over the years adopted a truncated waiver review procedure for two types of requests: a waiver request for an "enforcement procedure" rather than a "standard," and a waiver request that is "within the scope" of a preexisting waiver granted by EPA. An enforcement procedure is a regulatory requirement that does not directly reduce emissions but rather assists or reinforces a standard that directly reduces emissions. EPA usually considers modifications of an existing standard or incremental changes in emissions regulations that do not substantially affect its prior "protectiveness" finding to be within the scope of a previous waiver.

Issues Related to Waiver Implementation

An issue that arose during EPA's consideration of California's waiver request for its 1990 LEV standards was whether EPA could consider in its waiver decision the impact and implications of other states adopting the California standards under CAA section 177. EPA concluded that section 209(b) does not authorize the agency to consider the impacts of actions or potential actions taken by other states under section 177 in reviewing a waiver request (58 Fed. Reg. 4166 [Jan. 13, 1993]).

One shared concern of California, states seeking to adopt California standards under section 177, and manufacturers subject to such standards is that the EPA approval process for California preemption waiver requests takes too long. EPA often does not approve a California waiver until shortly before or, in some cases, even after the applicable regulations take effect. For example, the California Air Resources Board (CARB) submitted its waiver request for new onboard diagnostic system regulations for 1994 and subsequent model-year vehicles in August 1990, but EPA did not approve the waiver until October 1996, well after the regulations were in effect (61 Fed. Reg. 55371 [1996]). Similarly, California requested a waiver in December 1990 for its exhaust emissions standards and test procedures for utility and lawn and garden equipment engines for 1994 and subsequent calendar years, but EPA did not provide a final waiver approval of those regulations until July 1995 (60 Fed. Reg. 37440 [1995]). In October 1991, California submitted its waiver request for new emissions standards for low-emission medium-duty vehicles beginning in model-year 1998, but that waiver was not approved until April 1998 when the 1998 model year was already well under way (63 Fed. Reg. 18403 [1998]). EPA issued a waiver decision for California motor vehicle evaporative emissions standards for 1996 to 1998 model-year vehicles on August 5, 1999, well after the applicable model years were over (64 Fed. Reg. 42689 [1999]). EPA asked for public comment on December 31, 2003, on a "within the scope" waiver request for California heavy-duty diesel engine certification requirements and procedures that took effect 8 years earlier in model-year 1995 (68 Fed. Reg. 75500 [2003]).

Because neither California nor other states under section 177 can enforce California standards until they have received a federal waiver, the substantial delays in the waiver process can adversely affect states, as well as cause uncertainty for manufacturers. The delays in issuing a waiver decision are due to several factors. EPA must undertake a substantial technical analysis of the feasibility of the California standards for which a waiver is sought, even when waiver requests are relatively noncontroversial. EPA must also provide an opportunity for public comment and a public hearing and fully analyze and consider all comments and testimony submitted in those processes. The California standards are often amended or otherwise revised while the waiver request is pending; that requires EPA to redirect its analysis and, in some cases, to provide an additional opportunity for public comment. California also sometimes takes a long period of time before submitting a waiver request to EPA. Finally, the resources that EPA devotes to waiver requests might not be adequate.

EPA may be permitted to pursue more expedited waiver decisions under the existing statutory provision. Section 209(b) requires EPA to issue a waiver after "notice and opportunity for public hearing," but EPA has consistently held that its waiver decisions are not "rules" under the Administrative Procedure Act. As such, waiver decisions are not subject to many of the administrative formalities that apply to other regulatory actions such as the adoption of standards. Thus, under the existing statutory authority, EPA could probably adopt an expedited approval process by which it would quickly and provisionally approve those California waiver requests that appear to be noncontroversial, and such waivers would go into effect unless any party objects or requests a public hearing. For example, out of the past eight full waiver requests submitted by California, four received no important adverse comments, thus being the type of noncontroversial waiver that could be approved through an expedited process (59 Fed. Reg. 46978 [1994]; 69 Fed. Reg. 60995 [2004]; 70 Fed. Reg. 50322 [2005]; 71 Fed. Res. 335 [2006]). This process would be similar to direct final-rule-making, in which an agency publishes a rule without undergoing prior notice and comment and declares that the rule will take effect automatically unless the agency receives an objection within 30 days (Levin 1995). Using such a process to expedite the noncontroversial waiver requests would allow EPA to focus its resources on the waiver requests that raise more serious issues.

One possible disadvantage to such an approach is that any objection to the direct final waiver might be viewed by EPA's Office of General Counsel as potentially substantive because a court might find it substantive in any subsequent litigation. Upon receipt of an adverse comment, the Office of General Counsel would require that the full waiver process be performed and the resulting time lost could be substantial. However, as noted above, many past waivers received no important comments, and the agency is under no obligation to respond to frivolous or otherwise minor comments.

State Adoption of California Standards under Section 177 of the CAA Amendments

In contrast to section 209(b) in which Congress explicitly assigned EPA the role of approving waivers of federal preemption for California standards, in section 177, Congress did not assign EPA any role in approving adoption of California standards by other states. As EPA itself stated, "language requiring that other States request and receive authorization from EPA is noticeably absent. Indeed, the statutory text reads as authorizing States to adopt California standards on their own volition" (59 Fed. Reg. 36969, 36983 [1994]); Reilly 1993). Given the absence of EPA review, the courts have generally held that under section 177, states may enforce only California standards that have received a waiver.

Issues Related to Enforcement

The issue of whether section 177 states may adopt enforcement procedures that differ from California's has been controversial. In a 1991 letter to Congressman John Dingell, then the chairman of the House Committee on Energy and Commerce, EPA stated its position as follows:

> It is clear that states may not adopt additional new vehicle certification testing or enforcement procedures since such testing and enforcement procedures will be conducted under Federal and California laws. For example, a state may not establish a unique test procedure. Nor may it establish additional enforcement procedures for new vehicles. EPA believes, however, that a state may implement its own enforcement program which differs from California's enforcement program provided there is no undue burden imposed on manufacturers (Reilly 1991).

On the same issue, the Second Circuit Court stated that "although the 'piggyback' provision in section 177 requires states to adopt standards identical to those in place in California to avoid preemption, there is no such identicality requirement for the mechanism employed to enforce those standards" (MVMA v. NYDEC, 79 F.3d 1298, 1305 [2d Cir. 1996]).

Issues Related to the Identicality Requirement

Another issue litigated under section 177 is whether a state that adopts the California emissions standards must also adopt the California fuel requirements that may be associated with the vehicle standards in California. EPA and the courts agreed that a section 177 state had no duty to adopt California fuel requirements to satisfy the identicality requirement or third-vehicle prohibition of section 177 (59 Fed. Reg. 48664, 48690 [1994]). A key part of the reasoning in that determination is that a section 177 state is permitted to adopt only California emissions requirements that have been issued a federal preemption waiver by EPA, and California's fuel requirements are not included in its waiver requests under section 209(b) (59 Fed. Reg. 48664, 48690 [1994]).

EPA and the courts have rejected all claims by manufacturers that different conditions in a section 177 state (that is, different fuels than California that allegedly adversely affect emission-control equipment or cold temperatures that allegedly affect performance of battery-powered electric vehicles) will result in a prohibited "third vehicle." Two primary reasons are given for this rejection. First, EPA and the courts have concluded that any decision by a manufacturer to alter its vehicles for a state that adopts California standards are not compelled by that state but rather is a "marketing choice" voluntarily undertaken by that manufacturer (59 Fed. Reg. 48664, 48691 [1994]; MVMA v. NYSDEC, 79 F.3d 1298, 1307 [2d Cir. 1996]). Second, EPA and the courts held that not all changes in a vehicle would constitute a third vehicle, but that vehicle changes that imposed an "undue burden" on manufacturers would constitute a third vehicle (59 Fed. Reg. 48664, 48691 [1994]; MVMA v. NYSDEC, 79 F.3d 1298, 1308 [2d Cir. 1996]). EPA and the courts relied on floor statements by several senators in the debates on the 1990 CAA amendments to argue that the legislative history confirms that the third-vehicle prohibition does not require "physical identicality" of vehicles in the section 177 state with vehicles in California (59 Fed. Reg. 48664, 48691 [1994]; MVMA v. NYSDEC, 79 F.3d 1298, 1308 [2d Cir. 1996]). Thus, the Second Circuit Court opined that "a state can never violate the third vehicle prohibition so long as it satisfies the identicality requirement and 'does not administer or enforce the California emission standard in a more burdensome manner than what occurs in California'" (MVMA v. NYSDEC, 79 F.3d 1298, 1308 [2d Cir. 1996]).

Another issue is whether a state exercising its authority under section 177 must adopt all of California's motor vehicle emissions standards, or whether it can choose to adopt only a subset of the California standards. EPA has taken the position that a state need not adopt all of the California standards. For example, EPA determined that a state adopting California's LEV standards in the 1990s was not required to adopt California's ZEV mandate, which had been enacted as part of the LEV standards by California (59 Fed. Reg. 48664, 48667 [1994]). EPA initially based that determination on its 1993 determination that the ZEV mandate was not an emission standard under sections 209 and 177, as discussed previously in this chapter (Reilly 1993). Although EPA subsequently changed its position and concluded that the ZEV mandate was indeed a standard, the agency maintained its position that a section 177 state was not required to adopt the ZEV mandate. EPA stated that "adoption of the California LEV program does not require adoption of the California heavy-duty engine program," and similarly "section 177 does not require a state to promulgate standards that are clearly segregable from one another" (59 Fed. Reg. 48664, 48692 [1994]). EPA then concluded that the ZEV sales requirement is clearly segregable from the other parts of the LEV program (59 Fed. Reg. 48664, 48692 [1994]).

In contrast, EPA has suggested that the NMOG fleet-average requirement is not segregable from the LEV program and, in contrast to its evolving position on the status of the ZEV mandate, has always maintained that the NMOG fleet average is a standard under sections 209 and 177 (Reilly 1993). Although EPA has not officially ruled that a section 177 state adopting California standards must adopt and enforce an NMOG fleet average requirement that is identical to California's, it has strongly hinted that is its position. For example, in 1994, EPA stated that "the NMOG fleet average requirement is the heart of the California LEV program and is the central mechanism for ensuring reductions from the program, and any State implementing the LEV program should, and is probably compelled to, include enforceable NMOG fleet average requirements" (59 Fed. Reg. 21720, 21736 [1994]). EPA subsequently noted "the NMOG fleet average is the central provision of the LEV program to require manufacture of the low emission vehicles in the program and to obtain enforceable emission reductions from such vehicles" (59 Fed. Reg. 48664, 48692 [1994]). EPA suggested, however, that it might be appropriate for several adjacent northeastern states "to enforce the NMOG fleet average provisions through a region-wide averaging system" (59 Fed. Reg. 48664, 48693 [1994]). Some states adopting the NMOG fleet-average requirement (including Massachusetts and New York) announced that they would not initially enforce the NMOG fleet average. EPA was aware of, and did not disapprove of, this practice (Wilson 1992).

Issues Related to Lead Time

Finally, there is the issue of whether a state that is already implementing California standards is subject to the 2-year lead-time requirement again when it adopts revised California standards. The Second Circuit Court held that when California revises its regulations for which a waiver has been granted, section 177 requires a state that adopted California standards to revise its regulations to maintain the identicality requirement (American Auto. Mfrs. Ass'n v. Cahill, 152 F.3d 196, 200-01 [2d Cir. 1998]). In such circumstances, EPA has taken the position that the two-year lead time applies to each new set of California standards but not to minor revisions of the California standards that are within the scope of an existing waiver (Reilly 1993). A final timing issue is whether a section 177 state must wait for the California standards to receive a federal preemption waiver before adopting such standards, given that section 177 only permits a state to adopt and enforce California standards for which a waiver has been granted. EPA and the courts have consistently taken the position that a non-California state may adopt California standards that have not received a preemption waiver but that it cannot enforce such standards until EPA grants a preemption waiver (Reilly 1993).

FEDERAL (EPA) PROCESS FOR DEVELOPING MOBILE-SOURCE EMISSIONS STANDARDS

EPA is the agency responsible for developing federal mobile source emissions standards. Standards-setting activities are ultimately initiated by Congress through the CAA legislation. The 1990 CAA amendments direct EPA to conduct several types of activities. In some cases, Congress directs the agency to adopt numerical standards, such as the Tier 1 standards for motor vehicle emissions, or to study an issue such as the issue of nonroad engines and develop the standards (EPA 1991). For Tier 2 light-duty vehicle standards, Congress directed EPA to assess the adequacy of CAA numerical values for default Tier 2 standards and to provide alternatives if the agency deems such a course necessary.

The regulatory process for issuing standards is directed by a series of federal statues (Table 3-1) and executive orders (Table 3-2) that dictate how federal regulations are promulgated. In particular, all emission standard rules must be promulgated according to the procedures laid out in Section 307 of the CAA and, because of their important impacts, must be assessed and reviewed by the Office of Management and Budget (OMB) under EO 12866. Under Section 307, EPA must follow a notice and comment process that is similar to the procedure detailed in the Administrative Procedures Act. This process begins with a publication of a proposed rule in a Notice of Proposed Rulemaking (NPRM), which is published in the *Federal Register*. The NPRM calls for comments on the proposed rule to be provided to EPA over a specified time. Major rules often involve multiple public hearings to facilitate the collection of comments. Many major emissions standards also involve an earlier step, the publication of an advanced notice of proposed rule-making (ANPRM), to solicit early input from the public and key stakeholders. This step can add several months to the process. The whole process generally takes a

TABLE 3-1 Federal Legislation Controlling Regulation Development

- Administrative Procedures Act - Governs public notice and comment process
- Congressional Review Act - Before a rule may take effect, agencies need to submit the rule and other information to Congress
- National Technology Transfer and Advancement Act - Directs agencies to use voluntary consensus standards (e.g., ISO) where practical
- Paperwork Reduction Act - When information is required to be collected from the public, agencies must submit for OMB approval an Information Collection Request
- Regulatory Flexibility Act/SBREFA - Consider economic impacts on small entities
- Unfunded Mandates Reform Act - Addresses imposition of unfunded federal mandates on state and local governments

Source: France 2005.

TABLE 3-2 Executive Orders Controlling Regulation Development

- Regulatory Planning and Review (EO 12866) - Governs procedures for OMB review of "significant" federal rules
- Children's Health (EO 13045) - Addresses whether rule has a disproportionate effect on children
- Energy (EO 13211) - Addresses effects on the supply, distribution and use of energy
- Environmental Justice (EO 12898) - Addresses disproportionate health or environmental effects on minority and low-income populations
- Federalism (EO 13132) - Ensures input by state and local governments for rules that have substantial direct effects on states
- Tribal Governments (EO 13175) - Addresses implications on tribal governments

Source: France 2005.

year to develop the NPRM and a year to finish the final rule-making (France 2005).

Because regulations for emissions standards are typically deemed to be important, OMB has a major role in EPA's rule-making process. OMB's role includes a review of both the proposed and the final rules under EO 12866. OMB also issues directives and guidance, in the form of memoranda and circulars, on the types of regulatory analysis required. EPA is required under EO 12866 to prepare a regulatory impact analysis (RIA) of the proposed standards, which includes the agency's estimates of the likely costs, benefits, and impacts on industry. The RIA for EPA's Tier 2/gasoline sulfur proposal for light-duty vehicles, discussed in detail in Chapter 6, is typical of a RIA for a major rule. This RIA includes a discussion of health and welfare concerns about motor vehicle emissions; the impacts of the proposal on emissions, air quality, and vehicle and gasoline costs; the cost effectiveness and the cost and benefits of the proposal; and the impacts of the proposal on small businesses.

CALIFORNIA'S HISTORY AND PROCESS FOR DEVELOPING MOBILE-SOURCE EMISSIONS STANDARDS

California has played a central role in the control of emissions from mobile sources, especially the automobile. Environmental and human factors converged because California was susceptible to severe atmospheric inversions, population growth was rapid, and automobile use was greatly expanding. Before the discussion on the development of California's mobile-source emissions standards, a brief historical perspective is presented on California's role in mobile-source emissions controls. Aplet and Meade (1997) contains a more complete discussion of the history of mobile-source emissions regulation in California.

California's Pioneering Role in Light-Duty-Vehicle Emissions Control

Programs to address air pollution in the United States originated in the first half of the twentieth century in industrialized urban areas, such as Pittsburgh, Pennsylvania; Chattanooga, Tennessee; and Saint Louis, Missouri. World War II brought industrialization to Los Angeles to support the war effort; that was followed by a post-war population boom that resulted in a significant deterioration in air quality. The *Los Angeles Times* railed about the problem in the mid-1940s and sponsored a study led by Raymond R. Tucker in 1946. As shown in Figure 3-1, the results of Tucker's study were headline news in the January 19, 1947 edition of the *Los Angeles Times*. Among its conclusions was the finding that widespread collaboration was needed to address the growing air pollution problem. The state adopted the California Air Pollution Control Act in 1947, authorizing the creation of Air Pollution Control Districts by each county. Los Angeles, with its burgeoning air pollution problem, moved ahead of other counties by establishing the Los Angeles Air Pollution

Los Angeles Times

FINAL EDITION

'TIMES' EXPERT OFFERS SMOG PLAN

Talmadge Ready to Obey Courts

Closed Shop Frowned On

HUGE BIRDS KILL WHALE WITH BEAKS

U.S. Aid Used by Politicians,

SMOG SOURCES SUMMARIZED IN REPORT

Ban on Private Rubbish Burning and Analysis of Factory Fumes Proposed

FIGURE 3-1 *Los Angeles Times* sponsored a study by Raymond R. Tucker in 1946. Source: Los Angeles Times 1947.

Control District, the first such air pollution control district in the country and the forerunner to the current South Coast Air Quality Management District (SCAQMD). The Los Angeles district began programs to ban backyard incineration, to set smoke-stack standards, and to fund research into the cause of smog.

A critical issue was the characteristics of the air quality problem in Los Angeles. Through independent research, Dr. Arie J. Haagen-Smit, of the California Institute of Technology, determined the nature and causes of photochemical smog. Dr. Haagen-Smit and collaborators determined the atmospheric processes that led to the formation of ozone, the key component of smog, and implicated the automobile as a key source of the problem (Haagen-Smit and Fox 1954). The federal government took its first action to address air pollution in 1955 in the Federal Air Pollution Control Act, which provided funding to state and local governments to "protect the primary responsibilities and rights of the state and local governments in controlling air pollution…." In that same year, the Los Angeles County Motor Vehicle Pollution Control Laboratory was established. By 1959, California legislation directed the state Department of Public Health to establish air quality standards and necessary controls for motor vehicle emissions. Following California's lead, the federal government acted in 1960 directed the Surgeon General to study the "various substances discharged from the exhausts of motor vehicles…." (Public Law 86-493).

The decade of the 1960s proved to be a seminal time in the identification of the automobile as a significant source of urban air pollution and the establishment of some of the initial light-duty-vehicle control requirements. The pattern throughout the decade was consistent; California authorities would establish control requirements and the U.S. government would follow a few years later. For example, California required the control of crankcase emissions (now controlled with the positive crankcase ventilation [PCV] valve) in 1961. California acted again in 1964 by setting the first hydrocarbon (HC) and carbon monoxide (CO) emissions regulations in the nation, which applied to model-year 1966 vehicles. The federal government followed suit in the Motor Vehicle Air Pollution Control Act of 1965 when it adopted both the California crankcase and tailpipe emissions standards for 1968 model-year vehicles. In 1970, Congress moved ahead of California and the federal government process by amending the CAA to require the establishment of regulations to reduce motor vehicle emissions by 90% for model-years 1975 and 1976 vehicles. However, throughout most of the 1970s and the 1980s, California outpaced the federal regulatory process, for example, setting evaporative emissions standards for model-year 1970 vehicles and the first nitrogen oxide (NO_x) emission standards for model-year 1972 vehicles. In general, the federal process has continued to lag behind the California process by 1 or more years. Table 3-3 shows the chronology of federal and California exhaust emissions standards for new passenger vehicles.

California continued in the 1980s to set even stricter emissions standards years ahead of the federal government. Near the end of the 1980s, California took another pioneering step as it attempted to influence the fuels used in motor vehicles. That step led to work on methanol as a potential fuel in 1987, establishment of a ZEV mandate in 1990, and replacement of diesel fuel with natural gas. These pioneering efforts have not always had the anticipated results or been successful; however, the actions stimulated ancillary actions. For example, the Atlantic Richfield Company (ARCO) produced and advertised a cleaner burning gasoline to match methanol performance. General Motors launched an electric vehicle program followed by other manufacturers, and Honda and others produced experimental vehicles that burned gasoline but produced near-zero emissions. The current popular hybrid vehicle technology has benefited from the early electric vehicle work and the later near-zero-emission vehicle innovations.

California continued to outpace the federal government by establishing emissions limits for a broad range of vehicles and other mobile sources. It set standards for small off-road and utility engines in 1995, again before the federal government. It also set emissions standards for recreational marine engines. Table 3-4 compares the California vehicle regulatory initiatives with the federal vehicle regulatory process for light-

TABLE 3-3 California and Federal Exhaust Emissions Standards for Passenger Vehicles (grams/mile)[a]

	Federal			California		
Model Year	HC	CO	NO_x	HC	CO	NO_x
Uncontrolled	8.7	90	3.4	8.7	90	3.4
1966				4.3	44	
1967				4.3	44	
1968	4.1	34		4.3	44	
1969	4.1	34		4.3	44	
1970	4.1	34		2.2	23	
1971	4.1	34		2.2	23	
1972	3.0	28		1.5	23	3.0
1973	3.0	28	3.1	1.5	23	3.0
1974	3.0	28	3.1	1.5	23	2.0
1975	1.5	15	3.1	0.9	9	2.0
1976	1.5	15	3.1	0.9	9	2.0
1977	1.5	15	2.0	0.41	9	1.5
1978	1.5	15	2.0	0.41	9	1.5
1979	1.5	15	2.0	0.41	9	1.5
1980	0.41	7.0	2.0	0.41	9	1.0
1981	0.41	3.4	1.0	0.41	7	1.0
1982	0.41	3.4	1.0	0.41	7	0.4
1983	0.41	3.4	1.0	0.41	7	0.4
1984	0.41	3.4	1.0	0.41	7	0.4
1985	0.41	3.4	1.0	0.41	7	0.4
1986	0.41	3.4	1.0	0.41	7	0.4
1987	0.41	3.4	1.0	0.41	7	0.4
1988	0.41	3.4	1.0	0.41	7	0.4

1989	0.41	3.4	1.0	0.41	7	0.4
1990	0.41	3.4	1.0	0.41	7	0.4
1991	0.41	3.4	1.0	0.41	7	0.4
1992	0.41	3.4	1.0	0.41	7	0.4
1993	0.41	3.4	1.0	0.41	7	0.4
1994	0.41	3.4	0.4	0.25 [b]	1.7-3.4 [c]	0.2-0.4 [c]
1995	0.41	3.4	0.4	0.231 [b]	1.7-3.4	0.2-0.4
1996	0.41	3.4	0.4	0.225 [b]	1.7-3.4	0.2-0.4
1997	0.41	3.4	0.4	0.202 [b]	1.7-3.4	0.2-0.4
1998	0.41	3.4	0.4	0.157 [b]	1.7-3.4	0.2-0.4
1999	0.41	3.4	0.4	0.113 [b]	1.7-3.4	0.2-0.4
2000	0.41	3.4	0.4	0.073 [b]	1.7-3.4	0.2-0.4
2001	0.075 [b]	1.7-3.4 [c]	0.2-0.4 [c]	0.07 [b]	1.7-3.4	0.2-0.4
2002	0.075 [b]	1.7-3.4	0.2-0.4	0.068 [b]	1.7-3.4	0.2-0.4
2003	0.075 [b]	1.7-3.4	0.2-0.4	0.062 [b]	1.7-3.4	0.2-0.4

[a]There are subtle differences in the California and federal standards that cannot be presented in a table format. These differences involve certification methods and fleet averaging.

[b]Fleet average of nonmethane organic gases.

[c]Emissions standard varies depending on certification levels (for example, low-emission and ultralow-emission vehicles).

Source: Sperling et al. 2004a. Reprinted with permission by the authors; copyright 2004, University of California, Davis.

TABLE 3-4 Comparison of California with Federal Regulatory Initiatives for Light- and Medium-Duty On-Road Vehicles

Model Year	California Requirement	Result	HEW/EPA Model Year	HEW/EPA Comparison with California
1963	Reduced crankcase emissions	Reduced VOC emissions	1968	Adopt similar requirements (5 yr later)
1966	First VOC and CO exhaust emission standards	Leaner carburetors, air pumps, and retarded timing controls	1968	Adopt similar requirements (2 yr later)
1970	First evaporative emission standard	Reduced VOC emissions using carbon canisters	1971	Adopt similar requirements (1 yr later)
1974	NO_x exhaust emission standard set at 2.0 g/mi	Increased use of exhaust gas recirculation (EGR)	1977	Adopt similar requirement (3 yr later)
1975	VOC exhaust standard tightened by 72%, CO exhaust standard tightened by 77%	Use of oxidation catalysts	1975	VOC exhaust standard reduced by 56%, CO exhaust standard reduced by 62% (EPA action concurrent to CARB but standards reduced to lesser extent)
1977	VOC and NO_x exhaust standard tightened by 34%	Use of on-board computers and improved catalysts; some 3-way catalysts begin to appear	1980	Adopt similar requirements (3 yr later)
1980	Evaporative standard tightened by 67%	Reduced VOC emissions using larger carbon canisters	1981	Adopt similar requirement (1 yr later)
1980	NO_x standard tightened by 33%	Use of 3-way catalysts on all models	1981	Adopt similar requirement (1 yr later)

1982	NO_x standard tightened by 30%	Use of improved 3-way catalysts; fuel injection on some models, more durable vehicles	—	Did not adopt
1987	Alternative fuels strategy	Attempts to shift vehicle fleet to methanol fuel	—	Did not adopt
1988	First on-board diagnostic (OBD) systems	Alerted driver to failing emissions component	—	Did not adopt but some manufactures choose to sell CA OBD nationwide
1989	NO_x standard tightened by 43%	Universal use of electronic fuel injection systems	1994	Adopted similar requirement (5 yr later)
1990	ZEV requirement	Battery-powered vehicles	—	Did not adopt
1994	First LEV program		1999	Available in New England states in 1999 and nationwide in 2001 (6 yr later)
1994	Second-generation OBD systems	Alerted drivers to need for maintenance; also used as emissions inspection technology in inspector and maintenance (I/M) programs	1996	Adopted similar requirements but not applicable to medium-duty vehicles until 2004 (2-10 yr later)
1995	Real-time evaporative testing and standards	Improved control of fuel-vapor losses and lower fuel-vapor permeation materials	1996	Adopted similar requirement with additional test procedure (1 yr later)
1998 - 2005	ZEV mandate	Modified requirements in light of low consumer demand and ultralow-emitting vehicles	—	Did not adopt

(Continued)

TABLE 3-4 *Continued*

Model Year	California Requirement	Result	HEW/EPA Model Year	HEW/EPA Comparison with California
1999	Formation of California Fuel Cell Partnership	Public-private venture to demonstrate fuel-cell vehicles in California	—	No action
2004	Second LEV program	Required sport-utility vehicles and pickup trucks to meet same emissions limits as cars	2004	Adopted similar requirements but phased in more slowly with some higher emissions standards. Included heavier SUVs and light duty trucks within the standards. No advanced technology requirements.
2009	Greenhouse-gas emission standards	Projected to reduce greenhouse-gas emissions by 27% in 2030 (see Chapter 6)	—	Did not adopt

Source: Adapted from CARB, unpublished material, 2005.

duty vehicles over the past 40 years. Tables comparing California and federal standards for the full range of vehicles and nonroad engines are contained in Appendix C.

Process for Setting Mobile-Source Emissions Standards

The combination of the federal legislation and regulations and the California statutes establishes the general framework for air quality management in California. The process is derived from three critical pieces of legislation, the federal CAA, the California CAA, and the California Environmental Quality Act. The federal CAA and subsequent EPA policies control the setting of national priorities. These policies require California to develop a program to meet nationally determined air quality standards within specified times through state implementation plans, as discussed in the next major section. Since the 1950s, California has also adopted air quality legislation. One of the important amendments to the clean air laws was the California CAA of 1988, which introduced the low-emissions vehicle (LEV) program (Assembly Bill 2595, Sher). In addition, the California Environmental Quality Act adopted initially in the 1970s has an important role in setting requirements for the adoption of regulations in California.

The California Air Resources Board (CARB), which is a department within the California Environmental Protection Agency, is responsible for managing air quality in California. CARB bases its regulatory program on the combination of requirements specified in the federal and state statutes. It also adopts its own policies and procedures for adopting rules through board and executive officer directives. Additionally, California has 35 air pollution control districts or air quality management districts with authority to regulate stationary, indirect, and area sources of air pollution within a given county or region.

The California Clean Air Act

The California CAA provides a framework for air quality planning and regulation in the state. It provides air quality goals, planning mechanisms, regulatory strategies, and standards of progress. The California CAA requires attainment of state ambient air quality standards by the earliest practicable date. It also requires the development of air quality management plans for air districts in violation of the state ozone, CO,

TABLE 3-5 Important Features of the California CAA of 1988

- Air quality management is overseen by the CARB board of 11 members appointed by the governor and approved by the Senate (H&S39500,39510).
- CARB must classify California nonattainment areas into one of four classifications (moderate, serious, severe, extreme) (H&S40921.5).
- CARB must adopt mobile source emissions control to significantly reduce mobile source emissions by year 2001.
- Air quality management plans to achieve and maintain California ambient air quality standards must be developed in 1990 and revised at 3-year intervals thereafter (requirements for the South Coast Air Quality Management District) (H&S40911).
- Air quality management plans are required to
  - — Apply every "feasible emission reduction method" or at a minimum meet a 5% annual reduction in emissions (H&S40914).
  - — Consider the "cost effectiveness of their air quality programs, rules, regulations, and enforcement practices" and "strive to achieve the most efficient methods of air pollution control" (H&S40910).
- Air quality management plans must be considered at public meetings by the relevant local board followed by a public review by the state Air Board.

sulfur dioxide (SO_2), or nitrogen dioxide (NO_2) standards. Some highlights of the act are listed in Table 3-5.

The California Environmental Quality Act

The California Environmental Quality Act (CEQA) was designed to force evaluation of the overall impact of projects on the environment. The evaluations must include consideration of air pollution, water pollution, solid waste, biodiversity, fire safety, and other environmentally related issues. Major rules are considered projects by subsequent court interpretations, and all agencies in California who adopt rules are required to meet the CEQA requirements for their important rules.

A government agency can design its rule-making process to inherently meet CEQA requirements so that a separate specific CEQA document is not required. CARB and the South Coast Air Quality Management District (SCAQMD) are examples of California agencies that have designed processes that are deemed to be equivalent to CEQA. Extensive information on CEQA can be found at the website http://ceres.ca.gov.

CARB's Governing Board and Resources

Created in 1967, CARB is overseen by an independent governing board made up of a full-time chairperson as one of 11 members who appoint an executive officer to direct the day-to-day activities of the CARB staff. All regulations produced by CARB must be approved by the governing board. The members of the CARB governing board are appointed by the governor and approved by the California Senate. The governor must ensure that the governing board makeup meets the specifications given in Table 3-6.

The governor selects one of the board members to be the chairperson of the board. The governing board typically meets once a month or more. In 2000 through 2004, the governing board met 77 times for some or all of a day, an average of 15.4 meetings per year. The budget for CARB in the 2004-2005 fiscal year was $130 million, the SCAQMD budget for that period was $102 million (SCAQMD 2004), the Bay Area Air Quality Management District budget was $52 million, and the other 33 smaller California districts budgets averaged about $8 million per district. Thus, California annually spends about $550 million on government activities related to air quality management. As a comparison, the EPA air program budget is about $660 million (EPA 2005f).

TABLE 3-6 Composition of 11-Member CARB Governing Board

One member from each of the governing boards of the following districts:

- San Diego Air Pollution Control District
- Bay Area Air Pollution Control District (San Francisco region)
- San Joaquin Valley Air Pollution Control District
- South Coast Air Quality Management District (Los Angeles region)
- Any additional air quality management district in the state

One member having expertise in each of the following fields:

- Automotive engineering or a closely related field
- Agriculture, science, or law
- Medicine or health effects
- Air pollution control or an area similar to that indicated above

Two members having no special expertise may be selected from the public at large.

TABLE 3-7 Estimates of Emissions-Carrying Capacity Needed for Attainment of Ozone Standards

Area	Population (millions)	Carrying Capacity ($VOCs + NO_x$) Tons per Day	Carrying Capacity ($VOCs + NO_x$) Pounds per Person per Year
South Coast Basin	16.9	840	36
San Joaquin Valley	4.1	630	69
Houston	5.5	1360	181

Source: Witherspoon 2004.

Air Quality Management Planning and Subsequent Regulatory Processes

The basis for the development of air quality regulations in California is the need to meet state ambient air quality standards. CARB first issued ambient air quality standards for California in 1969 for total suspended particulates, photochemical oxidants, SO_2 , NO_2, and CO. Since the passage of the California CAA in 1988, nonattainment areas are required to address the steps needed to attain air quality standards through an air quality management plan (AQMP). In effect, the air quality management plan is the same type of plan needed to meet federal air quality standards described in the state implementation plan (SIP). However, the state AQMPs are designed to meet the state air quality standards (frequently more stringent) but do not contain mandatory attainment deadlines. Local air quality management districts develop AQMPs, which are sent to CARB for review. CARB may submit the plans to EPA if a SIP is required by federal rules.

The South Coast Air Basin (the region around Los Angeles) has traditionally had the worst air quality in the state and nation, although the San Joaquin Valley region of California is becoming an area of greater concern. Table 3-7 shows the emissions carrying capacity for the South Coast Air Basin and compares it with those of San Joaquin Valley and Houston, Texas. The carrying capacity refers to the upper limit of emissions in a region consistent with achieving air quality standards. This table shows how per capita emissions need to be much lower in the Los Angeles area compared with the other locations. In addition to its air quality problems, the South Coast Air Basin and environs houses about 17 million persons. This is 40% of the population of California and represents about 5% of Americans (more populous than 41 states). Thus, the South Coast Air Basin tends to drive regulatory decisions relative to both stationary and mobile sources in California and will be used as an example to discuss the California regulatory process.

At the beginning of a traditional planning cycle, SCAQMD and CARB officials ideally discuss reduction requirements and the appropriate emissions reductions to be obtained from stationary sources versus mobile sources and from consumer products for the next planning cycle. The SCAQMD is responsible for reducing the emissions associated with stationary sources in its jurisdiction, and CARB is responsible for reducing mobile-source and consumer-product-related emissions statewide. In addition to the local and state regulatory air agencies, the metropolitan planning organization in the area—the Southern California Association of Governments—participates in the AQMP development because of its responsibility for transportation control measures (TCMs) and for determining that the regional transportation plan (RTP) conforms to the AQMP. From the beginning, CARB and the SCAQMD have had to commit to achieving the highest emissions reductions possible from all source categories of under their jurisdiction to meet federal clean air goals. Keeping in mind the emission-reduction commitments, each agency begins the process of identifying cost, safety, and other factors that are part of the control options at its disposal for its area of responsibility. Federally preempted source-category rules are also added to this mix of strategies from state and local agencies to demonstrate attainment of the standards.

The discussions between the SCAQMD and CARB can be difficult and normally revolve around the relative cost effectiveness associated with various control approaches. Over the past decade, the regulatory responsibilities of the two agencies have become increasingly blurred. The AQMP produced by the SCAQMD also may not be consistent with the goals or views of CARB.

Development of California Emissions Standards

As part of the state regulatory activities undertaken to attain air quality standards, California uses stringent emissions standards for motor vehicles. The initiating activities can be by legislative directives or by CARB. For example, the California CAA mandated that CARB take whatever actions were necessary, cost effective, and feasible to achieve a 55% emission reduction of HCs and a 15% emission reduction of NO_x

from 1987 baseline motor vehicle emissions by 2001 (section 43018(b) California Health and Safety Code). As a result, CARB developed the LEV program (discussed in Chapter 6). The California legislature in 2002 passed the Pavley bill, which compelled CARB to issue motor vehicle greenhouse-gas standards. In many other cases, however, CARB has used its authority to regulate on-road and nonroad emissions sourcesin the absence of directions from the California legislature. Table 3-8 shows the steps in the CARB standards development process.

CARB and EPA exchange relevant information to the extent possible during their respective regulatory processes, and the efforts made after rule implementation are coordinated with EPA (for example, testing to determine whether a noncompliance determination and a recall are appropriate). In the end, the CARB regulation development process follows the same overall procedures as the EPA process, the exception being that the CARB final rule adoption process is a public process review by an independent board. As discussed later in Chapter 6 in reference to the ZEV mandate, CARB seems to be able to adjust its rules more often to meet changing technological developments than the federal process has demonstrated to date.

MOBILE-SOURCE REGULATION IN THE STATE IMPLEMENTATION PLAN

State Implementation Planning Process

SIPs[2] are the formally adopted air quality attainment plans required by the CAA. The SIP describes the combination of local, state, and federal actions and emissions controls that will be undertaken for an area to comply with and maintain compliance with the NAAQS. The SIP provides the basic link between state and local regulations, EPA oversight of state actions and enforcement, and federal contributions to controls. These plans are designed by the states that have NAAQS nonattainment areas and must submit attainment plans to EPA for approval. A SIP must

[2] Individual states have a single SIP that describes the state's plans for coming into attainment with all NAAQS. That SIP is amended as regulations related to NAAQSs are modified. An amendment, such as describing how a location will come into attainment with the new 8-hr ozone standard, is typically referred to as a SIP (the 8-hr ozone SIP), when in fact, it is a SIP amendment. In this report, we continue to refer to each SIP amendment as a SIP.

demonstrate how the state will attain and maintain the standard for each pollutant in the time lines established in the CAA. The SIP for each pollutant typically contains the elements listed in Table 3-9.

The emissions reductions outlined in a SIP may contain reductions achieved through federal, state, and local (municipal, county, and district) regulatory actions. In the case of state and local actions, the SIP must demonstrate that the state has the legal authority for the actions that they propose and that the proposed actions will achieve the needed reductions. The attainment demonstrations must use approaches and models accepted by the EPA. In some instances, EPA has accepted nonregulatory actions for small (2-4%) portions of needed emissions reductions in nonattainment areas. Due to the difficulty of achieving air quality standards in many locations, successive amendments to the CAA have extended attainment deadlines. Table 3-10 indicates the design values[3] and attainment deadlines for the 1-hr ozone NAAQS associated with successive CAA amendments and standard promulgation. Though the implementation schedule for the new 8-hr ozone standard and transition from the 1-hr ozone standard is still being developed, attainment dates vary from 3 years for marginal areas to 20 years for extreme areas after an area's status has been designated by EPA.

Considerable technical and policy debate is typically associated with the first seven SIP elements described in Table 3-9. EPA attempts to minimize the debate by establishing in advance the requirements for developing emissions inventories, the acceptable emissions and ambient air quality models, and the acceptable emission reductions for some key types of control measures, such as vehicle I/M. Even with those safeguards, many issues and inconsistencies evolve. For example, in the early 1980s, during the development of the 1987 SIPs, EPA produced a new mobile-source emissions model. Some states preferred the older emissions model that predicted greater emissions reductions and argued that the timing of the process did not allow use of the newer model. EPA reacted by allowing the use of both models, creating an inconsistency in the predictions in various SIPs that had used different models.

---

[3] The design value is the monitored reading used by EPA to determine an area's air quality status; e.g., for the 1-hour ozone standard, the fourth highest reading measured over the most recent three years is the design value.

TABLE 3-8 Steps in CARB Standards Development

- Assessment of potential emissions reductions from specific source category typically involve the following:
  - — Review of technical papers.
  - — Discussions with control technology developers.
  - — Identification of one or more technologies and reformulation of a fuel or solvent.
- Technology assessment or approach further identified
  - — More thorough evaluation of its potential to reduce emissions.
  - — Preliminary assessment of cost and other impacts.
- If management decides to proceed with regulatory proposal, staff may
  - — Undertake testing to demonstrate viability and safety.
  - — Meet with companies that might be affected by the proposed regulation.
- Public workshops held to gather information from a broader set of stakeholders, including users of engines or equipment.
- Staff develops proposed regulation, including stringency of a standard, lead time, compliance and enforcement provisions.
  - — Regulatory requirements to assess flexibility, safety issues, cost of the regulation, impact on businesses
  - — Continued information gathering through expert consultants, additional workshops, and peer review by the University of California
- Action by CARB Governing Board
  - — Regulatory proposal is published at least 45 days before hearing.
  - — Governing board meets to hear the staff's proposal and testimony from concerned persons.
  - — Governing board discusses the proposal and the testimony and votes on whether to adopt the regulation.
  - — Governing board may make changes to the regulation or postpone adoption.
  - — If new information is needed, the hearing may be continued for a month or more.
  - — Frequently the governing board directs staff to provide periodic updates, which are presented at subsequent public hearing.

- Completion of regulatory process
    - — additional public comment period on any changes made by the governing board.
    - — written response to all comments made in written or oral testimony.
    - — review by the State Office of Administrative Law.
- Additional review prior to and during implementation that can include the following:
    - — Formal periodic reviews for governing board.
    - — Additional testing.
    - — Modifications to the adopted regulations.

TABLE 3-9 Elements of State Implementation Plans

- An initiation of an emissions inventory for all sources of each pollutant under consideration.
- A projection of emissions on the attainment date for all sources of each pollutant under consideration assuming no control intervention and expected population growth.
- A description of the control programs that the state will use emission for each pollution source under consideration and an indication that the authority exists for the state to achieve the indicated control intervention—the description of control problems includes those carried out by the federal government independent of the state.
- A commitment to adopt the indicated state control interventions and a schedule of adoption.
- A projection of emissions on the attainment date that includes the anticipated reductions achieved by the implementation of the control interventions superimposed over anticipated growth.
- A calculation of ambient air quality based on the projected reduced emissions using the approved model.
- A set of contingency measures that will be applied if the committed control interventions do not achieve the desired emission reductions.
- A public review process allowing stakeholders the opportunity to review and comment on the proposed plans.

TABLE 3-10 Attainment Years Set in Various CAA Amendments for the 1-hr Ozone Standard[a]

	Attainment Year by Severity of the Problem				
CAA Amendments	Marginal Area (design value = 0.121-0.138 ppm)	Moderate Area (design value = 0.138-0.160 ppm)	Serious Area (design value = 0.160-0.180 ppm)	Severe Area (design value = 0.180-0.280 ppm)	Extreme Area (design value = 0.280 ppm and above)
1970	1975 (all areas)				
1977	1982 (all areas)		1987[a]		
1990	1993	1996	1999	2005	2010

[a]The terms *marginal, moderate, serious, severe,* and *extreme* did not come into use until the 1990 CAA amendments. The 1977 CAA amendments required attainment in 1982 but allowed an extension to 1987 for those areas that could demonstrate that it was not possible to meet the 1982 date, which had the effect of giving serious, severe, and extreme areas an additional 5 years. This classification scheme was devised before EPA promulgated a new 8-hr standard and, thus, is related to the old 1-hr form of the standards.

Evaluation of California Standards as a Potential SIP Control Measure

The decision to adopt California emissions standards is best based on a comprehensive technical evaluation of emissions reductions needed, as well as an evaluation and comparison of alternative control measures. A comprehensive analysis includes the following components:

- Estimation of emissions reductions needed by pollutant. For example, in an ozone nonattainment area, performing air quality modeling to see what percent reductions in NO_x and HC emissions would be required to bring the area into attainment.
- Emissions inventory evaluation to assess the relative contributions of source categories to the pollutants of interest.
- Comprehensive evaluation of a broad range of control measures in all source categories (stationary point, area, on-road, and off-road). Such an evaluation would include estimating emissions reductions from specific control measures, costs for implementation of control measures, cost-effectiveness, and implementation issues.
- Modeling the effects of evaluated control measures individually or as a group.

Many nonattainment areas follow the comprehensive approach described above. For example, control-measure evaluations have been conducted in ozone nonattainment and near-nonattainment areas in Texas (for example, Coulter-Burke et al. 2002; ENVIRON 2000) and are continuing; Sacramento, California (SMAQMD 2004), the Upper Midwest States (MACTEC 2005) and are continuing; and Tennessee (University of Tennessee 2003). However, such an analysis is resource-intensive and might be prohibitively expensive for some air quality planning agencies, who may rely in part on analyses performed by other agencies. For example, NESCAUM's analyses of the California LEV program have been used by several states to justify adoption of the program.

Example of a State Implementation Plan

SIPs contain an array of emissions-reduction strategies for mobile, area, and stationary sources. Table 3-11 lists the controls and emission reduction estimates projected in the 1-hr ozone attainment SIP using

TABLE 3-11 SIP Control Strategies for NO_x for the Dallas-Fort Worth Area

EPA-issued Rules	Estimated NO_x Reductions in 2007 (tons per day)
Federal on-road measures:	93
Federal phase II reformulated gasoline (RFG)	
Tier 2 vehicle emission standards and federal low-sulfur gasoline	
National low emissions vehicle program	
Heavy-duty diesel standards	
Federal off-road measures:	48
Lawn and garden equipment	
Tier 3 heavy-duty diesel equipment	
Locomotives	
Compression ignition standards for vehicles and equipment	
Spark ignition standards for vehicles and equipment	
Recreational marine standards	

Texas Council Environmental Quality–issued Rules	Estimated NO_x Reductions in 2007 (tons per day)
Major point source NO_x reductions in four counties	129
Light-duty-vehicle I/M	54
Senate Bill 5 Voluntary Incentive Program	16
Airport ground service equipment electrification in four counties	6
Heavy-equipment fleets—gasoline in nine counties	2
Gas-fired water heaters, small boilers, and process heaters (statewide rule)	1

Other State measures:	1
Energy efficiencies	

Dallas-Fort Worth Local Initiatives	Estimated NO_x Reductions in 2007 (tons per day)
Speed limit reduction in nine counties	5
Voluntary mobile emissions reduction program (VMEP) in nine counties	2 - 5
Transportation control measures (TCM) in four counties	5

Source: TCEQ 2004.

federal, state, and local initiatives in the Dallas-Fort Worth (DFW) area. This plan relies heavily on NO_x controls. Reductions in mobile-source emissions in the plan come from tightening emissions standards for on-road and nonroad sources, as well as reducing in-use emissions through motor vehicle emissions I/M. As seen in the table, the DFW area relies heavily on federal mobile-source emissions standards to reduce emissions. Box 3-1 discusses issues related to using I/M programs versus federal emissions standards to control motor vehicle emissions.

Box 3-1 New Vehicle Emissions Standards and Controls on In-Use High-Emitting Vehicles

Motor vehicle emissions standards are part of the array of programs designed to reduce emissions and improve air quality. Air quality programs at the state and local level are typically a mix of programs as described in the adjoining section on the state implementation plan for the Dallas-Fort Worth (DFW) area. A prominent method for reducing motor vehicle emissions is by inspecting vehicles and requiring repairs when they fail emissions testing. Emissions from vehicles with malfunctioning emission-control equipment, referred to as high emitters, contribute disproportionately to overall emissions. An earlier NRC report (2001) discussing vehicle emissions inspection and maintenance (I/M) programs quoted estimates that typically less than 10% of vehicles typically contributed over 50% of total light-duty-vehicle pollutant emissions for any given pollutant. Given the significant fraction of emissions that comes from such vehicles, it is prudent to question whether strategies designed to reduce the emissions from high emitters would be more effective than tightening new emissions standards. The committee concludes that new vehicle emissions standards contribute to reducing motor vehicle emissions in ways not matched by in-use emissions controls. Specifically, new vehicle emissions standards have driven continual improvements in the ability to monitor and control in-use emissions and in the durability of emission-control equipment.

By the early 1980s many elements of modern emission-control equipment, including the automobile catalyst and onboard computers, were introduced. The further tightening of vehicle standards over the past 2 decades has inspired significant refinements in catalyst design and materials. These standards have also resulted in the introduction of sophisticated onboard diagnostic equipment to improve the emissions performance and to alert motorists and repair technicians to problems with emission-control equipment. Other standards have resulted in systems to reduce evaporative emissions and in improvements in the durability of emission-control systems.

Improvements in emission-control technologies have had a positive impact on in-use emissions, including emissions from high emitters. As shown in Chapter 2, on-road emissions of CO have greatly decreased since 1970, most of that reduction coming from light-duty vehicles. Pokharel et al. (2003) discussed how improvements in emission-control technologies have affected the fraction of vehicles defined as gross emitters of CO. Gross emitters are defined in their analysis as those highest emitting vehicles that together cause half of the total light-duty vehicle on-road CO emissions. Using remote-sensing emissions measurements of on-road vehicles in Denver, Colorado, described in Bishop and Stedman (1990) and Pokharel et al. (2002), Pokharel et al. (2003) showed that over the 1989-2001 time period the fraction of vehicles considered gross emitters and the average emissions from these vehicles had declined. Pokharel et al. (2003) also found that new vehicles manufactured in 2000 emitted less CO, HC, and NO_x than new vehicles manufactured in 1998, even while standards for new vehicles remained constant, a result the authors attributed to continual improvements made by manufacturers to meet standards and improve durability. The trend for new vehicles to show decreasing emissions during successive years of measurements are shown for CO in Figure 3-2. This finding has been observed in other remote-sensing emissions data obtained in Chicago, Illinois (Bishop et al. 2003). Improvements in emission-control technologies might also have caused a slowing in emissions deterioration, the increase in emissions that occurs as vehicles age (Wenzel et al. 1997). Thus, improvements in emission-control technologies spurred by emissions standards have affected in-use emissions, including emissions from high emitters.

Finally, previous efforts to decrease contributions from high-emitting vehicles have been problematic. I/M programs are the most prominent method of controlling in-use emissions from vehicles with malfunctioning emission-control equipment, including high emitters. The NRC report (2001) on I/M programs recommended more focused efforts be given to high emitters. However, the report noted some of the difficulties in reducing the number of high emitters that have been encountered in I/M programs. Studies California and elsewhere have shown that many vehicles on the road have failed I/M programs and have not been retested and others have never shown up for initial screening (Ando et al. 2000; Wenzel 2001). Temporary repairs, tampering, and other possible methods of avoiding program requirement have also been used (Lawson 1993; Stedman et al. 1998). Solving some of these persistent problems is a continuing issue with I/M programs.

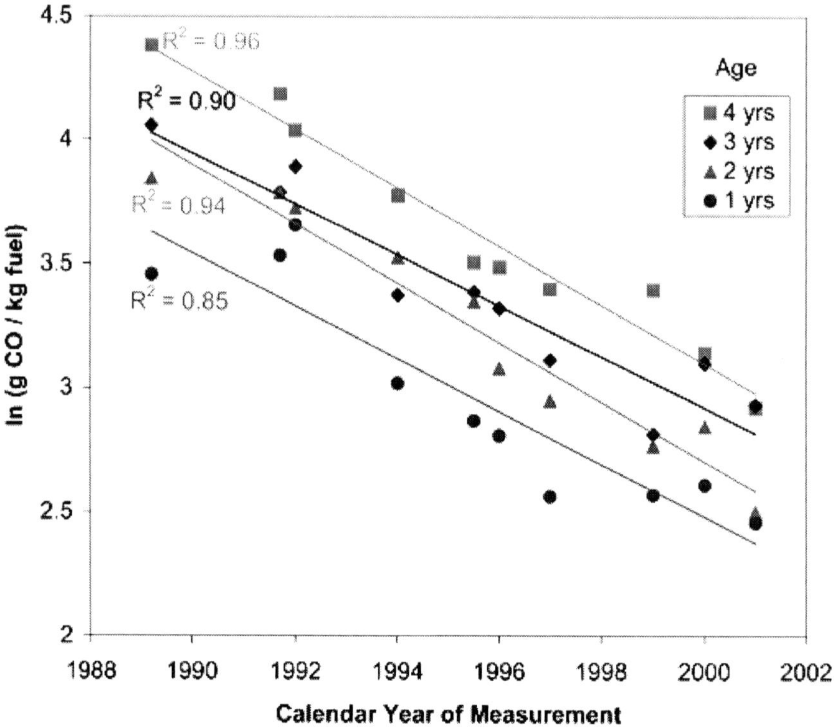

FIGURE 3-2 Decreasing CO emissions from new vehicles. The natural log of grams CO emissions per kilogram of fuel used is plotted by year of measurement for vehicles aged 1-4 years measured between 1989 and 2001 in Denver. Age of vehicle is calculated by subtracting model year from measurement year. Regression lines for each age are drawn and R^2 values are included. Source: Pokharel et al. 2003. Reprinted with permission; copyright 2004, American Chemical Society.

It is also clear from this SIP that federal emissions standards are only part of the emission-control strategies. Various state initiatives are implemented in the DFW SIP to increase total reductions. One major source of reduction is point-source NO_X reduction. This method alone accounts for 35% of the NO_x reductions. I/M of light-duty vehicles contributes 15% to NO_X reductions. Local initiatives also include transportation control measures, which are designed to achieve on-road mobile-source emissions reductions and are included as control measures in the SIP. Examples in the DFW area include intersection improvements, signal improvements, high-occupancy-vehicle (HOV) lanes, freeway corridor management, park and ride lots, pedestrian and bike facilities, rail, and van pools. The DFW SIP also includes controls on small-engine equipment, such as gas-fired water heaters, small boilers, and process heaters, and relies on regulating smaller equipment to contribute to total NO_X emissions reductions. Table 3-11 includes a proposal for speed limit reductions, which were not adopted by the state legislature.

CONCLUSIONS

Regulation of mobile-source emissions through standards on new vehicles and engines is an important component of air quality management. The CAA gave the state of California the authority to set mobile-source emissions standards that differ from the federal standards as long as they are as protective in the aggregate as federal standards. Later amendments to the CAA allowed other states to adopt California standards. The history of California mobile-source emissions standards is told through a succession of scientific findings, legislative actions, and administrative and judicial interpretations that are summarized in this chapter. Several conclusions come from this examination.

- Over the history of mobile-source regulation to date, California has typically led EPA in tightening emissions standards on light-duty vehicles and become a laboratory for emission-control innovations. CARB's regulatory process is supportive of this laboratory role in that California emissions standards can be amended rapidly in the face of changing market and technological conditions.
- Each time CARB sets or substantially revises a California mobile-source emission standard, it must seek a waiver from EPA. The waiver review process can take several years and significant EPA resources to complete, even for relatively straightforward and uncontroversial waivers. In some cases, waivers have been approved after vehicles and engines that meet the standards are in the market. This process creates uncertainty for California, other states considering adopting California standards, and manufacturers.
- States first began using their authority under section 177 of the CAA in the early 1990s when New York and Massachusetts adopted California emissions standards for new light-duty vehicles. The primary reason that these states adopted California emissions standards was to obtain additional emissions reductions to help attain and maintain the NAAQS.

4

Co-evolution of Technology and Emissions Standards

Emissions-control technology for mobile sources has developed in a series of interactive steps with the promulgation of emissions standards for new vehicles and engines and for fuel regulation. For light-duty vehicles, emissions-control hardware has changed greatly over the past 50 years to reflect the changes in emissions standards, vehicle design, fuel-efficiency standards, and technological capabilities. The efforts of motor vehicle manufacturers, the manufacturers of emissions controls and other equipment, the California Air Resources Board (CARB), and the U.S. Environmental Protection Agency (EPA) have made vehicles much cleaner and more durable. Although the relationship among the parties has not always been harmonious, it has produced benefits not only for the United States but also for the world. On-highway diesel vehicles have had emissions-control standards since the 1970s while EPA's emissions-control activities have been authorized for nonroad sources only since the passage of the 1990 Clean Air Act (CAA) amendments.

This chapter discusses the basic elements of mobile-source emissions control, emphasizing the interaction of emissions-control research and new-equipment emissions standards. Chapters 6 and 7 contain a detailed discussion of several emissions standards developed by CARB and EPA. The descriptions of emissions controls in this report are not intended to be comprehensive. Heywood (1988) provided a comprehensive summary of the primary technical issues related to pollutant formation and control methods for light-duty vehicles before 1990. More recent publications discussing current research in combustion and pollution control are available from many publications, including those by the Society of Automotive Engineers (SAE). In particular, the SAE Progress in Technology (Johnson 2002a,b; Johnson 2005a,b) series contains collections of SAE technical papers relating to specific technologies, including combustion and pollutant control in internal combustion engines. The Manufacturers of Emissions Control Association (www.meca.org) is another source for more comprehensive information on mobile-source emissions controls.

TECHNOLOGY-FORCING STANDARDS

A central concept of the standards-setting process for mobile-source emissions by EPA and CARB is "technology forcing." The committee defines a technology forcing standard to be the establishment by a regulatory agency of a requirement to achieve an emissions limit, within a specified time frame, that can be reached through use of unspecified technology or technologies that have not yet been developed for widespread commercial applications and have been shown to be feasible on an experimental or pilot-demonstration basis. The use of technology-forcing standards in the United States contrasts with the standards-setting process in Europe where new emissions-control technologies on mobile sources are required only after they have succeeded in the U.S. market (Faiz et al. 1996).

When controls on vehicle emissions were first considered in Los Angeles in the 1950s, requiring of devices that were not commercially available was not accepted as a viable approach. Los Angeles County Supervisor Kenneth Hahn was denied permission to require the installation of emissions-control devices on vehicles sold in the county by a county legal counsel opinion that stated, "Any such requirement would be "arbitrary, capricious, and void" until "a satisfactory device is perfected and available on the market" (Krier and Ursin 1977, p. 98, as cited in Lents et al. 2000, p. II-10). Lack of progress on air pollution control in California and elsewhere in the United States prompted Congress in the 1970 CAA amendments to feature ambient air quality standards with deadlines for their attainment and a technology-forcing program to reduce emissions from vehicles (Muskie 1990). Chief Senate sponsor, Senator Edmund Muskie, stated that the CAA was not "to be limited by what is or appears to be technologically or economically feasible" but "to establish what the public interest requires to protect the public health of persons" even if that means that "industries will be asked to do what

TABLE 4-1 Major Control Technologies of New Light-Duty Vehicles Sold in the United States

Model Year	Noncatalyst	Oxidizing Catalyst	Three-Way Catalyst and/or Fuel Injection
1974	100%	—	—
1975	16.7	80.7	2.4
1976	16.2	81.7	2.1
1977	10.8	85.3	3.9
1978	8.4	88.0	3.6
1979	8.1	89.6	2.2
1980	0.0	94.4	4.6
1981	0.0	31.0	69.0

Sources: Bresnahan and Yao 1985; Gerard and Lave 2002.

seems to be impossible at the present time." (Committee Report Compiled for the Senate Committee on Public Works by the Library of Congress, Ser. No. 93-18, p. 227, 1974, 116 Cong. Rec. 32901-32902 [1970]). A central component of the CAA was the requirement that new-vehicle emissions standards would be reduced by 90% for carbon monoxide (CO), nitrogen oxides (NO_X), and hydrocarbons (HCs) by 1975, although it was not clear in 1970 how such a reduction would be achieved. The act allowed for any type of technology to be used. Although the date to attain that level of reductions was subsequently delayed, the approach resulted in manufacturers introducing the two-way catalyst to control HCs and CO in 1975 and the three-way catalyst to control HCs, CO, and NO_X in 1981. Table 4-1 shows the penetration of the catalyst into the light-duty vehicle fleet. Using the 1970 requirements for light-duty vehicles, Gerard and Lave (2005) discuss some factors critical to implementing technology-forcing policies.

Technology-forcing standards are not used only for light-duty vehicles. For example, the 1990 CAA amendments recommend that technology-forcing emissions standards be adopted for evaporative emissions and nonroad engines. Section 202(k) requires EPA to develop standards for evaporative emissions control that "shall take effect as expeditiously as possible and shall require the greatest degree of emission reduction achievable by means reasonably expected to be available for production during any model year to which the regulations apply, giving appropriate consideration to fuel volatility, and to cost, energy, and safety factors." Section 213 of the 1990 CAA amendments requires that nonroad emissions standards, which at that time were uncontrolled, "achieve the greatest degree of emission reduction achievable through the application of technology which the Administrator determines will be available for the engines or vehicles to which such standards apply, giving appropriate consideration to the cost of applying such technology within the period of time available to manufacturers and to noise, energy, and safety factors associated with the application of such technology. In determining what degree of reduction will be available, the Administrator shall first consider standards equivalent in stringency to standards for comparable motor vehicles or engines (if any) regulated under section 202."

TECHNOLOGIES TO CONTROL LIGHT-DUTY-VEHICLE EMISSIONS

Emissions-control hardware on motor vehicles can be grouped into three basic types: engine and exhaust controls, evaporative controls, and diagnostic controls. Brief descriptions of these emissions-control technologies are given below. A more complete description of control technology may be found in many sources, including Heywood (1988), Mondt (2000), and NRC (2001). In addition, Chapter 6 contains a discussion of emissions standards adopted since 1990, including the California low emissions vehicle (LEV) standards and the federal standards, and the technologies used to meet those standards. Box 4-1 describes the process of certification and enforcement of emissions standards for light-duty cars and trucks.

Engine and Exhaust Controls

Table 3-3 in Chapter 3 displays a chronology of federal and California exhaust emissions standards for new passenger vehicles. Allowable emissions for light-duty trucks have been reduced at a similar pace, although emissions standards for these vehicles were not as low as those for passenger vehicles until recently. Engine emissions are regulated by performance-based standards (as opposed to technology-based standards),[1] but the standards typically result in certain classes of hardware

---

[1] A performance-based standard is generally technology-neutral and sets an upper limit for emissions coming from a source; a technology-based standard dictates the specific control technology. See Glossary for complete definitions.

being applied for compliance. The first nationwide emissions-reduction requirements were mandated for model-year 1968 and consisted of crankcase and engine controls. As shown in Table 4-2, these standards led to the use of positive crankcase ventilation (PCV) valves, higher air-

Box 4-1 Certification of Light-Duty Vehicles to Emissions Standards

Certification and enforcement of mobile-source emissions standards ensure their effective implementation. The general process of certification and enforcement is described here for light-duty cars and trucks.

The requirements for light-duty-vehicle certification for EPA and CARB are similar because of collaborative efforts between the agencies and industry in the 1990s aimed at harmonizing EPA and CARB procedures to the greatest extent possible, streamlining the process for manufacturers, and shifting the focus to in-use testing of vehicles. These efforts resulted in EPA's adoption of the Compliance Assurance Program 2000 (CAP2000) described further in its 1998 notice of proposed rule-making (NPRM) (63 Fed. Reg. 39654 [1998]) and 1999 Final Rule (64 Fed. Reg. 23906 [1999]), and CARB's adoption of the elements of the CAP2000 program with its LEV II rule-making. Certification and enforcement occur in two phases: pre-production certification and in-use testing (enforcement), which are described in the following paragraphs based on information in EPA's Final Rule (64 FR 23906) and CARB's LEV II staff paper "Initial Statement of Reasons" (CARB 1998a). A separate issue relates to the enforcement of sales of California-certified vehicles in California and states that have adopted those standards. Such enforcement is connected to vehicle registration, since California-certified vehicles can be identified by the vehicle identification number (VIN). The committee did not investigate sales enforcement practices.

Before to selling a vehicle in the California or national market, manufacturers must submit to CARB or EPA, respectively, test-data that demonstrate the vehicle will meet corresponding emissions standards. Data must be submitted yearly for each engine family (group of vehicles with the same engine and emissions-control characteristics). The agencies then approve the vehicle for sale by issuing a certificate of conformity (EPA) or executive order (CARB). Preproduction testing must demonstrate that vehicles will comply with emissions standards at the end of their useful life, which is 120,000 mi for present-day light-duty vehicles. Compliance is demonstrated by determining the deterioration characteristics of emissions-control equipment (durability demonstration) and then showing that production vehicles will meet emissions standards using this same equipment and configuration (certification testing). Manufacturers have some flexibility in durability demonstrations, although their methods must be approved by the agencies. Durability demonstrations are performed on one prototype representing a broad group of vehicles with similar deterioration characteristics (durability groups) rather than on each engine family or vehicle model. Bench-aging of catalysts is one method for testing durability of emissions-control systems, as opposed to accumulating full useful-life mileage on an actual vehicle. Once the useful-life emissions of an emissions control system/configuration have been determined, it must be shown to result in emissions below the certification levels on production vehicles. One prototype vehicle from each test group (vehicles of the same emissions rates and certification levels) is used to demonstrate compliance, for example, by attaching a bench-aged catalyst and measuring the emissions according the defined federal test procedure (FTP). California and EPA certify vehicles in this same framework, but manufacturers must certify separately with each agency because certification categories are different in the California and federal programs (see Chapter 6 for more information on the LEV II and Tier 2 programs). Furthermore, there are differences between CARB and EPA in test procedures—for example, the ambient temperatures at which vehicles must be certified.

In-use testing comprises the second part of the certification and enforcement program. Both agencies procure vehicles and test their emissions in-use to detect those that might not be achieving emissions levels at their certification levels. Manufacturers are also required to procure vehicles and test them in-use for emissions at low (10,000 mi), medium (50,000 mi), and high mileage (75,000-105,000 mi). These vehicles are tested as is; they are not screened for proper maintenance. If the vehicles do not meet emissions requirements, the manufacturers must test other in-use vehicles of this model to ascertain whether the emissions-control equipment functions properly. Testing data are provided to the agencies to help them target other vehicle models that might not comply with emissions standards. If in-use testing shows that certain vehicle models are not meeting their certification levels, remedial action may be required. CARB and EPA may order a vehicle recall at the manufacturers' expense. A secondary benefit of in-use testing is that it provides EPA and CARB with in-use emissions data to be used in accurately assessing and predicting vehicle emissions. The in-use testing data also helps manufacturers to evaluate how well their durability demonstrations simulate actual deterioration.

TABLE 4-2 Summary of Emissions Controls Used to Meet Federal Standards for Light-Duty Vehicles

Engine Emissions Controls

- Crankcase controls (1961 to present)
  —PCV used to recycle hydrocarbons that leak past the piston rings into the crankcase back into the combustion process.
- Engine adjustments (1968 to present)
  —Primary control consisted of modifications to mixture strength and spark timing as well as EGR.
- Oxidizing catalysts (1975 to present)
  —Lean mixtures and oxidization catalysts were used for HC and CO control. EGR was used to control NO_x. Air-pump or pulse-air valves were incorporated.
- Unleaded gasoline phase-in (enabling technology for catalysts) (1975 to present)
- Closed loop three-way catalysts (1981 to present)
  —Precise mixture control and three-way catalysts control HCs, CO, and NO_x; >90% removal of each.

Electronic Controls and Onboard Diagnostic (OBD) Systems

- Onboard computers, oxygen sensors (1981 to present)
- Preregulatory OBD Systems (1981 to 1993)
  —GM and Ford had OBD systems starting on 1981 models.
- OBDI (1994 to 1995)
- OBDII (1996 to present)

Evaporative Emissions Controls

- Early trap test technology (1971 to 1977)
  —Tank and carburetor bowl were vented to a small carbon canister.
- Early sealed housing for evaporative determination (SHED) test technology (1978 to 1995)
  —Material in the detail seals on the carburetor is claimed for reduced permeation and increased purge.
- Enhanced evaporative emissions controls (1996 to present)
  —Three-day diurnals, measuring running losses, high-temperature hot soaks, and 10-year life required larger canisters and more permeation control. Refueling controls were added to cars starting in 1998.

Source: NRC 2001.

to-fuel ratios, spark-timing modifications, and external exhaust-gas recirculation (EGR). As described in Chapter 3, these standards were modeled after standards mandated for California vehicles a few years earlier. Emissions-control systems used on the precatalyst vehicles of 1968-1974 relied on combustion controls, and some aspects of these controls compromised the performance of the engine, resulting in degraded drivability and fuel economy and promoting the practice of tampering with the emissions-control system (Bresnahan and Yao 1985; NRC 2001).

In contrast, EGR systems developed during this precatalyst period retained most of the vehicle performance while lowering tailpipe NO_x. Such systems are still in use today. Figure 4-1 shows a diagram of a modern EGR system. Higher combustion temperatures result in greater NO_x formation. The EGR valve acts to recirculate exhaust gas in controlled amounts to dilute the air-fuel mixture, which lowers the peak combustion temperature and NO_x formation. EGR is also used to reduce NO_x emissions from diesel engines.

Exhaust emissions-control devices consist of catalytic converters and air injection systems. The first generation of catalytic converters added to model-year 1975 vehicles promoted the oxidation of HCs and CO by passing the exhaust over a bed containing small amounts of platinum, palladium, and rhodium. These were known as two-way or oxidation catalysts. A critical regulatory requirement for technology was the use of unleaded gasoline since the combustion by-products of tetra-ethyl-lead, an additive used to increase gasoline octane quality, were found to reduce catalyst conversion efficiency.

Stricter model-year 1981 standards provided another challenge. A development known as the three-way catalyst, which provides control of NO_x in addition to CO and HCs, came into use and continues to be a central component of emissions controls. Figure 4-2 shows a diagram of a three-way catalyst. Another key technological development needed for the three-way catalyst is the adoption of electronic controls to tightly maintain the air-to-fuel ratio[2] through the metering of fuel. Over the years since the three-way catalyst was first introduced, important im-

---

[2] The air-to-fuel ratio is the ratio of the weight of air to gasoline entering the intake in a gasoline engine. The ideal ratio for complete combustion is 14.7. This ratio is called the stoichiometric air-to-fuel ratio. Air-to-fuel ratios less than 14.7 are called "rich" and contain excess fuel for complete combustion; air-to-fuel ratios greater than 14.7 are called "lean" and contain more air than is required for complete combustion.

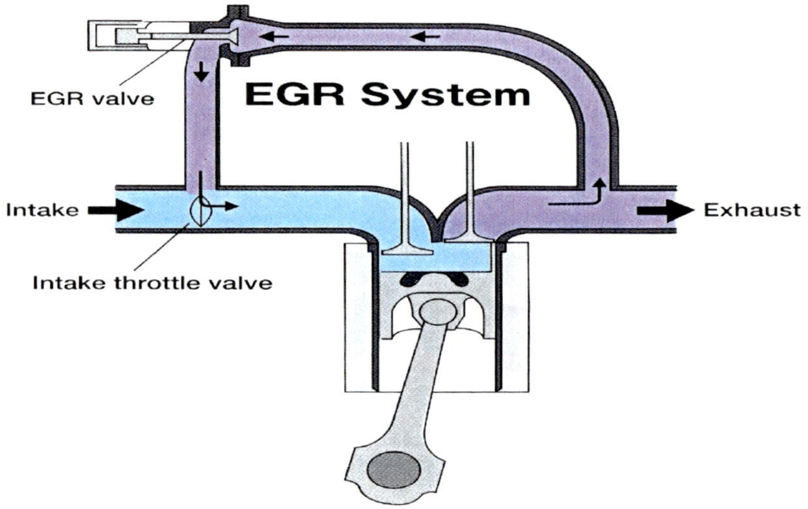

FIGURE 4-1 Exhaust gas recirculation. Source: UTI 2005. Reprinted with permission; copyright 2005, Isuzu Motors, Japan.

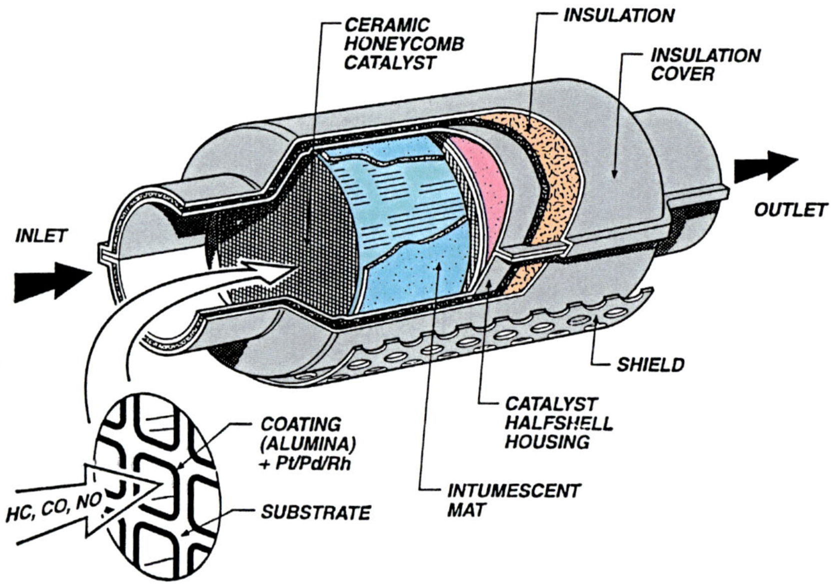

FIGURE 4-2 Schematic of a three-way catalyst. Source: MECA 2003. Reprinted with permission; copyright 2004, Manufacturers of Emissions Control Equipment.

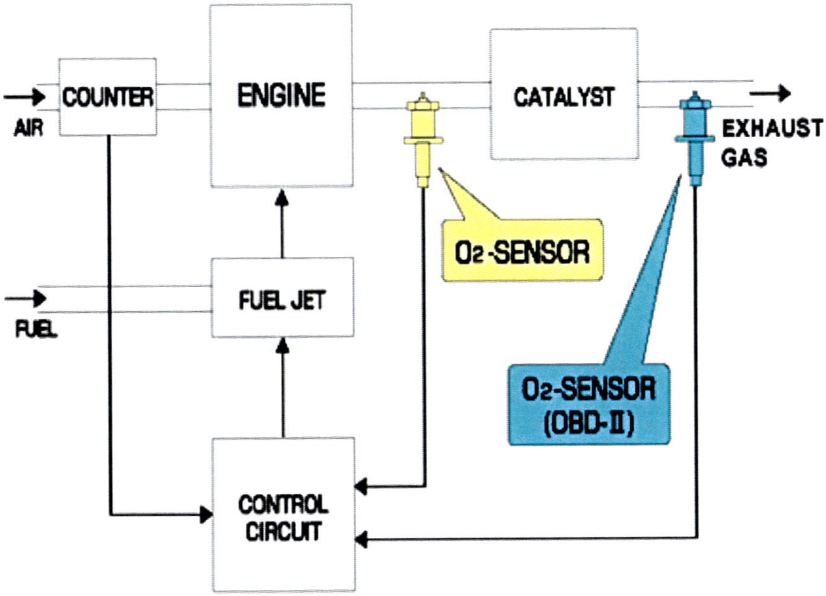

FIGURE 4-3 Schematic of closed-loop controls. Source: Kyocera 2006. Reprinted with permission from Kyocera; copyright 2006, Kyocera International Inc.

provements have been made in catalyst efficiency and life and exhaust gas sensors. The scope of improvement has allowed the three-way system to achieve the progressively stricter California and federal standards.

Electronic Controls and Onboard Diagnostics

Electronic controls on light-duty vehicles involve the use of on-board computers, sometimes known as engine-control units, and oxygen sensors and enable the adoption of closed-loop fuel control. Closed-loop control consists of the use of oxygen sensors in the exhaust and fuel-rate adjustment capability in the carburetor or fuel injection system. Figure 4-3 shows a schematic representation of the closed-loop system. The front (or upstream) oxygen sensor monitors the efficiency of combustion in the engine and allows the creation of a self-adjusting fuel metering system that maintains the air-to-fuel ratio within a very narrow range. As shown in Figure 4-4, the ability to decrease NO_x, CO, and HCs simultaneously in a three-way catalyst depends on the accurate control of the air-to-fuel

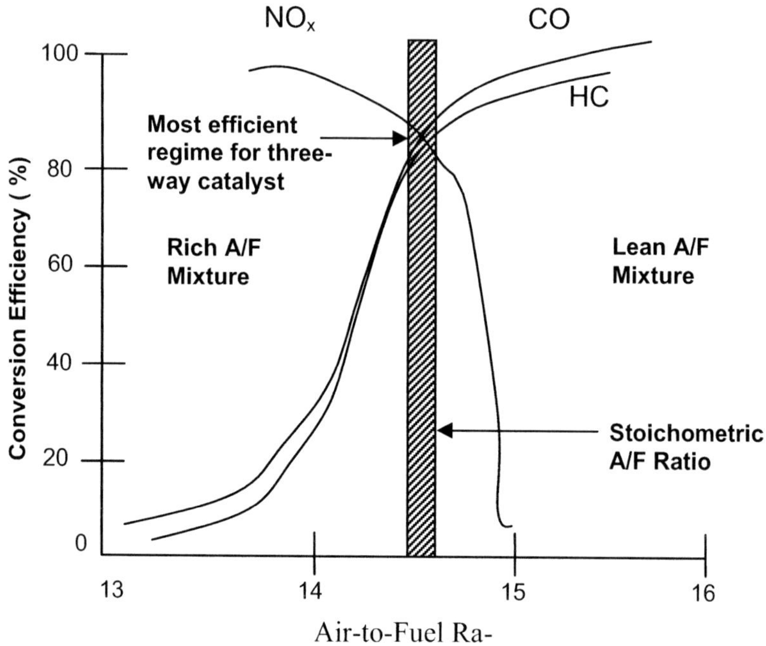

FIGURE 4-4 Catalyst conversion efficiency as a function of air-to-fuel (A/F) mixture ratio. Source: Adapted from Canale et al. 1978. Reprinted with permission; copyright 1978, SAE International.

ratio. The rear (or downstream) oxygen sensor monitors the efficiency of the catalytic converter. Onboard computers are also able to control ignition timing, transmission gear changes, and in a few engines, even valve timing.

Onboard diagnostic (OBD) systems are incorporated into the computers of vehicles to monitor the performance of the emissions controls. OBD hardware and software do not directly control emissions but are a vital part of emissions-control systems by monitoring various engine functions, including the emissions-control system. Some manufacturers incorporated OBD on a voluntary basis in model-year 1981 to help with the service and reliability of their vehicles (Grimm et al. 1980; Gumbleton and Bowler 1982). The OBD system is made up of the sensors and actuators used to monitor specific components as well as the diagnostic software in the onboard computer. California regulators recognized the potential of the OBD system, expanded the scope, and required it on new vehicles starting with a 1988 model-year phase-in. California and EPA expanded the scope and coverage of diagnostics with the OBD II regulations, which were phased in beginning with the model-year 1994 vehicles. All light-duty vehicles built after 1996 (with a few exceptions) are equipped with the OBD II system. OBD II periodically checks many emissions-control functions, including the following components: catalysts, oxygen sensors, evaporative canister purge system, fuel-tank leak check, misfire detection, and onboard computers. As indicated in Figure 4-3, the rear oxygen sensor is required as a part of the OBD II—OBD I did not require the monitoring of the performance of the catalytic converter.

Cold-Start Emissions Control

HC and CO emissions are higher during starting and the first few minutes of vehicle operation. Under cold-start conditions, the engine computer commands the fuel injectors to add excess fuel to the intake air to ensure that enough fuel evaporates to yield a flammable mixture in the engine. A typical engine-computer strategy injects excess fuel during the first engine start using a fixed fueling schedule to reach idling conditions. This open-loop operation, before the catalyst reaches peak efficiency, can continue for several minutes at low ambient temperatures. During this period, the engines in properly operating modern vehicles have the highest emissions rates of CO, air toxics, and unburned HCs. Typical heat-up times under mild ambient conditions (70-80°F) can be about 1 minute (min) for modern catalysts and even as short as a few seconds for modern close-coupled catalysts (catalysts close to the engine). When ambient temperatures are −20°F or lower, however, catalyst and engine warm-up times can exceed 5 min (Sierra Research 1999). The long warm-up time can result in a substantial increase in emissions, as shown in Figures 4-5 and 4-6. In some locations, cold-start emissions can also be a large fraction of the emissions inventory. For example, in Fairbanks, Alaska, winter cold-start and initial-idle emissions contributed an estimated 45% of overall on-road CO emissions (NRC 2002a). Because of the importance of cold-start CO emissions, new cars and the lightest category of light-duty trucks (LDT1) have been required since 1994 to meet a CO limit of 10 g/mi on certification tests conducted at 20°F.

To meet stricter emissions standards, including the cold-start CO standards and federal Tier 2 and California LEV II standards, vehicle

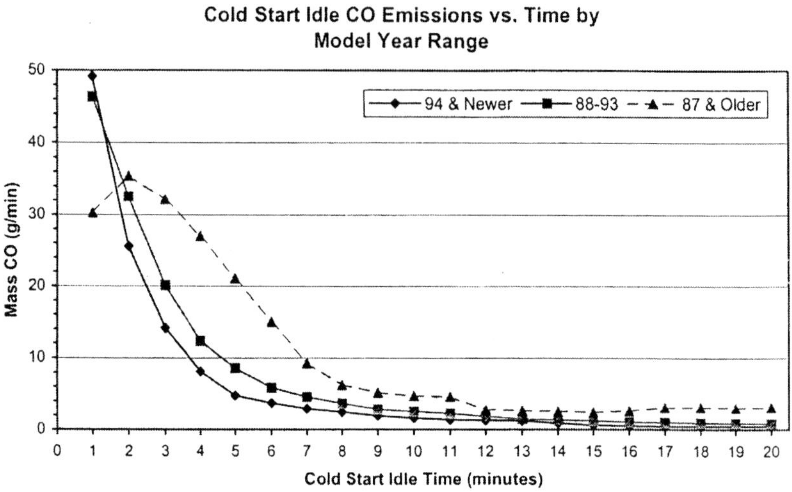

FIGURE 4-5 Average CO emissions for a 111-vehicle test sample taken in Fairbanks and Anchorage, Alaska, under temperatures ranging from −34° F to 14° F ("94 & Newer" applies to model years 1994-1998). Source: Sierra Research 2000. Reprinted with permission; copyright 2000, Sierra Research, Inc.

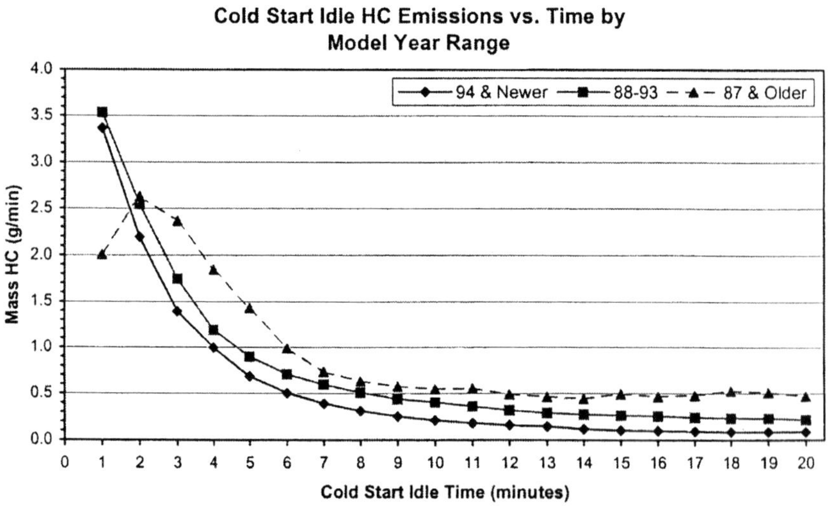

FIGURE 4-6 Average HC emissions for a 111-vehicle test sample taken in Fairbanks and Anchorage, Alaska, under temperatures ranging from −34° F to 14° F ("94 & Newer" applies to model years 1994-1998). Source: Sierra Research 2000. Reprinted with permission; copyright 2000, Sierra Research, Inc.

manufacturers and suppliers focused much effort on reducing cold-start emissions. This has been accomplished with low thermal inertia manifolds, ignition timing changes, and leaner air-fuel mixtures during cold start along with low-heat-capacity catalysts with higher geometric surface area and improved catalytic layer design.

Evaporative Emissions Control

Evaporative emissions are the HCs that escape from the vehicle that do not come from the tailpipe. Originally, losses due to the evaporation of residual fuel in the fuel metering system and diurnal fuel tank losses were of concern. Running losses (evaporative emissions during vehicle operation) also have been found to contribute to vehicle fuel emissions. Evaporative emissions are evaluated using the Sealed Housing Evaporative Determination (SHED) test developed by General Motors.

The uncertainties in estimating the evaporative contribution to mobile-source HC emissions are large. Evaporative emissions were first controlled nationwide in model-year 1971. Residual gasoline fuel vapors in the carburetor and fuel tank were routed to a small (about 1 liter) container of activated carbon for temporary storage and eventual use by the engine. Figure 4-7 shows the design of a carbon canister. The basic design of evaporative control hardware has not changed much since 1971, but as standards on evaporative emissions became more stringent with the conclusion of running losses, control effectiveness increased greatly due to improvements in materials, understanding, and measurement techniques. For example, Babik (2005) reported that General Motors would use less-permeable materials to construct fuel tanks, reduce the number of connections in the fuel lines, increase the size and efficiency of the carbon canister, and make other modifications to meet new California evaporative emissions standards.

Fuel Composition and Emissions-Control Technologies

Fuel composition and quality are intrinsic in the design, development, and performance of vehicle systems to meet the tailpipe and evaporative emissions regulations. Some regulations can be considered stand-alone programs, such as those for reformulated fuels or oxygenated fuels. These are outside the scope of this report. Other properties affect

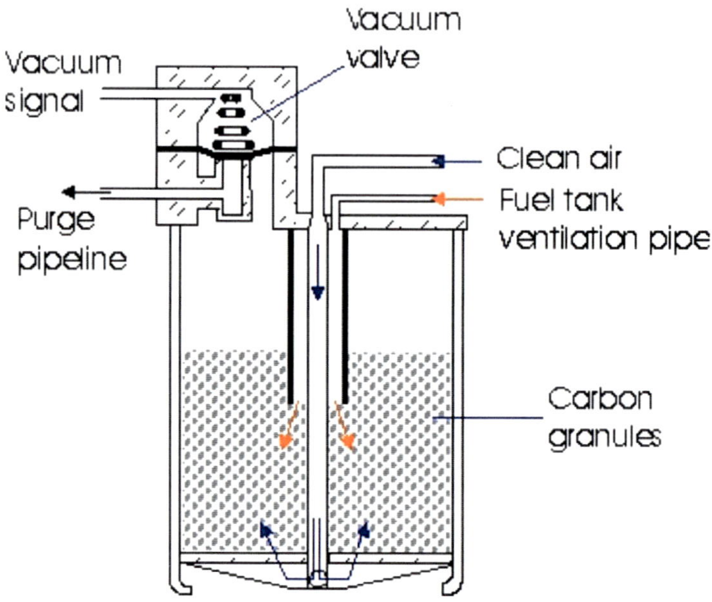

FIGURE 4-7 Cross-section of a carbon canister. Source: TTM 2004. Reprinted with permission; copyright 2004, Tightrope Technologies Motors, Inc.

the operation of emissions-control equipment. An important example is the removal of lead from gasoline. The removal of lead from gasoline, beginning in 1975 and completed by 1992, enabled the widespread adoption of catalysts. Other advantages came with the adoption of catalysts and the use of unleaded gasoline, including increased spark-plug and engine life, longer exhaust-system life, and extended oil-change intervals. In addition, the health effects of airborne lead were reduced. The sulfur content of fuel has also been recognized to adversely affect the performance of catalyst technology (MECA 1998). This was a key finding of the Auto-Oil Project (Benson et al. 1991), an industry-funded multiyear multicompany research initiative that helped to bring about the introduction of reformulated gasoline. Among this project's conclusions was that reducing sulfur concentrations from 450 to 50 ppm would result in over a 10% decrease in CO and HC exhaust emissions in 1990 model year vehicles. Although the impacts of sulfur are not as severe as the impacts of lead in gasoline, sulfur in fuel is converted during combustion to various sulfur-containing compounds that react with the catalyst surface and inhibit the removal of other target pollutants. To address concerns about the increased sensitivity of the newer technology in vehicles to sulfur poisoning, EPA included new fuel standards that require refiners to meet an average sulfur concentration of 30 ppm beginning by January 1, 2006, in its Tier 2 rule (65 Fed. Reg. 6697 [2000]). Low-sulfur fuel provided by the Tier 2 regulation is expected to improve catalyst performance of many on-road light-duty vehicles by up to 30% for HCs, CO, and NO_x. Meeting an average sulfur content of 30 ppm has been required in California gasoline since 1996.

TECHNOLOGIES TO CONTROL HEAVY-DUTY-VEHICLE EMISSIONS

Current federal regulations do not require certification of complete heavy-duty diesel vehicles, requiring instead certification of only the engines. This is because of the difficulty in devising per mile limits for the broad range of vehicles covered and the difficulty in developing a practical chassis dynamometer test. Consequently, the basic standards are expressed in grams per brake horsepower-hour (g/bhp·hr) instead of grams per mile, the unit used for cars and light trucks.

Early regulations on heavy-duty diesel engines began in 1968 when the National Air Pollution Control Administration, the agency that set emissions standards before the founding of EPA, issued regulations to limit visible smoke emissions from diesel engines used in on-road trucks and buses. The technology used to achieve less smoke involved increasing the air-to-fuel ratio by turbocharging, intercooling (cooling the engine intake air to lower NO_x emissions), and on some models, limiting the fuel rate on acceleration and adjusting the engine timing. On-road heavy-duty diesel vehicle emissions standards were implemented in California in 1973 and the rest of the United States in 1974, the standards were harmonized in 1988 (Lloyd and Cackette 2001). It was not until the 1977 CAA amendments that technology-forcing requirements for diesel particulates and NO_x were adopted, calling for heavy-duty diesel engines to achieve the greatest emissions reduction achievable consistent with consideration of costs, technology feasibility, and other factors. These standards became more stringent throughout the 1990s, including more stringent particulate matter standards for urban buses. Diesel engine and combustion technology during 1988-2000 was vastly improved, more than was achieved in the first 100 years of the diesel engine. Through model-year 2006, the standards were met primarily through engine-operation modifications, including fuel injection, electronic engine controls, combustion chamber design, air handling, and reduced oil consumption (MECA 1997). In particular, the standards were attained through improved diesel combustion systems using high-pressure electrically actuated fuel injectors. Variable injection timing and multiple injections during a combustion event were possible through computer control. EGR is also becoming popular on diesel engines as a means to reduce NO_x.

In early 2001, EPA issued new, more stringent regulations on emissions from heavy-duty vehicles (65 Fed. Reg. 59896 [2000]; 66 Fed. Reg. 1535 [2001]). These regulations tighten emissions standards and require a decrease in the fuel sulfur content, strategies similar to those adopted for light-duty vehicles. Low sulfur fuel is an important prerequisite for developing technologies to lower PM emissions. The regulations that will be phased in for 2007 to 2010 model-year vehicles will reduce PM and NO_x emissions by at least 90% compared with current standards. The use of new exhaust emissions-control technologies will be required to meet the more stringent standards for diesel engines. These standards force examination of a range of technologies, including diesel particle filters (DPF), selective catalytic reduction (SCR) of NO_x with ammonia, and NO_x absorber catalysts. Figure 4-8 shows a diagram of an SCR system. Chapter 7 contains a discussion of these standards and the technologies that will be used to meet these standards.

Gasoline heavy-duty engines, like those used in passenger vehicles, have benefited from the application of high-energy ignition (HEI), positive crankcase ventilation (PCV), exhaust gas recirculation (EGR), and oxidation catalyst technologies. Subsequently, more stringent heavy-duty gasoline engine regulations were met using a heavy-duty version of the three-way catalyst and closed-loop oxygen-sensor fuel-metering system proven in passenger cars.

TECHNOLOGIES TO CONTROL NONROAD SOURCES FROM LAWNMOWERS TO LOCOMOTIVES

As emissions from on-road sources were reduced, emissions from nonroad sources became a more important issue. Nonroad sources include the following engine categories:

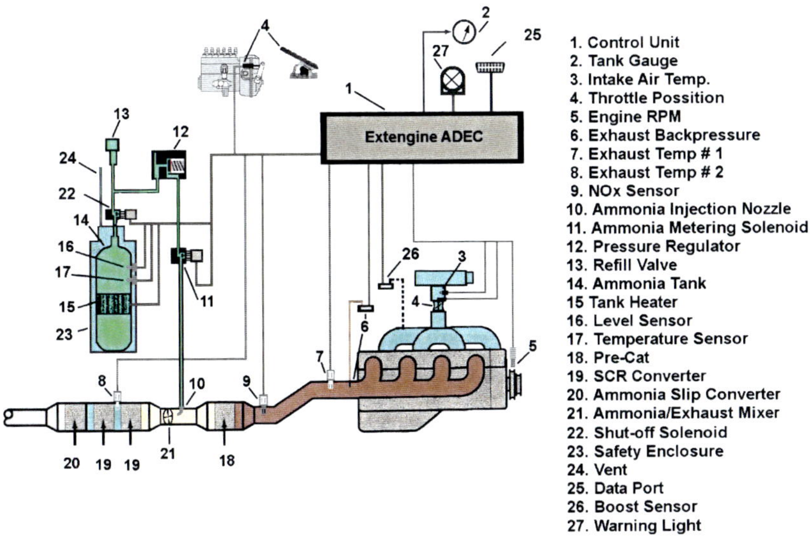

FIGURE 4-8 Schematic of selective catalytic reduction (SCR) system for NO_x reduction. Source: Omnitek 2006. Reprinted with permission; copyright 2006, Omnitek Engineering Corporation.

- Compression-ignition-equipment engines (construction and mining, agriculture).
- Small spark-ignition engines (lawn mowers, chain saws)
- Large spark-ignition engines (forklifts, generators)
- Marine compression-ignition vessel engines (commercial, recreational inboard)
- Marine spark-ignition vessel engines (jet skis, personal watercraft)
- Recreational vehicles (all-terrain vehicles [ATVs], snowmobiles, motorcycles)

The 1990 CAA amendments directed EPA to prepare a study of the scope and sources of nonroad emissions and to regulate them if they were found to make a substantial contribution to nonattainment of ozone or CO ambient air quality standards. The EPA report did not make a formal determination of a significant effect, but it contained an inventory of emissions from nonroad sources and concluded, "because nonroad sources are among the few remaining uncontrolled sources of pollution, their emissions appear large in comparison to the emissions from sources that are already subject to substantial emissions-control requirements" (EPA 1991). Space does not permit a full description of all regulatory actions by EPA and CARB. A few categories of such sources and their emissions-control technologies are provided. Chapter 7 contains a discussion of recent emissions standards for small nonroad spark-ignition engines and personal watercraft.

Nonroad emissions are the new frontier for mobile-source emissions control for EPA and CARB. Table 4-3 lists some of the possible emissions-control technologies for nonroad source. The multiple engine types in nonroad sources and their variety of uses are likely to require multiple emissions standards. Different nonroad sources will need to have sets of standards and procedures for demonstrating compliance. Although some control technology, such as the oxidation catalyst, is well understood, the challenge will be in applying this technology to specific engine applications. Transfer of technology to farm and construction equipment from heavy-duty diesel and gasoline engines involves different duty cycles, environment, and durability needs. Nowhere is the coop-

TABLE 4-3 Possible Emissions-Control Technologies for Nonroad Mobile Sources

- Spark-Ignition Engines
  - Fuel injection and feedback control systems
  - Exhaust gas recirculation systems
  - Three-way catalysts and advanced catalyst systems
  - High-energy ignition
  - Hybrid electric systems
  - Advanced combustion system design and control

- Compression-Ignition Engines
  - Turbocharging
  - Intercooling
  - Cooled exhaust gas recirculation systems
  - Oxidation catalysts
  - Selective catalyst reduction system
  - Lean NO_x catalysts
  - NO_x storage catalysts
  - Catalyzed particulate filters
  - Hybrid electric systems

erative evolution of technology and standards more appropriate than that for nonroad sources.

Nonroad Diesels and Locomotives

As described in Chapter 3, locomotives and nonroad engines used in farm and construction equipment with engines smaller than 175 horsepower (hp) have been and continue to be exempt from state regulation. Locomotive engines operate on diesel fuel and can be either two- or four-stroke design engines. EPA has regulated locomotives nationwide since 2000. Engines smaller than 175 hp used in farm and construction equipment can be either gasoline or diesel powered and are primarily four-stroke design. Nearly all large farm and construction equipment is diesel powered and can share control technology with heavy-duty trucks. Spark-ignition engines in farm and construction applications may use versions of gasoline heavy-duty-truck emissions controls.

Handheld Engine Applications

Handheld engine applications can be two- and four-stroke gasoline-fueled engines. String trimmers, leaf blowers, and chain saws rely on the low weight, compact high power of two-stroke gasoline engines. As discussed in Chapter 7, two-stroke engines emit large amounts of unburned fuel and therefore emit more HC emissions than four-stroke engines (see also Boyle 2002). Substituting a larger, heavier four-stroke engine poses challenges in weight, cost, and product performance, yet many four-stroke engines are being developed with lower exhaust emissions. Some applications (for example, European chain saws) include some emissions-control technology. However, the capability of such control technology to achieve EPA and state chain saw emissions regulations and durability remains to be determined by manufacturers.

Lawn Mowers

Most lawn mowers are powered by four-stroke spark ignition engines, although a few are powered by electric motors. Most gasoline-powered lawn mowers have high HC and CO emissions during operation and high evaporative losses during refueling, operation, and storage. Catalyst use to reduce exhaust emissions presents some complications for this equipment. The heat release when HCs and CO are oxidized can result in high exhaust temperatures and hot catalytic-device surfaces. The application of catalytic exhaust systems can increase the risk of fire during operation, refueling, and storage. Improved four-stroke engine designs and catalytic-device shielding might reduce the risk of fires. New technologies, such as stratified scavenging, have been found to improve two-stroke engine emissions performance. These technologies create a separation of the exhaust flow and the intake flow by creating pressurized pockets that prevent the flow of fresh fuel charge into the combustion chamber until the exhaust cycle is complete. Unburned fuel (HCs) is thereby reduced. The most recent CARB standards for these engines, including fuel evaporative controls, are discussed more in Chapter 7.

Watercraft and Related Two-Stroke Engine Applications

Personal watercrafts, such as jet skis, are powered by two-stroke engines for their superior power-to-weight ratio. Lower emissions and direct-injected two- and four-stroke engines are being developed for these applications and for snowmobiles. Even though the four-stroke engine emits much less HC than a two-stroke engine, emissions-control equipment to reduce engine-out emissions of CO and HCs may be necessary to meet regulations. Chapter 7 contains more information on recent emissions standards for jet skis.

CONCLUSIONS

The basic elements of mobile-source emissions control result from the co-evolution of emissions-control research and the promulgation of vehicle and engine emissions standards. A central concept of the standards-setting process for mobile-source emissions is technology forcing. A technology-forcing emissions standard requires a new vehicle or engine to achieve an emissions limit through use of unspecified technology or technologies that have not yet been developed for widespread commercial applications. A review of emissions-control technologies emphasizes two general conclusions.

- The concept of technology forcing is central to the standards-setting process for CARB and EPA. It has been applied to a wide range of sources, including light-duty vehicles, on-road diesel engines, and nonroad engines, for the control of CO, HCs, and NO_x.
- Over the almost 50 years of mobile-source emissions regulations, controls have evolved from the use of simple technologies to control light-duty vehicles to today's sophisticated integration of engine, fuels, and emissions-control technologies to control emissions from an array of mobile sources. Compared with emissions rates of 1967 model-year light-duty vehicles, the rates of new, properly operating vehicles decreased by 95-99%.

5

Assessment of Different Approaches to Setting Mobile-Source Standards

The current system for setting mobile-source emissions standards permits California to set more stringent standards than the federal U.S. Environmental Protection Agency (EPA) and allows other states to choose between the two. This system is an attempt to create a compromise between two legitimate, yet competing interests—that of state governments attempting to tailor regulations to meet air quality objectives and that of manufacturers attempting to keep costs of equipment and distribution down.

In this chapter, the committee elaborates on the rationales for uniform versus two standards to clarify the key factors favoring one or the other. The committee also discusses the harmonization and possible impacts of different approaches in setting emissions standards. The chapter concludes by outlining a framework for assessing the costs and benefits to provide a basis for comparing the methods used by EPA, California Air Resources Board (CARB), and other states that choose to adopt California's emissions standards.

RATIONALES FOR DIFFERENT APPROACHES TO SETTING MOBILE-SOURCE STANDARDS

The current system for establishing mobile-source emissions standards in the United States is a compromise between two conflicting interests. From the perspective of a mobile-source manufacturer, the advantages of uniform national emissions standards are clear and compelling. Dual standards complicate equipment design, magnify certification costs and risks, erode economies of scale, and complicate supply-chain logistics. Uniform national standards provide substantial economies in production throughout the supply chain. By helping to keep costs down, uniform standards benefit consumers. Dual standards will mean higher prices for mobile sources in some or all regions, depending on how higher costs are passed on to consumers.

From the perspective of a state government given the responsibility of protecting the public's health by meeting the National Ambient Air Quality Standards (NAAQS), the need to have commensurate authority to regulate emissions by all possible means is equally clear and compelling. If one option for reducing emissions from a major contributor such as a mobile source is preempted, the task of meeting air quality requirements might be more difficult. Both perspectives are valid, and the current system under which California may set its own standards and other states may adopt those standards reflects a particular compromise between the two competing interests. In this section, the arguments for having uniform mobile-source emissions standards and for having dual standards are examined.

Arguments in Favor of Uniform Emissions Standards

From the earliest days of regulation of pollution sources, some argued that pollution standards for new mobile sources should be set at a uniform level in all parts of the country. In fact, as this report describes in Chapter 3, the federal government preempted the ability of individual states to set separate mobile-source emissions standards and allowed only California to obtain a waiver from the preemption.

The early legislative debate brought out a number of justifications for setting uniform standards. On the basis of arguments presented by vehicle manufacturers and others, Congress found that allowing states to set different standards might cause economic disruption in vehicle markets and increased cost to consumers. Because of the substantial time frame needed to modify and manufacture vehicles to different emissions specifications, a particular concern was having consistent and fixed standards for manufacturers to take advantage of economies in productions. An additional argument in favor of uniform standards was that mobile emissions sources often cross state lines; therefore, the impacts of different state standards on air quality and human health may be diluted and difficult to discern. For example, many heavy-duty diesel trucks routinely cross state lines.

In general, a uniform standard is thought to lower overall costs of bringing mobile-source products to market. These arguments are drawn from early congressional debates, vehicle manufacturers, and economic principles. A uniform standard would accomplish the following:

- Lower costs as a result of economies of scale in production. Large-scale production also allows fixed costs of meeting new emissions standards to be spread across more vehicles.
- Keep the costs of distributing vehicles low. Different standards require mobile- source products to be matched to particular states.
- Avoid duplicating inventory, certification, and testing costs that might result when there is more than one standard.
- Avoid boundary and enforcement problems. Mobile sources cross state lines and enforcing different state standards would add to costs.

Economies of Scale

Economies of scale in the production of mobile sources occurs when the cost per unit declines with the number of units produced. Large-scale production to meet a single standard could affect costs in several ways. First, fixed costs imposed by a new standard include research and development (R&D) expenditures, redesign and retooling of plants, and certification of the vehicles or other mobile-source products. Those costs would be reduced on a per unit basis with a larger volume of production. For more than one standard, those fixed costs would be higher but spread over approximately the same number of vehicles, raising the cost per vehicle.

Pure economies of scale in production often occurs up to a certain output level. Due to specialization and production efficiencies, costs per unit fall as production volume increases along an assembly line or across an entire production facility. It has long been argued that scale economies continue for very large volumes of vehicle production (Maxcy and Silberston 1959). However, automakers argue that it is increasingly difficult today to take advantage of scale economies in production because of the rapid pace of technical change, which causes models to be outdated more quickly than in the past. Different regulatory requirements may also contribute to this trend. Although lower costs through larger production on a given assembly line still occur, manufacturers often cannot take full advantage of scale economics (J. German, Honda, personal commun., May 18, 2005; R. Babik, General Motors, personal commun., June 21, 2005).

There is evidence, however, that manufacturers are attempting to find and exploit economies of scale in production whenever possible. Truett and Truett (2003) find evidence of economies of scale for Fiat in the Italian auto industry. The authors argue that for Fiat to be competitive with other manufacturers in Europe, they will need to increase volume to lower costs. Also, Chrysler is moving toward a world engine for all its vehicles, which would allow Chrysler to achieve further scale economies and reduce engine prices (Fletcher 2004). General Motors has moved to purchase other companies, including Saab, in part to increase the potential for lower costs. It is integrating production facilities to achieve savings through scale economies and through common administrative offices in finance, information systems, and fleet management. In a similar approach, Ford and PSA Peugeot Citroën are integrating design and production of diesel engines to achieve scale economies for lower-cost engines that will be used in each companies' different product lines. The consolidation of vehicle manufacturers also points to the increased globalization that is occurring among manufacturers of mobile sources.

Scale economies can reduce the cost of meeting any standard. However, the extent of these economies can vary a great deal for different engine types, vehicle types, and design changes. Compared with light-duty-vehicle volumes, heavy-duty engine and vehicle sales volumes are lower overall, so increasing sales volume with a uniform standard might be particularly important for manufacturers of heavy-duty engines and vehicles to take advantage of scale economies (French 2004).

In general, the committee could not obtain data from individual manufacturers on the effects of specific emissions standards on fixed costs or production costs. Manufacturers were reluctant to disclose such proprietary information. Therefore, it is difficult to say that the extent to which scale economies would be affected by a different standard in one or more regions as compared with a uniform standard.

Inventory Costs

Uniform standards are most likely to keep inventory costs at the lowest possible level for holding stocks of vehicles. Production levels are always subject to uncertainty about demand conditions. To the extent that production does not exactly match demand, inventories will rise or fall. The larger the number of types of products that must be delivered to market, the greater the inventory holdings must be to meet fluctuations and uncertainties in demand. Gruenspecht (2002) explained this issue for the fuels industry in meeting multiple fuel standards and provided some evidence about how large the effect could be. The increase in costs from the need to hold larger inventories depends on the extent of the tailoring of standards and the size of the different markets.

Automobile manufacturers have argued that higher inventory costs of meeting two standards will result in higher vehicle prices and, therefore, fewer sales (R. Babik, General Motors, personal commun., June 21, 2005). Manufacturers have to produce more vehicles to meet uncertain demand for several types of vehicles. Dealers may end up holding more inventory or having to sell at a discount to prevent inventory buildup. However, the committee could not obtain data from manufacturers to evaluate the effects of two emissions standards on inventory costs.

Distribution Costs

Mobile-source distribution costs are most likely to increase if there is more than one standard. There are logistical costs of getting different types of vehicles or other products to different locations. In a presentation to the committee, Babik (2005) stated that General Motors has full-time staff ensuring that fleet-average emissions within an individual state meet the fleet-average emissions requirements specified for the California low-emission vehicle (LEV) II program. Honda has argued that distribution costs are one of the most serious problems it faces when there is more than one standard. There is a great deal of uncertainty about whether the right vehicles will get to the right markets—dealers can swap over state lines, and the logistical difficulties in allocating vehicles are substantial. To avoid dealing with these distributional issues, Honda tries to certify at least some vehicles to meet a California and a federal standard (J. German, Honda, personal commun., May 18, 2005). Manufacturers of handheld lawn and garden equipment argued that distribution costs would be higher with two standards even if they produced the exact same product because they would have to use different stock-keeping-unit (SKU) identifiers on the same product. Different SKUs would have to be scanned as a way to ensure that products sold in a state are complying with certification required by the state. Chapter 7 discusses some distribution complications associated with small-engine equipment.

Economies of scale can also occur in distribution. For example, if a vehicle is sold in a small state, it might be sold in all surrounding states to take advantage of the scale economies in distribution. There is evidence of that occurring in the LEV II program in northeastern states. For example, Maine did not adopt the zero-emission-vehicle (ZEV) component of the California LEV II program, but the surrounding states did by 2004. There is evidence that advanced-technology partial zero-emission vehicles (AT-PZEVs), which are required under the ZEV mandate, have been sold in Maine, even though Maine is not required to sell them.

Enforcement Costs

Because mobile-source emissions sources are numerous and can cross state lines, separate standards in different jurisdictions must be enforced. In California, new vehicles cannot be sold or imported into the state unless they show proof of California certification. There is a civil penalty of $5,000 per vehicle for vehicles offered for sale or brought into the state that do not meet the standards. California has had a fair amount of success in enforcing their light-duty vehicle standards, mostly through required registration, and through the inspection and maintenance (I/M) program. In the last few years, California and EPA have worked together to produce a uniform standard for heavy-duty trucks, many of which are frequently driven across state lines. Even without separate standards, California has found that trucks need to be inspected to ensure they have in operation vehicle emissions-control components required based on their model-year standards. In general, if vehicle I/M and centralized registration programs are already in place and enforced, the costs of enforcing a separate standard will be reduced. In contrast, nonroad sources do not have registration or inspection procedures in place and enforcement must be done at the point of sale. The CARB annual reports of enforcement activities (for example, CARB 2004a) detail the process used to enforce mobile-source emissions standards in the state.

Enforcement costs are likely to vary a great deal with the type of product and the technology for enforcement (Harrington et al. 1996). In some industries, the production of the engine is separate from assembly of the product. Heavy-duty diesel and small-engine manufacturers operate under that type of nonintegrated manufacturing system, unlike automobile manufacturing, which has fully integrated production of the vehicle. Manufacturers using the nonintegrated system face a variety of issues to comply with emissions standards, as discussed for small engines in Chapter 7. It is much harder for engine manufacturers to achieve compliance with different standards if the engines cannot be tracked easily through assembly and distribution.

Arguments in Favor of Dual Standards

Laboratory for Emissions Control

A key rationale for allowing one state to set the most stringent technology-forcing emissions standards is that it could limit the risks of failure. The principle that air quality standards are to be established at levels that protect the public's health implies that emissions standards might be too stringent for the capabilities of existing technologies. Such technology-forcing standards inevitably entail the risk that it might not be possible to develop technologies that can achieve the standard at acceptable costs. The risks of setting technology-forcing standards are of three types: (1) expenditures on research and development (R&D) do not result in an acceptable technological solution, and the emission standard is subsequently relaxed; (2) if the standard is implemented and technologies are implemented that are later found to be ineffective, capital invested in production facilities and labor expended in production are also wasted; and (3) if the modification of vehicles or equipment to meet the emission standard results in unacceptable deterioration of the quality of the product or other serious malfunctions, additional damages or recall and retrofit costs might occur. Allowing more states to adopt the alternative standard can increase these risks. Allowing one state to function as a laboratory is unlikely to lower R&D costs substantially. However, the potential costs of derivative damages and remedies for technologies that prove to be ineffectual or defective should be reduced because fewer vehicles or pieces of equipment will have been produced.

If the "laboratory" state's technology-forcing emissions standards result in cost-effective technological breakthroughs, the rest of the country can take advantage of the new technologies by adopting equally stringent standards without risk, although with some delay. By reducing the risks of technology-forcing standards, the laboratory state model allows the rest of the nation to enjoy cleaner air sooner than would be achievable otherwise. As discussed in previous chapters, much of the progress made in reducing emissions from mobile sources has come via regulations that set emissions standards beyond levels that could be achieved with commercially available control techniques. Such technology-forcing standards have often produced impressive advances in emissions control that have enormous air quality and public health benefits. Giving the responsibility for setting alternative emissions standards to the state with the most difficult air quality problems has ensured that the alternative standards would challenge the limits of available technology. It is also not surprising that the most interest in having a second tighter standard has historically been in California and the Northeast, which have been struggling with reaching ozone attainment for the longest periods.

As the experience of having California and federal mobile-source emissions standards has increased, other advantages have appeared. The CARB regulatory process allows California standards to be amended more rapidly without the federal regulatory review in the face of changing market and technological conditions. This process aids in the implementation of technology-forcing regulations. Having a second authority establish emissions standards occasionally allows a division of labor in research and analysis. For example, EPA specializes in developing standards for heavy-duty vehicles, and CARB specializes in developing standards for light-duty vehicles. Their coordination allows the leading agency to do most of the work in the area in which it specializes and allows the other agency to avoid duplicating the effort.

Matching Strategies to Conditions

A second rationale for having different emissions standards is the potential efficiency gained by states when they have a greater ability to match control strategies to the local conditions that determine air quality. The severity and nature of air quality problems vary substantially across geographic regions, as shown in detail in Chapter 2. The factors that determine air quality vary from place to place and time to time. The quantities of air pollutants vary with such factors as the regional population, intensity of vehicle traffic, number and types of industries, types of fuels used to generate electricity, and the rate of biogenic emissions in the region. In addition, the factors that turn pollution into violations of NAAQS depend on atmospheric chemistry, wind and air circulation patterns, and many other climatic and topographical characteristics, all of which vary substantially across the United States.

Regional differences in the factors that determine the quantities of pollutant emissions and the processes that transform pollutants into air quality problems and consequently into public health and environmental damage imply that uniform national emissions standards could be economically inefficient in low-pollution areas and potentially ineffective for meeting NAAQS in regions with the most extreme air quality problems.

Illustrative Example for Dual Standards

To illustrate this economic efficiency argument, assume there are two hypothetical regions attempting to reduce ozone pollution, one that has greater sensitivity to nitrogen oxide (NO_x) emission reductions and the other that has greater sensitivity to hydrocarbon (HC) emission reductions. Figure 5-1a shows the marginal costs and the marginal benefits of reducing NO_x emissions in both regions. Assume in this example that NO_x reductions are most important for reducing ozone in Region 2. Therefore, the damage prevented by greater NO_x reduction in Region 2 (the marginal benefit of compliance, labeled, $MB_{Region2}$) is above that prevented by greater NO_x reduction in Region 1 ($MB_{Region1}$). Assume further that the additional costs of achieving successively greater NO_x reduction (the marginal cost of compliance, labeled MC_{NOx}) are identical in both Regions 1 and 2. If the emission standard for both regions is set at Q1—the efficient standard for Region 1 where marginal costs of control equals marginal benefits—Region 2 will incur excessive damage in that its marginal benefit from reducing NO_x will be greater than its marginal costs. Region 2 would be willing to incur additional costs up to C2 to reduce NO_x emissions up to Q2. The single uniform standard represented by Q1 is economically inefficient because it does not allow Region 2 to equate marginal control costs to marginal benefits.

Similarly, in Figure 5-1b, the marginal costs and benefits of HC emission reductions are illustrated for the two regions. Assume that Region 1 has ozone levels that respond more to HC emission changes and that Region 2 has ozone levels that respond more to NO_x changes and less to HC emission changes. Once again, the standard has been set at the efficient standard for Region 1, namely q1. However, this level of control is excessive for Region 2, which would prefer control level q2 at the lower cost of c2. In this example, Region 2 has too little NO_x control, as shown in Figure 5-1a, and too much HC control. Not only is Region 2 incurring excessive damage in that its marginal benefit from reducing NO_x is greater than its marginal costs, it is also wasting money on unnecessary HC controls. In this example, allowing each region to tailor their emission reductions to their needs (Q1 in Region 1 and Q2 in Region 2

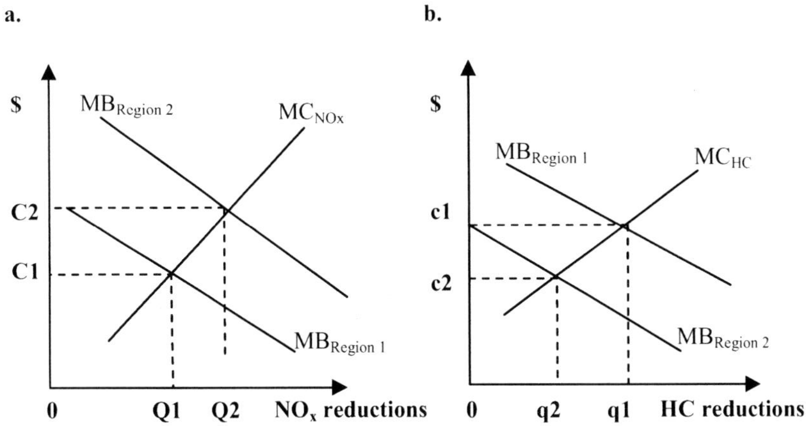

FIGURE 5-1 Costs and benefits of (a) NO_x reduction and (b) HC reduction in Regions 1 and 2. The lines represent marginal costs (MC) and marginal benefits (MB) for NO_x and HC controls in each region. Vertical axes represent the costs and benefits for NO_x (C) and HC (c) controls. Horizontal axes represent NO_x (Q) and HC (q) emissions reductions.

for NO_x and q1 in Region 1 and q2 in Region 2 for HC) allows them to meet air quality goals in the most economically efficient manner.

Additional Complexities

Emissions and their air quality effects are more complex than shown in this simple figure. NO_x is a precursor to particulate matter (PM) as well as ozone, so the effect of NO_x standards in PM markets should also be considered. HC controls also have an effect on the reduction of hazardous air pollutants. Another complication is that a single region is not necessarily homogenous. For example, NARSTO (2000) concluded that local HC emission reductions may be effective in reducing ozone in urban centers, while NO_x emissions reductions are more effective at distances removed from urban centers.

Other qualifications to the simple model are also important. The argument as illustrated in Figure 5-1 is based on the assumption that states can be expected to make the correct choice between two available options, one focused on HC control and one focused on NO_x control. Policy choices are not always so clear, however. The model also does not address the possibility that a state might choose a set of standards that adversely affects air quality in a downwind state.

For many reasons, the single emission-standard system is economically inefficient and often ineffective in achieving air quality goals. However, setting mobile-source standards for each region is not a feasible option and not necessarily optimal for many of the reasons argued above. Having two standards to select from, the California standard or the federal standard, offers regions at least a choice and may provide states with opportunities to develop more efficient levels of control.

HARMONIZATION OF STANDARDS AND PROCEDURES

General Context

Harmonization refers to the practice of aligning regulatory standards and procedures of different jurisdictions to relieve regulatory burdens on industry and consumers by allowing for greater scale economies in production and distribution. Most nations have adopted one of three types of emissions standards for mobile sources: (1) the United States federal emissions standards, (2) the European Union (EU) standards, or (3) the Japanese emissions standards (Jorgensen 2005; Walsh 2005). Most countries outside North America are adopting EU emissions standards rather than developing their own. For example, China and India have adopted the EU standards for both light- and heavy-duty vehicles, whereas Korea and Brazil have adopted EU standards for heavy-duty vehicles and U.S. or CARB standards for gasoline-powered light-duty vehicles (Walsh 2005). Regulators from the three authorities also meet frequently to find ways to make their emissions requirements and certification procedures similar.

Their efforts have had some success but have not achieved complete unification of standards and procedures. Different driving characteristics and atmospheric conditions make the nature of emissions and the severity of pollution problems vary across the globe. For example, differing driving conditions in Japan and the United States contribute to differences in standard driving cycles for certifying vehicles. As within the United States, there are economic interests that seek harmonization, and there are air quality needs and other characteristics that are better served by differences in standards and certification tests. Harmonization need not imply that standards are to be the same. Many countries are adopting older EU or U.S. standards first and then progressively tighten them in the same manner that has historically occurred in the EU and the United States, although often at an accelerated pace.

Global Operations of Manufacturers

The increasing globalization of manufacturing of mobile-source equipment is having an impact on improving harmonization on an international basis. Increased globalization means that foreign manufacturers are producing and selling their products within the United States, and domestic manufacturers are producing and selling globally. As discussed above under Economies of Scale, major automobile manufactures now have alliances with foreign competitors. In addition, many of the major automobile manufactures, both U.S. and foreign, have manufacturing subsidiaries around the world. Globalization also has an impact on the small-engine manufacturing sector. Importing and exporting in the small-engine industry has increased steadily since 1990. Since 1995, imports have been rising and rose to $359 million in 2000, according to the U.S. International Trade Commission. Exports of lawn and garden equipment have also increased steadily to a level of $867 million. Canada and France have continued to be the largest market for U.S. manufactured outdoor equipment. Both countries accounted for over $300 million in exports for U.S. companies in 2000.

Harmonization of Standards and Certification

Harmonization involves the levels of pollutant emissions and the test procedures used to measure emissions. Harmonization of certification procedures can be as important as harmonization of emissions levels because driving cycles and operating conditions (such as cold versus hot start and ambient temperatures) can have a strong impact on emissions levels. Different certification regimes may not only require repeat testing of the same vehicle and separate testing facilities but may also impose different technology and design requirements. Driving behavior, traffic conditions, and the ambient environment can vary substantially from one region to another. As a result, full harmonization is not always the most desirable situation. It may be preferable and more cost-effective to allow both standards and procedures to differ in accordance with local needs and conditions. Nonetheless, even where different standards are appropriate, there may be opportunities to harmonize technical aspects of requirements and procedures with meaningful benefits.

In the United States, California and the EPA have jointly sought opportunities to harmonize mobile-source emissions standards. A single nationwide standard can have strong advantages, such as successful harmonization of heavy-duty diesel engine (HDDE) standards and test procedures. The following excerpt is from the California Air Resources Board digest (CARB 2005d):

> On October 25, 2001, the Board considered and adopted, without modifications, the amendments to California's current HDDE standards and test procedures that staff proposed. To harmonize federal and California requirements for 2007 and subsequent HDDEs, the Board adopted more stringent emission standards and slight changes to the supplemental test requirements used in the California certification process for 2007 and subsequent model year HDDEs. The adopted requirements are identical to those adopted by the U.S. EPA.

Balancing Harmonization

Recognizing the needs of some states to adopt more stringent mobile-source emissions standards to help meet air quality goals, a desirable objective is harmonization of CARB's and EPA's certification procedures. Although meaningful differences in standards can be important in achieving clean air, superficial differences, in areas such as certification procedures, can be wasteful. As technological advances create the ability to manufacture extremely low-emissions vehicles at a reasonable cost, regulators should make a determined effort to harmonize not only the levels of emissions required but also the procedures for testing and certification to the maximum extent possible without sacrificing the primary goal of achieving clean air.

POTENTIAL IMPACTS OF EMISSIONS-CONTROL TECHNOLOGY COSTS

The pricing, employment, and production effects of standards are complex and will depend on consumer preferences, overall vehicle and machine characteristics, and the ability of firms to innovate in the face of

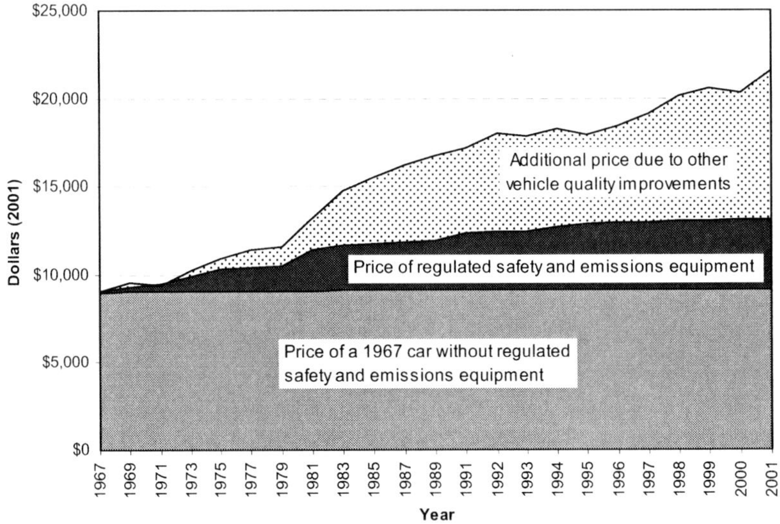

FIGURE 5-2 Impacts of safety and emissions equipment and other quality improvements on average vehicle price from 1967 to 2001. Source: Sperling et al. 2004b. Reprinted with permission; copyright 2004, University of California, Davis.

regulation. In general, emissions-control regulations can be assumed to result in higher prices paid by consumers, other things being the same.[1] There is evidence from the literature that vehicle prices have been higher through time than they would have been without environmental regulation (White 1982; Crandall et al. 1986). For example, Figure 5-2, uses information provided in Ward's Automotive Yearbook to estimate that one-third of the price increase on new cars between 1967 and 2001 could be attributed to safety and emissions regulations. However, prices charged in markets do not always reflect the full cost of producing the product; pricing decisions also depend on other considerations, such as profits and sales. The assumption here is that manufacturers will try to pass on the higher costs of emissions controls in the form of higher prices. Under such conditions, higher prices will reduce unit sales and, depending on the price elasticity of demand, may increase or decrease

[1] The exception to this rule would be when technological change induced by emissions regulations leads to a serendipitous discovery that lowers overall costs. While recognizing that this phenomenon can occur, we do not discuss it here because it represents an unanticipated outcome where lower emissions come with lower costs (a win-win situation).

manufacturers' revenues, which will affect profits. Employment is likely to fall in some parts of the vehicle and engine manufacturing industries as a result of fewer products being produced. Employment is likely to rise in other industries, such as manufacturers of emissions-control equipment as more or different pollution-control components are needed. Despite requests, the committee has been unable to obtain detailed quantitative information on how vehicles and other mobile sources are priced and what the production and employment effects of different standards have been in the past.

In general, if emissions standards are necessary to protect public health as defined by the Clean Air Act, then the possibilities of reductions in profits and employment are trade-offs that society has implicitly accepted. There are still, however, the distributional effects when one region sets or adopts stricter standards. If all the negative impacts occur within the state, it can be assumed that the state will take account of them, but there can be both costs and benefits that fall outside the adopting state. When making the decision to adopt, California and other states usually do not consider impacts outside their own borders.

An argument has also been made that the imposition of emissions standards will cause domestic manufacturers of small engines to move production from the United States to a lower-cost location. If a firm is competitive in the U.S. today and does not have sufficient incentive to move offshore, this argument asserts that the additional cost of the emissions-control equipment tips the scales in favor of relocation. The cost of capital, such as that for major pollution control requirements, could trigger a relocation decision. For example, it might be less expensive for a company to retool and relocate to a new lower-cost facility in another country than to retool and re-outfit an existing manufacturing plant in the United States. The elasticity of demand for products affects the ability to pass costs on and therefore to recover the investment of capital. A decision to relocate could possibly be the only option to remain competitive.

The relocation decision can also be affected by the product and market diversity of a company. A company that is solely in the small-engine or small-equipment industry would not have the same capability to absorb capital costs and fixed and variable overhead that a company that has small nonroad engines and equipment plus motorcycles, automobiles, marine, agricultural or industrial products. Because the small nonroad engine and equipment industry is diverse, having large diversified companies, large undiversified companies and many small companies, located in the United States and overseas, the impacts of state standards may vary widely. Some may choose to move production to a lower-cost location to remain competitive whereas others may choose to not offer a product in a particular market. Some might have the flexibility to absorb the cost of regulations, and some might have to take drastic measures to maintain their competitive position in the market. In other cases, firms might already have an economic incentive to move production offshore and may do so even if the emissions regulation is not enacted.

One important issue is when and whether the emissions regulations would trigger or accelerate a move out of the country. The committee lacks sufficient information to evaluate substantively such issues with respect to domestic small-engine manufacturers. However, the question raises a number of important policy issues. Given the wide range of factors that determine whether domestic manufacturers are at a competitive disadvantage compared to foreign manufacturers, these issues are often resolved in the political arena.

AN OVERVIEW OF THE POSSIBLE STANDARD-SETTING OUTCOMES

The committee summarizes the general benefits and costs of allowing an additional state standard compared with a uniform standard. Table 5-1 surveys the benefits and costs with some indication of the distribution of the effects in different regions. This table does not identify all effects of the alternative regulatory policies. It summarizes benefits by describing only the direction of the effect on benefits from improved air quality resulting from probable changes in emissions. The real benefits come from health and welfare improvements resulting from changes in air quality and emissions. More issues in estimating benefits and costs are discussed in greater detail in the following section.

Separate Standards Set Only by California

The second column of Table 5-1 compares the cases of having a separate standard in California and having a national uniform standard. Under the current waiver process, California may set a separate and more stringent standard than the rest of the country. The emissions reductions, air quality benefits, and health improvements of a stricter standard will occur in California, and the state is likely to weigh those benefits in assessing the need for the stricter standard. In addition, other jurisdictions

TABLE 5-1 Costs and Benefits of Possible Outcomes of Allowing Separate State Standards

Benefits and Costs	Effects of Allowing California to Set Standards Different from National Standards	Effects of Allowing Other States to Adopt California Standards
Benefits to health and welfare from emissions and air quality changes	Increased benefits primarily in California; possible spillover effects to other states	Increased benefits in opt-in states; extent of benefits depends on meteorological and geographical conditions and overall mix of emission sources in the opt-in states; possible spillover effects to nearby states
Costs	Higher costs of stricter standards; possible higher costs of mobile-source products (how costs are passed on depends on pricing strategy of manufacturers)	Somewhat higher costs in opt-in states, although full costs of standards might not be reflected in prices (how costs are passed on depends on pricing strategy of manufacturers)
Fixed costs: R&D, certification, retooling	Higher R&D, retooling, and certification costs because of two standards	Possible higher fixed costs than if stricter standard only in California because of possible need for additional retooling for larger volume or for additional certification
Equipment and production	Higher equipment costs per vehicle; vehicle-price change depends on pricing strategy; possible higher production costs due to smaller scale in California market and all other state markets; depends on scale economies and learning	Higher equipment costs per vehicle; vehicle-price change depends on pricing strategy; production-cost changes depend on scale economies in two separate markets and depends on technology and learning over time in production
Inventory and distribution	Higher inventory and distribution costs for manufacturers because of two markets	Higher inventory and distribution costs than if stricter standard only in California because of regional spread of two markets

Enforcement	Higher enforcement costs	Higher enforcement costs
Employment effects	Possible regional employment effects in California and other areas, depending on the mobile-source type, the standard, and the location of production	Employment effects increase to the extent of changes from uniform-standard employment because of larger market for stricter standard
Overall distributional effects	Higher costs and higher benefits, although full costs most likely not paid by California motorists; possible spillover effects to other states	Higher benefits and higher costs, although not all costs paid by opt-in states; possible spillover effects to other states

might be affected by the stricter standards, depending on the mobile source and the pollutant. For example, air quality benefits might spill over into adjacent states, depending on wind patterns, such as ozone-reduction policies that can affect air quality in parts of Arizona and Nevada and relocation of cleaner California units to neighboring states. Current California law requires that the standard-setting process in the state not consider effects elsewhere in the country. This law is unlikely to cause any objections if there are only spillover benefits. However, costs that affect other regions do raise objections, as discussed below.

Direct Costs

The costs of a separate standard in California are the resource costs of developing and producing a different product for the California market. These costs include the higher costs to the manufacturers in fixed costs (R&D, retooling, and certification costs) and production costs. Inventory costs are also likely to be higher if inventory holdings have to be higher with two markets. The costs of distributing a different vehicle to California might be somewhat higher as well. Higher costs to the manufacturers are likely to be passed on at least partially to buyers in California, but higher prices also mean fewer products sold. It is difficult to determine who pays for these higher costs, because pricing strategies by mobile-source producers often depend not only on the cost of the vehicle but also on other marketing and profit strategies. As Sperling et al. (2004b) point out, "the relationship between pricing and costs is quite complex." Finally, to maintain compliance with the separate standard, California probably has some administrative and enforcement costs.

Distribution of Costs

Some of the costs of California's separate standards probably fall on jurisdictions other than California. As discussed in Chapter 3, these costs are not typically considered in the formal standard-setting process in California. Production and employment effects could be either positive or negative in other parts of the country as a result of the separate standard. Production sites might need to be different for a product with a new standard, or a new standard may mean retooling or relocation of existing plants, as discussed earlier in this chapter. For example, Hanz and Hotz (2005) argued that a more stringent standard in California for Briggs and Stratton Corporation might require sufficient retooling to cause it to relocate plants outside the United States. The committee did not obtain any information to determine whether relocation was a plausible result of the imposition of a second emissions standard on any industry. The case study is discussed more in Chapter 7.

There are also changes in employment and sales for the emissions-control industries when there are additional stricter standards. Because new technology must be developed and sold, new products will be developed, and the product mix of emissions-control parts will be changed. Emissions-control industries will expand, although sales of some equipment may not be as robust in the absence of a national market that is available under a uniform standard.

CARB (2001a) discussed how the board found that the costs of one emission standard was passed on to consumers. CARB concluded that (1) manufacturers might not pass on all the costs associated with the zero-emission vehicle (ZEV) mandate and might absorb them internally; (2) automakers do not mark up cars sold in California to reflect the increased cost of emissions controls there; and (3) costs associated with the ZEV program and other California vehicle emissions-control initiatives might be spread across all car purchases in the United States, not just in California. Vehicle manufacturers also claim that they usually cannot pass emissions-control costs to consumers through higher prices if vehicles are similar in other respects, although views differ on whether the increased costs are only to California consumers or are spread across the country (J. German, Honda, personal commun., May 18, 2005; R. Babik, General Motors, personal commun., June 21, 2005).

Impacts on Federal Standards

A final point about allowing California to set more stringent standards than the rest of the country is that this policy has sometimes led to more stringent standards nationwide. One of the major rationales described above for allowing California to explore when and how to set a more stringent standard is that it will serve as a laboratory for emissions control in the rest of the nation. When California acts as a laboratory, additional federal standards might have higher benefits and lower costs than would otherwise occur. California standards as they evolve over time might help to identify ways to more cost-effectively reduce emissions. The federal government might then adopt stricter standards already proved to be effective in California. In that case, inventory and distribution costs will depend on whether the new standard is harmonized with California's standard or whether it is different. In some cases, the federal government has adopted a stricter standard at least in part because of California's more stringent standard. For example, the federal Tier II program for light-duty vehicles is stricter than earlier standards but is different from California's LEV II program. In other cases, such as HDDE standards after 2007, the standards are identical.

Some have argued that when CARB sets a stricter standard that is adopted by EPA, California in effect sets standards for the entire nation. Since CARB bases its standards on the needs of the Los Angeles area, the most polluted regions of the country are likely to benefit the most, but these standards are unlikely to be efficient for all regions.

Opt-in Provision for Other States to Adopt California Standards

The third column of Table 5-1 compares (1) the benefits and costs of allowing states to adopt (also called opt-in) the California standards with (2) the benefits and costs of not allowing states to adopt the California standards. There are effectively two standards—one in California and opt-in states and one in the rest of the country.

When states are allowed to adopt the California standard, as some do now under the section 177 provisions of the Clean Air Act, there are a range of benefits and costs. The opt-in states will identify improvements in health and welfare within the state. Emissions reductions from stricter standards may also spill over into other states that do not adopt, allowing for air quality and health benefits in other jurisdictions. These spillover benefits probably will not be considered during the state decision process, because few decision-making bodies have any regional jurisdiction over environmental issues. In the case of the northeastern states, the Northeast States for Coordinated Air Use Management (NESCAUM) provides a regional perspective on mobile-source emissions standards policy.

The costs of the stricter standards for the opt-in states are difficult to determine. The full costs of additional emissions controls will not be reflected in the prices of individual vehicles. As mentioned above, vehicle pricing by manufacturers is complex and reflects many factors, including marketing strategies for different vehicles. Vehicle manufacturers argue that they usually cannot pass emissions-control costs to consumers through higher prices if vehicles are similar in other respects. If emissions-control costs are not included in vehicle prices in opt-in states, the costs will probably be spread across the vehicle fleet and perhaps the nation. In that case, the opt-in region might benefit, but the costs are spread nationwide. If prices of vehicles in the opt-in states reflect cost differences, consumers might try to cross states lines to buy vehicles, and regions would have to make greater enforcement efforts.

Overall, there will be benefits and higher costs in the opt-in states. The higher costs might not reflect the full costs of the standards, but the committee did not have the data needed to evaluate that issue. Some spillover effects from opt-in states can occur from changes in emissions and air quality, sale of products in regional markets, and employment effects in states that do not opt-in.

There is some evidence that allowing a separate standard in California and all states to opt-in might induce manufacturers to produce uniform products for the nation at the more stringent standard. For example, automobile manufacturers have stated their intention to certify all their vehicles according to the stricter California evaporative emissions standards rather than produce some vehicles certified according to federal standards and some certified according to California standards. The increased costs of having a uniform standard, even a stricter standard, may be lower than the costs of complying with two separate standards. Various manufacturers have expressed a willingness to comply with some stricter standards as long as they are uniform (Hanz and Hotz 2005; Jorgensen 2005; J. German, Honda, personal commun., May 18, 2005). Manufacturers' preference for one standard suggests that the scale economies and inventory and distribution costs of different standards are high enough to offset any additional costs of meeting a stricter uniform standard.

ISSUES IN ANALYSIS OF BENEFITS AND COSTS OF STANDARDS

The next two chapters review mobile-source emissions standards for a number of specific cases. Before the committee examines those cases in detail, we review here some of the important overall issues related to costs and benefits of standards in the analyses of these regulatory activities. OMB (2003) and EPA (2000b) provide specific guidance on the preparation of economic analysis for EPA rule-makings.

Benefits: Importance of Health and Welfare Effects

The effects of a regulatory policy on human health and welfare are the key elements that define the benefits of any mobile-source regulation. Further, in a full benefit-cost analysis of a regulation, both benefits and costs are estimated in dollars. Dollar values are particularly difficult to estimate on the benefit side because emissions changes must be tracked through their effects on air quality to their health consequences and finally to the value of various health outcomes. Attempts at such estimates are increasingly frequent, however, in regulatory assessments. The last link to dollar benefits is important because of the range of different pollutants affected by a mobile-source regulation and the variation in the severity of health outcomes that occur under those regulations. Aggregating emissions changes, as is done in cost-effectiveness analysis (see below), usually does not account for the different effects of pollutants on human health and welfare. A thorough analysis of estimating the benefits of air pollution regulations can be found in NRC (2002b).

The data are often not available for a full analysis of the benefits, and the data needed for such an analysis are often expensive to collect, particularly for multiple small regions. Nevertheless, the link of the regulatory change to air quality and to health is important and often gets lost in the regulatory analyses. One can argue that the goal of the emissions reductions targeted in a state implementation plan (SIP) is to provide an indirect link to air quality and health benefits. Thus, regions often evaluate policies in terms of the emissions reductions since the NAAQS are designed to protect public health.

Cost-Effectiveness Analysis

Cost-effectiveness analysis can be a useful way to compare regulatory policies when only the emission changes resulting from the policy and the costs are known. A cost-effectiveness measure is calculated as the full cost of the policy in relation to the resulting emissions reductions. The costs and emissions-reductions outcomes over time must be measured relative to a baseline set of outcomes. For example, one cost-effectiveness comparison of the on-road diesel standards would be the costs and emissions resulting from those standards compared with the emissions allowed under the previous standards. With SIP emissions goals in place, this comparison of standards is the most frequent policy tool used by state and local governments to consider policy alternatives.

Several issues arise in the implementation of cost-effectiveness analysis. Looking only at the average costs and emissions reductions from a specific regulatory policy does not reveal whether that policy should be adopted. One aspect of adoption is environmental equity: costs may be borne by a segment of the population that gets little of the benefit or vice versa. The policy must be compared with alternative policies that attain similar emissions reductions at lower costs or with policies that attain greater reductions at the same cost. When it is possible, incremental analysis of policy components may be more appropriate than analysis of the average cost-effectiveness of an entire package of policies or standards. Parts of the policy may be cost-effective but other parts costly relative to alternatives.

Finally, policies to reduce mobile-source emissions often result in the reduction of multiple pollutants because a single pollutant, such as NO_x, might contribute to increased concentrations of multiple other pollutants, such as ozone and PM. If the denominator of the cost-effectiveness calculation is in tons, then the effects of different pollutants on air quality and human health will not be distinguished. For example, if the effect of NO_x reductions on ozone and therefore on health differ between Massachusetts and Texas (for example, because of interactions among pollutants), those different effects will not show up in cost-effectiveness analysis unless the tons of reductions from the different pollutants are somehow weighted by their relative impacts on air quality and health.

Estimating Costs

Estimating costs is more complex than simply assessing the costs of the emissions-control technologies. Costs should include the full opportunity costs, including not only private costs but also public costs. As described above, the costs of many mobile-source standards involve not only additional equipment costs but also fixed costs, such as R&D, distribution, and certification costs, associated with a new standard. The public costs of enforcing a stricter standard include testing and monitoring products in the region where they are being sold. Cost savings from the regulation, if any, should also be included, and direct and indirect costs should be considered.

There are also complex incentives for the various groups that produce cost estimates to inflate or understate costs, making independent analysis difficult. A frequent argument is that a regulated industry has an incentive to overstate the costs of complying with regulations, in part to convince regulatory authorities to make regulations less stringent. The regulatory agencies might have the opposite incentive, to understate costs, if their goal is to enforce regulations that will have the greatest effect on emissions and air quality. For example, Anderson and Sherwood (2002) found that EPA cost estimates for regulations on highway vehicles and their fuels were generally significantly lower than estimates from other stakeholders. Thus, ex ante estimates of the costs of regulation are not only predictions based on best estimates of future technology, design, and behavior, they may also be subject to bias based on the perspective of the estimator. Goodstein and Hodges (1997) and OTA (1995) found that ex ante costs tend to be overestimated, compared with ex post costs, because new technologies not anticipated in the ex ante analyses are often discovered in response to the regulation. Squitieri (1998) and Harrington et al. (2000) find the issue of assessment of costs to be more complex than these early studies suggest. First, it matters who is doing both the ex ante and ex post studies, since the biases discussed above can be inherent in both. Second, it is important to assess costs in combination with emissions reductions. For example, if total costs are accurately estimated but the emissions are only half of those originally predicted, then the costs of achieving the emissions goal are actually higher than estimated. When these issues are taken in account, the accuracy of ex ante cost estimates are somewhat more mixed than the early studies suggested (Harrington et al. 2000).

Estimating Indirect Benefits and Costs

Indirect effects should be examined, and any that are significant should be included in any regulatory analysis. For example, one presentation to the committee (Austin 2005) included an analysis of the effect of the cost of stricter new light-duty-vehicle standards and the effects of the higher price of new vehicles on fleet turnover. Higher new vehicle prices may slow new vehicles purchases. Austin argued that the emissions reductions of stricter standards are not as high as the direct estimates show because the slowing of fleet turnover tends to increase fleet emissions. Another example is that if stricter standards are adopted in one state, residents there might cross-over to adjacent states to buy vehicles. The magnitude of this behavioral change can be estimated to determine how much benefits would be reduced. Recent rules that treat fuels and engines as a system, such as the federal Tier 2/low sulfur standards discussed in the following chapter, add to the complexity of estimating impacts since such standards produce non-emissions benefits such as improved engine life, longer spark plug life, greater oil change intervals, and reduced maintenance. Other indirect impacts include impacts on safety, which are discussed in more detail in Chapter 7.

Accounting for Learning Over Time in Cost Estimates

The tendency for manufacturing costs to decrease as experience is gained through cumulative production increases the difficulty of estimating the costs of future emissions controls. For most manufacturing processes, including vehicle production, the unit costs of production have been shown to fall over time as producers gain experience in producing the product. A number of studies of vehicular and other products show that there is a distribution of this type of learning effect on costs. The amount of learning will depend on the technology and production process involved. However, the most common value for this "progress ratio" found in the empirical literature is about 0.8, meaning that as production increases by 100%, costs fall to 80% of their previous value (Argote and Epple 1990; Manson et al. 2002). Rubin (2005) found evidence that costs for installed vehicle control devices have fallen at slower rate, closer to 93% as production volumes double. Although the committee is not in the position to judge the evidence for the presence of such a progress ratio for reducing costs, it does note that such learning curves have been used by EPA to estimate the long-term cost of regulations (more than 5 years after implementation) for the Tier II tailpipe standards (EPA 1999a), for the heavy-duty diesel rule (EPA 2000c), and for the Phase II final rule on handheld spark-ignition engines (EPA 2000d).

Comparing Cost-Benefit Estimates with a Baseline

Cost-benefit estimates will also vary depending on the assumptions made about conditions in the absence of new emissions standards. It is important that costs and benefits of emissions reductions be compared in a consistent way with an assumed or forecast state of the world in the absence of the proposed standards. The baseline can include changes in external factors or other policies that might occur over time if the policy under consideration is not adopted. The baseline will have a time profile to which the costs and emissions reductions of the proposed standards can be compared as they vary over time.

Comparing Cost-Benefit Estimates with Alternatives

An evaluation of the costs and benefits of adopting a new standard should include a comparison of the costs and benefits of feasible alternative policies. Alternative design or implementation of a standard should consider whether the same air quality goals could be met by a new standard or by a more flexible set of standards. It could also consider whether the costs and benefits of a specific set of standards can compare to alternative ways to reduce emissions from the same sources or from different sources.

Accounting for Uncertainty

Estimating the future emissions reductions (or when possible, the benefits) and costs of a stricter standard are going to be highly uncertain. Point estimates of emissions reductions and costs often convey precision in the estimates that is not warranted. Some knowledge of the extent of uncertainty is necessary to decide whether a new policy would be cost-effective.

There are various ways to reflect the extent of uncertainty. The most accurate way is to use statistical techniques that show the underlying probability distributions of different outcomes. Probabilistic analyses of health outcomes are being done for federal standard setting. The most recent example is the Regulatory Impact Analysis done by EPA (2004e) for the federal rules for nonroad heavy-duty engines.

For mobile-source standard setting, particularly at the state level, there are not enough data on either costs or benefits to do this type of statistical analysis. Some sensitivity analysis on costs and benefits would be useful but is rarely done. The emissions reductions are often estimated using the MOBILE or EMFAC models, which have assumptions incorporated about future changes. Uncertainty must be included by identifying the assumptions that have the greatest uncertainty and by varying those assumptions in plausible and transparent ways. For example, if there is uncertainty about the effectiveness of a particular technology, that can be reflected by showing a range of results. Similarly for costs, areas of uncertainty can be identified and plausible alternative assumptions can be used to generate a range of cost estimates.

Distributional Effects

It is important to identify the groups who are likely to be affected by emissions reductions from a proposed standard and who are likely to bear those costs. The benefits and costs may differ across geographic areas within the region considering the standard or they may be distributed in other regions or throughout the country. As discussed above, in many cases, vehicle pricing policies result in the same price for the same model vehicle regardless of whether the standards are stricter, indicating that the costs of adopting stricter standards in one region are paid throughout the country.

CONCLUSIONS

The current system of regulating mobile-source emissions that allows a second set of emissions standards is a compromise between the interests of manufacturers, who prefer a single set of emissions standards to reduce compliance costs, and state air quality managers, who prefer standards tailored to air quality objectives. The desire to harmonize emissions standards and certification testing procedures stems from the need to improve regulatory efficiency. Several conclusions come from this chapter.

- Emissions-control regulations can be assumed to result in higher prices paid by consumers, other things being the same. Impacts from a second set of standards include changes to equipment and certification costs as well as costs related to distribution and enforcement. Quantifying the impacts of a second set of standards, however, is difficult. Costs are difficult to assess. Pricing, employment, and production effects are complex and depend on consumer preferences, overall vehicle or machine characteristics, and the ability of firms to innovate in the face of regulation. The committee has been unable to obtain detailed quantitative information on the pricing of vehicles and other mobile sources, and the production and employment effects of different standards in the past.

- Recognizing the needs of some states to adopt more stringent mobile-source emissions standards to help improve air quality, a desirable objective is to harmonize CARB's and EPA's certification procedures. Although meaningful differences in standards can be important in achieving clean air, superficial differences, in such areas as certification procedures, can be wasteful. Harmonization of standards and testing procedures also has the global context of foreign manufacturers producing and selling their products within the United States and domestic manufacturers producing and selling their products globally.

6

Light-Duty-Vehicle Emissions Standards

On-road light-duty vehicles (LDVs), such as cars, vans, pickup trucks, and sport utility vehicles (SUVs), have the longest history of mobile-source emissions regulation in the United States. Because of their numbers and activity, these vehicles have historically contributed the most to total mobile-source emissions. As discussed in earlier chapters, California's low-emission-vehicle (LEV) program, introduced in 1990, was an important milestone that helped define today's California and federal on-road emissions standards. The LEV program is the primary California mobile-source emissions standard adopted by other states. This case-study chapter presents an overview of the LEV program and compares the standard-setting practices of the California Resources Board (CARB) and the U.S. Environmental Protection Agency (EPA), including the practices used to develop the California LEV II and federal Tier 2 standards now in place. The chapter also discusses the practices used by other states in adopting California's LEV emissions standards.

THE LOW-EMISSION-VEHICLE PROGRAM

The California legislature enacted the California Clean Air Act of 1988, which instructed CARB to "achieve the maximum degree of emission reduction possible from vehicular and other mobile sources" (Cal. Health & Safety Code § 43018(a)). In response to this new legislative mandate, CARB approved an ambitious new rule-making in 1990 to regulate vehicle emissions. The LEV program consisted of several regulations to reduce emissions substantially from light- and medium-duty vehicles beginning in model-year 1994. These regulations included stringent new exhaust emissions standards for nonmethane organic gas (NMOG), nitrogen oxides (NO_x), carbon monoxide (CO), particulate matter (PM), and formaldehyde.

Rather than requiring every vehicle to meet the same emission standard, the LEV program featured a fleet-based approach, which allows manufacturers the flexibility to meet new emissions standards averaged across their entire product line. This format of the standard, which reduces overall compliance costs, allowed manufacturers a longer development time for vehicles that are the most difficult to control. For LDVs, CARB defined a set of four categories of emissions standards and allowed each manufacturer to certify its vehicle models to any mix of the available standards, provided that the sales-weighted fleet of the manufacturer met the applicable average emission level for that model year. The four available emissions standards in order of increasing stringency were for the transitional low-emission vehicle (TLEV), low-emission vehicle (LEV), ultra-low-emission vehicle (ULEV), and zero-emission vehicle (ZEV). The fleet-average requirement was based on NMOG emissions, and this average became progressively more stringent each model year from 1994 through 2003 (see Table 6-1).

A second feature of the LEV program is that it sought to regulate the vehicle and its fuel as an integrated system. CARB determined that the proposed regulations would encourage vehicle and fuel manufactures to work together to develop LEVs and clean fuels (CARB 1990). In addition to California's ultra-clean reformulated gasoline, the slate of clean fuels a manufacturer could choose from included methanol, ethanol, liquified petroleum gas (LPG), and compressed natural gas (CNG). Under the regulations, vehicle manufacturers were required to notify CARB 2 years in advance if they intended to certify a LEV vehicle using such an alternative fuel (13 California Code of Regulations [CCR] § 2303). If vehicle manufacturers announced an intention to market a combined total of at least 20,000 vehicles operating on a given clean fuel, then CARB would mandate the availability of such a fuel at California service stations (CARB 1990).

A third feature of the LEV program is that it included a mandate for ZEVs. CARB defined ZEVs as vehicles that have no exhaust or evaporative emissions of any regulated pollutant (CARB 1990). As initially adopted, this ZEV sales mandate required 2% of the passenger cars and light-duty trucks produced and delivered for sale in California by each

TABLE 6-1 LEV Fleet Average NMOG Standard in Grams per Mile

	1994	1995	1996	1997	1998	1999	2000	2001	2002	2003
PC, LDT, 0-3,750 lb	0.250	0.231	0.225	0.202	0.157	0.113	0.073	0.070	0.068	0.062
LDT, 3,751-5,750 lb	0.320	0.295	0.287	0.260	0.205	0.150	0.099	0.098	0.095	0.093

Abbreviations: PC, passenger car; LDT, light-duty truck.
Source: CARB 1998a.

large-volume[1] manufacturer to be ZEVs in the 1998 model year. This requirement increased to 5% in 2001 and 10% in 2003. Although the ZEV mandate was technologically neutral in that it did not specify the technology required to meet the ZEV standard, CARB noted that, at the time, only battery-powered electric vehicles were candidates to be ZEVs. CARB also stated that other technologies (such as fuel cells) could be developed in the future to meet the standard (CARB 1990).

The ZEV mandate was a notable departure of the LEV program from the flexibility of allowing manufacturers to choose how to comply with the NMOG fleet-average requirement. Nevertheless, CARB concluded that such a mandate was necessary because a "significant penetration of ZEVs is crucial to long-term attainment of the ambient standards in the South Coast, and there is no assurance that ZEVs will be developed without the limited, measured ZEV sales requirements in the regulations" (CARB 1991). An important aspect of a ZEV is the additional emissions benefits that are gained because there is no deterioration of emissions-control equipment over time. Although the primary objective of the ZEV mandate was to reduce vehicle emissions, CARB identified secondary benefits, including the investment by industry and communities in batteries and infrastructure for ZEVs (CARB 1991), the potential to "contribute to national and state energy diversity and security," and the potential "to revitalize California's economy through job creation and growth in an emerging industry" (CARB 1994a). (Although ZEVs have no emissions, electricity produced from fossil fuels to charge ZEVs results in air pollutant emissions, and these emissions, either inside or outside of California, affect air quality.)

The LEV program included a credit program to give manufacturers additional flexibility to meet the standards. Manufacturers could earn NMOG fleet-average credits in any model year by achieving a sales-weighted fleet-average emissions level lower than the applicable fleet-average standard for that model year. The credits could be sold to other manufacturers or applied to help the manufacturer achieve compliance in future model years, although the credits are substantially discounted with time. A second credit program pertaining to the ZEV mandate allowed manufacturers to obtain credits for complying with the ZEV mandate

[1] Generally, small-volume manufacturers in California sell fewer than 4,500 vehicles and engines per year, intermediate-volume manufacturers sell 4,501-60,000 per year, and large-volume manufacturers sell the remainder of the vehicles sold in California. See Title 13, California Code of Regulations, Section 1900 for exact definitions.

early. Like the NMOG credits, ZEV credits could be banked internally for future use or sold to other manufacturers. CARB initially rejected suggestions that hybrid electric vehicles (HEVs) should be eligible for partial credits under the ZEV mandate (CARB 1991) but provided some additional NMOG credits for HEVs that achieved certain performance goals.

A final feature of the California LEV program is its built-in process for periodic review of the program and revision, if necessary. Because of the far-reaching and long-term nature of the LEV program, CARB committed to a biennial review of its LEV program to monitor manufacturer compliance plans and to identify any problems with the feasibility of its demanding program. In response to this mandate, CARB has produced several reviews (CARB 1994b, 2000a,b).

The Zero-Emission-Vehicle Mandate

Although the overall LEV program was widely considered successful at reducing vehicle emissions and promoting advanced emissions-control technologies, the ZEV experiment has fallen short of its original expectations to promote the widespread use of electric vehicles. This requirement, which was premised on the availability of electric vehicles by model-year 1998, is an example of misjudgment by CARB that the required and expected ZEV technology would be feasible. CARB has revised its original ZEV mandate with four sets of successive revisions, resulting in a much diluted requirement today that no longer emphasizes electric vehicles and affecting CARB's credibility in the program itself. General Motors invested an estimated $1 billion over several years to develop ZEVs, which now do not appear to be headed for widespread use in society (GM 2005). To put this number in context, General Motors spent approximately 6.5 billion on total research and development in 2004 (Hira and Goldstein 2005).

The problems of the ZEV mandate are reflected in CARB's shifting estimates on the feasibility and costs of using electric vehicles. CARB initially thought in 1990 that, by 2000, electric vehicles would be comparable in cost to conventional vehicles plus an estimated $1,350 per vehicle cost for the batteries (CARB 1990). In 1994, CARB increased its estimate of the incremental additional costs of an electric vehicle to $5,000-$10,000 more than a conventional gasoline-fueled vehicle (CARB 1994b). In its 2000 review of the ZEV mandate, CARB staff estimated that the incremental cost of a freeway-capable ZEV would be approximately $20,000 more than a conventional vehicle (CARB 2000a).

The biennial reviews of the LEV program that CARB promised in adopting the program have largely focused on the ZEV mandate from the second biennial review in 1994. In the 1994 review, vehicle manufacturers emphasized concerns about the feasibility of the ZEV mandate and requested that CARB relax the ZEV mandate to permit ZEV credits for such vehicles as HEVs with extremely low emissions (CARB 1994b). CARB concluded at the end of a 2-day public hearing on the subject that the ZEV mandate was an important part of the LEV program, and no revisions were necessary at that time. CARB also instructed the staff to reconsider the role of HEVs within the framework of the ZEV mandate.

In preparation for the biennial review in 1996, CARB established an independent panel of experts to evaluate the readiness of electric vehicle battery technology to meet the ZEV mandate in the 1998 model year. The expert panel report concluded that even under "a complete success scenario," with no delays or unforeseen obstacles, "electric vehicles with commercial-production advanced batteries could become available in 2000 or 2001" at the earliest (Kalhammer et al. 1995). The panel also found that lead-acid batteries limited electric vehicles using such batteries to a small "niche" market. Based on the basis of the expert panel's recommendations, CARB voted to repeal the California ZEV mandate for model-years 1998 through 2002 to provide "time necessary for advanced technology battery developers to achieve commercialization" (CARB 1996).

In 1996, CARB also negotiated separate memoranda of agreement (MOA) with each of the seven manufacturers initially subject to the ZEV mandate. The MOAs required the manufacturers to place over 1,800 advanced-battery electric vehicles into operation in demonstration programs in California between 1998 and 2000. Manufacturers were also required to offset the emissions benefits that would have been achieved by the ZEV mandate from 1998 through 2002 by agreeing to the nationwide introduction of LEVs several years before such vehicles could be mandated under the federal CAA (CAA). (This manufacturer obligation was implemented through the National Low Emission Vehicle [NLEV] Program described below in the section Influences of the LEV Program on National Mobile-Source Emissions Standards.) The MOAs also obligated CARB to work with state and local governments to help develop ZEV infrastructure and remove other barriers to ZEV implementation.

Following the biennial review in 1998, CARB further relaxed the 2003 mandate in its LEV II rule-making (see below) by allowing ZEV credits to be earned by vehicles with near-zero emissions, referred to as partial ZEVs (PZEVs). Intermediate-volume manufacturers were permitted to meet their ZEV mandate requirements entirely with PZEV credits, and large-volume manufacturers were permitted to meet up to 60% of their ZEV mandate requirements with PZEV credits. A manufacturer could obtain from 0.2 to 1.0 ZEV credit for each PZEV sold, depending on the vehicle's characteristics.

In the 2000 biennial review, CARB again appointed an advisory panel of battery experts to evaluate the availability and cost of electric vehicle batteries. The expert panel concluded that advanced technology batteries with reasonable cost and performance characteristics would not be available in time to meet the 2003 mandate. The panel found that the most promising advanced battery was a nickel-metal hydride battery but that production of such batteries in quantities to meet the 2003 ZEV mandate would be \$9,500 to \$13,000 per battery, many thousands of dollars above the cost that could be commercially viable. Moreover, such batteries would produce a range of only 70-100 miles, which was below the expectations of most potential customers. The battery panel concurred with the manufacturers "that EVs with the battery costs and limitations anticipated for the foreseeable future will find only very limited markets, well below the numbers of vehicles called for by the ZEV regulatory provisions beginning in 2003" (Anderman et al. 2000).

Moreover, as part of its analysis for the 2000 biennial review, CARB determined that the PZEV standard would be "extremely challenging" for many vehicles to meet and that most manufacturers would not be capable of complying with the ZEV mandate in 2003 by using PZEV credits (CARB 2000a). CARB also recognized that in the initial years of the ZEV mandate, manufacturers would not be able to recover the full cost of ZEV production by using only price. Unless the state and local air districts could provide the substantial funds that would be needed to subsidize these vehicles, manufacturers would have to either absorb the economic losses internally or pass on the costs to purchasers of other vehicles (CARB 2000a).

Despite those findings, CARB voted unanimously at its September 2000 meeting to affirm the mandate as an essential component of California's long-term air quality strategy and to retain the basic ZEV requirements (CARB 2000c). CARB nevertheless instructed staff to develop some regulatory fine-tuning to address the challenges associated with the successful long-term implementation of the ZEV program. A month later, CARB staff expressed concern that the biennial review process and resulting revisions to the ZEV mandate "has interfered with the orderly growth of the ZEV market, because of the uncertainty it introduces into planning and implementation activities on the part of manufacturers, government agencies, and other parties" (CARB 2000d).[2]

At its next meeting to review the ZEV mandate in January 2001, CARB further relaxed the ZEV mandate. The 2001 amendment allowed large-volume manufacturers to meet another 20% of their ZEV obligation with partial credits from advanced-technology vehicles known as AT-PZEVs. AT-PZEVs include gasoline HEVs that meet specific criteria. Several other refinements and additions to the ZEV credit structure were enacted in this same rule-making.

In June 2002, a California federal judge issued a preliminary injunction against implementation of the ZEV mandate in a lawsuit brought by vehicle manufacturers and dealers. The lawsuit contended that the ZEV mandate as modified in 2001 was preempted by the federal fuel economy standards (Central Valley Chrysler-Plymouth, Inc., et al. v. Witherspoon, Case No. CIV F-02-05017 REC SMS [E.D. Cal.]). As part of a settlement of that litigation in April 2003, CARB amended the ZEV mandate to provide an Alternative Compliance Plan (ACP) option in which large-volume manufacturers could meet much of their ZEV requirement by producing their sales-weighted share of approximately 250 fuel-cell vehicles by 2008. The required number of fuel-cell vehicles would increase to 2,500 from 2009 to 2011, 25,000 from 2012 to 2014, and 50,000 from 2015 to 2017.

CARB recognized that other states might adopt the California ZEV program and ACP under section 177 of the federal 1990 CAA and might require production of their own fuel-cell vehicles, which would require manufacturers to produce more fuel-cell vehicles (CARB 2004b). To address that problem, CARB allowed a fuel-cell vehicle placed in any state that had adopted the California ZEV program to count toward California's ZEV requirement and conversely allowed any fuel-cell vehicle placed in California to count toward another state's ZEV requirement. The start date for those requirements was delayed from 2003 to 2005.

The 2003 amendments to the ZEV mandate are the most important yet, and they strongly signaled CARB's recognition that electric vehicles are not a promising technology to achieve the zero-emission goal to which the agency still strongly adheres (CARB 2004b). As part of this

[2] Public Workshop to Discuss Issues Related to the Zero Emission Vehicle Regulations: Agenda and Background Material, Oct. 16, 2000.

rule-making, CARB undertook another evaluation of the feasibility of battery electric vehicles and concluded that "the cost and performance characteristics of advanced batteries have not meaningfully changed" since the 2000 battery panel report, and that "even at substantially increased production levels full function EVs would not be cost competitive with conventional vehicles, and that there does not appear to be a viable path that will result in commercialization for general markets." (CARB 2004b). In contrast, "manufacturers appear to believe there is a business case for fuel cell development. Staff concurs that the technology shows great promise and fully expects fuel cell development to proceed to commercialization" (CARB 2004b). CARB estimated that vehicle manufacturers had already invested "several billion dollars to date in developing fuel cell technology and have publicly stated plans to continue heavy investment in the next decade" (CARB 2004b). CARB concluded that despite this effort, "fuel-cell ZEVs are clearly not ready for volume production" at this time (CARB 2004b). In the meantime, there have been "rapid advances" in HEVs and similar extremely clean vehicles certified according to the PZEV and AT-PZEV standards (CARB 2004b). "Under these circumstances, CARB has concluded that the best course of action is to take full advantage of the near-term possibilities afforded by PZEVs and AT-PZEVs, and adopt a stepwise approach toward pure ZEV commercialization that takes into account the desire of vehicle manufacturers to devote their entire 'gold' vehicle [pure ZEVs] focus to fuel cell ZEVs" (CARB 2004b).

Despite the numerous revisions and delays in implementing the ZEV mandate, CARB claims that its ZEV mandate has been "instrumental in promoting battery, fuel cell, component and vehicle research and development" (CARB 2000b). A study commissioned by CARB found that the California ZEV mandate had produced important secondary benefits in such categories as new economic development in California, advanced vehicle development, vehicle emissions reductions outside of California, and nonelectric vehicle applications of advanced batteries (Burke et al. 2000). Similarly, the 2000 CARB battery panel (Anderman et al. 2000) found that the battery research motivated by the California ZEV mandate undoubtedly benefited the development of better batteries for use in HEVs. These batteries have much less stringent technical requirements than do totally electric vehicle batteries. Although the ZEV mandate has had some indirect beneficial impacts in motivating research into hybrid electric and fuel-cell vehicles, these benefits cannot obscure the fact that the ZEV mandate forced manufacturers to devote considerable resources to develop electric vehicles, which appear at the present to be economically unviable. CARB's chairman was quoted in the press, "We have put a lot of faith in battery electric vehicles to meet the [zero-emission vehicle] mandate, but in spite of significant efforts batteries have inherent limitations. We're not giving up on the goal of the zero-emission vehicle, but we have to be realistic. No matter how you cut it, it is disappointing" (Polakovic 2002).

This history of the LEV and ZEV demonstrates the benefits of using California as a laboratory to experiment with aggressive, high-risk strategies. The technology-forcing requirements that CARB imposes can result in major breakthroughs in emissions controls. When CARB adopted its LEV standards, vehicle manufacturers claimed that the standards were not technologically feasible within the available lead time (CARB 1991). Under the pressure of the LEV regulations, vehicle manufacturers were able to exceed expectations in reducing emissions from gasoline-powered vehicles to near-zero levels. As discussed below, the success of the LEV program in California benefited emissions-control strategies across the nation and was primarily responsible for making the new federal standards for model-year 2004 more stringent than they otherwise would have been.

An inevitable consequence of a high-risk strategy is the likelihood that some policies will fail, and the electric vehicles envisioned under the California ZEV mandate appear to have failed. Manufacturers had a good record of complying with performance-based emissions standards. In contrast, the ZEV mandate was a technology-specific regulation in its early stages and proved to be too challenging. Although the costs and disruptions of the electric vehicle mandate were substantial, they were limited to a relatively small segment of the national market. The results of the CARB ZEV experiment support the idea of having one state (California) serve as a laboratory for experimentation in emissions control.

Impacts of LEV Program on LDV Emissions-Control Technologies

The LEV program provides a good example of California's role as a laboratory for innovation and technology to reduce mobile-source emissions. The primary success of the LEV program has been in achieving near-zero levels of tailpipe emissions from gasoline-fueled vehicles, exceeding the expectations of experts in both industry and government. One of the reasons that CARB initiated the low-emission vehicle/clean fuels program was that it thought that alternative fuels were likely to be needed to get vehicle emissions down to the ULEV standard or lower. In fact, because of the combination of technology-forcing and flexibility in the LEV program, manufacturers have achieved much greater progress in reducing emissions from conventional vehicles than was believed possible in 1990. In the words of CARB's chairman, Dr. Alan Lloyd, under the LEV program, "We've seen the near impossible accomplished with gasoline vehicles: zero evaporative emissions, exceedingly clean exhaust—cleaner, in some cases, than the outside air entering the cabin for ventilation purposes and emission control systems that are twice as durable as their conventional forebearers, forecasted to last an astonishing 150,000 miles" (CARB 2003b).

The technologies that enabled this enormous progress are a combination of improved catalyst technology, better on-board diagnostic systems, and cleaner reformulated gasoline (Ehlmann and Wolff 2005). For example, in adopting the LEV program, CARB expected that manufacturers would comply with the LEV and ULEV standards primarily by installing electrically heated catalysts. Electrically heated catalysts reduce cold-start emissions, which account for a large majority of the remaining emissions from modern motor vehicles (CARB 1991). However, manufacturers achieved the ULEV standard and beyond in gasoline-fueled vehicles through improved materials that allowed the catalyst to be more heat resistant, resulting in faster warm-up by allowing them to be placed much closer to the engine. The elimination of the electrically heated catalysts to comply with the LEV standards reduced the cost and simplified integration into the vehicle design.

Indeed, compliance with California standards creates its own subset of automobile engineering research featured in such journals as *Topics in Catalysts and Applied Catalysis* and publications of the Society for Automotive Engineers. Examples of such research for the LEV program are Summers et al. (1993), Smaling et al. (1996), and Truex (1998). More recent examples include technologies to comply with aspects of the LEV II program, such as McKinnon et al. (1999), Heck and Farrauto (2001), and Kim et al. (2001).

The impact of California standards is characterized several times in this report as either a success due to CARB's role in looking at fuels and engines as an integrated unit or as a failure due to its inability to bring about widespread use of electric vehicle technology. In reality, it is too simplistic to characterize the outcomes as simple successes and failures. For example, the impact of the ZEV on the introduction of popular hybrid technologies is not straightforward. The pursuit of many low-emissions technologies, including hybrid technologies, came in response to CARB standards for lower and zero emissions vehicles. However, CARB initially did not favor the HEVs as a part of the ZEV mandate. Sorting out the full relationship between the ZEV mandate and HEVs requires an in-depth understanding of many factors, including strategic business strategies, beyond the purview of this committee.

Influence of the LEV Program on National Mobile-Source Emissions Standards

Many northeastern states actively pursued adoption of California's LEV program in the 1990s to achieve emissions reductions that would help to meet their air quality goals. Thirteen northeastern states that were the members of the Ozone Transport Commission (OTC), created by the 1990 CAA amendments, pledged in October 1991 to adopt the California LEV standards. Massachusetts, New York, and Maine adopted LEV emissions standards between 1991 and 1993. As a result of those state efforts, manufacturers proposed in the fall of 1993 that they would voluntarily provide low-emitting cars that exceed the federal standards to the 49 non-California states in return for the northeastern states abandoning the California LEV program, particularly its ZEV mandate (Jolish 1999). This initial proposal resulted in over 4 years of active negotiations among the states, vehicle manufacturers, and other interested parties under the active supervision of EPA. The resulting NLEV program could provide stricter standards before EPA since the 1990 CAA amendments specified that the next set of federal emissions standards (Tier 2), if necessary, were not to begin until 2004 model-year vehicles.

In October 1995, EPA published a public notice of its plans to implement a voluntary NLEV program (60 Fed. Reg. 52733 [1995]). EPA determined that the NLEV program would result in equivalent or better emissions reductions in the Northeast Ozone Transport Region than would be achieved by having state-by-state adoption of the California LEV program (including the ZEV mandate) (60 Fed. Reg. 52737 [1995]). EPA also concluded that NLEV would reduce states' costs of improving air quality by avoiding separate, duplicative state programs (60 Fed. Reg. 52736 [1995]). It took another 2 years until the NLEV program was finalized by EPA in early 1998 (63 Fed. Reg. 926 [1998]). Under this voluntary program, vehicle manufacturers would provide low-emitting vehicles to all northeastern states beginning in 1999 and to all 49 non-California states beginning in 2001, and the states would forego implementation of the California LEV program. The program would remain in effect until the 2006 model year. Four states (Massachusetts, New York, Maine, and Vermont) refused to adopt the NLEV program because they were unwilling to forego the ZEV mandate and the California standards for medium-duty vehicles, which were not included in the NLEV program. Despite the four holdout states, EPA calculated that the NLEV program would achieve greater emissions reduction in the Northeast region than either regionwide adoption of California LEV by the OTC or state-by-state adoption of the LEV program. The greater emissions reduction would be largely due to the reduced emissions from permanent or short-term migration of cars into the Northeast from other parts of the nation and to the implementation the NLEV program at an earlier time than would be possible for adoption of the LEV program in all northeastern states. EPA described the NLEV program as "cleaner, smarter, cheaper" than the California LEV program for reducing motor vehicle pollution in the Northeast states (62 Fed. Reg. 31192 [1997]).

The NLEV program provides two major lessons. First, the technology-forcing nature of California standards can benefit not only California but also the rest of the country. The NLEV program, which resulted in substantial reductions of vehicle emissions across the nation beyond what was required under the 1990 CAA amendments, could exist only because of California's leadership in forcing stricter emissions controls and the rights of other states to adopt those standards under section 177 of the federal CAA. Without the emissions-control technology resulting from the California program and the legal authority of other states to adopt the program, vehicle manufacturers would have had no incentive to enter into the NLEV program. The second lesson of NLEV is that vehicle manufacturers are prepared to expend substantial resources to avoid a patchwork of state emissions requirements. The NLEV commitment by manufacturers demonstrates with actions rather than words that two standards do impose a substantial burden on manufacturers.

CURRENT CALIFORNIA AND FEDERAL EMISSIONS STANDARDS

In 1998, CARB proposed a rule-making for LEV II emissions standards to take effect with model-year 2004 vehicles. The technologies used by auto manufacturers to meet the original LEV emissions standards, particularly catalysts, created the potential for emissions control beyond the LEV standards. The LEV II program kept the key features of the LEV program and introduced several new features. The LEV II program included a restructuring of the light-duty-truck (LDT) classifications, and many pickup trucks and SUVs are now subject to the same emissions standards as passenger cars. Since the start of the LEV program, the fraction of pickup trucks and SUVs in the fleet had increased substantially (see discussion in Chapter 3). Many of those vehicles were previously classified as LDT (3,751 to 5,750 lb loaded vehicle weight [LVW[3]]) or medium-duty vehicle (MDV) (various categories from 3,751 to 14,000 lb test weight [TW]) under the LEV program and were subject to less stringent emissions standards than passenger cars and the smallest LDTs (those with a LVW of 3,750 lb or less). In the LEV II program, the LDT 2 category includes trucks weighing 3,751 to 8,500 lb GVW, which includes many of the models of pick-ups, SUVs, and vans that are now commonly used as passenger vehicles. All passenger cars and LDTs less than 8,500 lb GVW are subject to the same emissions standards under the LEV II program.

In addition to subjecting more LDTs to the most stringent standards, the LEV II program certified each vehicle according to standards at 50,000 and 120,000 miles. The LEV II program introduced another new emissions class called super-ultra-low emission vehicle (SULEV) and removed the transitional LEV (TLEV) category. SULEV emissions levels are even lower than ULEV levels and have no intermediate life certification. Furthermore, the permissible level of NO_x emissions in the LEV and ULEV certification categories at 50,000 miles was reduced from 0.2 to 0.05 grams/mile (g/mi). Under LEV II, the NMOG fleet average will continue to decline yearly from 0.062 g/mi in 2003 to 0.035 g/mi in 2010 for passenger cars and LDT 1 and from 0.085 g/mi in 2004 to 0.043 g/mi in 2010 for LDT 2. (These and further specifications for the LEV II program are discussed in CARB [1998a, 1999].)

The federal 1990 CAA amendments included new national LDV emissions standards to begin with model-year 1994, known as Tier 1 standards. Although these levels were more stringent than the pre-1990

---

[3] Vehicle weight is defined a number of ways and is different for different vehicle classes. Loaded vehicle weight (LVW) is the curb weight plus 300 lb. Gross-vehicle weight (GVW) is the curb weight plus a full payload. Test weight (TW) is used for CARB's medium-duty vehicles and is the average of a vehicle's curb weight and GVW.

emissions standards, Tier 1 standards did not include the features of California's LEV program—for example, flexibility of vehicle classes, fuel requirements, and a ZEV sales mandate. As discussed above, NLEV standards based on California's LEV program replaced Tier 1 standards in 1998 and aligned federal and California vehicle emissions standards, although the NLEV did not include a ZEV component. The CAA, however, specified that Tier 2 emissions standards for LDVs and MDVs were to begin with model-year 2004. The CAA directed EPA to assess the need for Tier 2 standards and to adjust the levels of the standards in light of air quality needs, technical feasibility, and cost-effectiveness. Logically, EPA set Tier 2 standards at levels that were commensurate with the great advances in emissions-control technologies introduced in the 1990s, in part, as a result of the LEV program. Tier 2 also incorporated some features of the LEV program, such as the flexibility of a fleetwide-average standard, the ability for manufacturers to certify according to different emissions classes (called bins), and the reliance on low-sulfur fuel to achieve the prescribed standard. Because of concerns about the increased sensitivity of the new-technology vehicles to sulfur poisoning, the Tier 2 standards required refiners to meet an average sulfur concentration of 30 ppm beginning January 1, 2006. Low-sulfur gasoline was introduced in California in the 1990s. Another important similarity between Tier 2 and LEV II is that LDTs up to 8,500 lb GVW are subject to the same stringent emissions standards as passenger cars after a phase-in period. In contrast to LEV II, Tier 2 includes a fleet-average NO_x emissions standard, which does not decline with time after the program is phased in. Table 6-2 summarizes the features of LEV II and Tier 2.

Emissions reductions over the fleet of vehicles subject to Tier 2 and LEV II programs in coming years will depend on many factors and assumptions about the future mix of vehicles in each fleet. Given the greater than 95% emissions reductions achieved by both programs, California and federally certified passenger vehicles are considered by some to produce "near-zero" emissions of NO_x, volatile organic compounds (VOCs), and carbon monoxide (CO) (Ehlmann and Wolff 2005). The committee heard arguments that the two programs are now practically equivalent in terms of emissions benefits (Dana 2004). As discussed in later sections, however, some parties consider the difference in emissions between the programs to be appreciable, especially due to the ZEV mandate (NESCAUM 2003). Figure 6-1 compares the California and EPA tailpipe standards by model year in grams per mile. These projected rates were provided by CARB and are based on the assumption that manufac-

TABLE 6-2 Comparison of the Features of the LEV II and Tier 2 Programs

Feature	LEV II	Tier 2
Regulated vehicle types	PC, all weights LDT1, 0-3,750 lb GVW LDT2, 3,751-8,500 lb GVW MDV, 8,500-10,000 lb GVW MDV, 10,000-14,000 lb GVW	LDV, 0-8,500 lb LLDT, 0-6,000 lb GVW HLDT, 6,000-8,500 lb GVW MDPV, 8,500-10,000 lb GVW
Certification categories	ZEV, AT-PZEV, PZEV, SULEV, ULEV, LEV (see certification levels in Table 6-3)	Bin 1 through Bin 11; Bin 10 and 11 removed after phase-in; Bin 11 for MDPV only (see certification levels in Table 6-4)
Phase-in	For PC, LDT1 and LDT2 at least 25%, 50%, 75%, and 100% resp. in 2004, 2005, 2006, 2007 must be certified to LEV II standards. Alternate phase-in is possible if equivalent NOx emissions reductions are achieved.	For LDV/LLDT, at least 25%, 50%, 75%, 100% in 2004, 2005, 2006, 2007 respectively, must be Tier 2; for HLDT/ MDPV, at least 50% in 2008 and 100% in 2009 must be Tier 2
Fleet-average exhaust emissions requirements (corporate average)	Based on NMOG emissions (all values in g/mile NMOG) Model Year / PC/LDT1 / LDT2 2004 / 0.053 g/mi / 0.085 g/mi 2005 / 0.049 g/mi / 0.076 g/mi 2006 / 0.046 g/mi / 0.062 g/mi 2007 / 0.043 g/mi / 0.055 g/mi 2008 / 0.040 g/mi / 0.050 g/mi 2009 / 0.038 g/mi / 0.047 g/mi 2010+ / 0.035 g/mi / 0.043 g/mi	Based on NO_x emissions During phase-in: 0.30 g/mi for all non-Tier 2 LDV/LLDT 0.20 g/mi all non-Tier 2 HLDT/MDPV 0.07 g/mi for all Tier 2 vehicles After phase-in: 0.07 g/mi for all LDV/LLDT after 2007 for all HLDT/ MDPV after 2009

Certification age	Intermediate life, 50,000 mi; full useful life, 120,000 mi; optional, 150,000 mi for credits toward NMOG fleet average	Intermediate life, 50,000 mi; full useful life, 120,000 mi
Sales mandate	ZEV (with PZEV, partial credits and ACP plans)	No sales mandate
Fuel requirements	Sulfur at 15 ppm	Sulfur at average 30 ppm by 2006

Abbreviations: ZEV, zero-emission vehicle; PZEV, partial-ZEV; AT-PZEV, advanced-technology-PZEV; LEV, low-emission vehicle; SULEV, super-ultra-LEV; ULEV, ultra-LEV; PC, passenger car; LDT, light-duty truck; LLDT, light-LDT; HLDT, heavy-LDT; MDV, medium-duty vehicle; LDV, light-duty vehicle; MDPV, medium-duty passenger vehicle; ACP, alternative compliance plan.

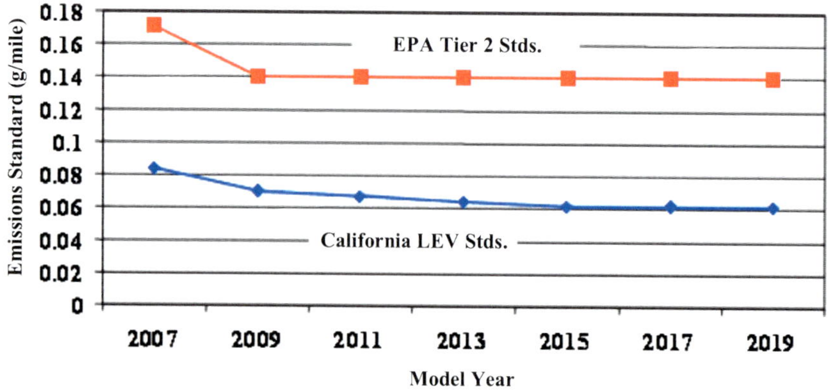

FIGURE 6-1 Comparison of California and federal emissions standards (NO_x + NMOG) in g/mile estimated by CARB. Only vehicles less than 8,500 lb GVW are included. Source: Adapted from CARB 2005e.

TABLE 6-3 LEV II Full Useful Life (120,000 mi) Exhaust Mass Emissions Standards in grams/mile

Type	Category	NO_x	NMOG	CO	HCHO	PM
PC and LDT1	SULEV[a]	0.02	0.010	1.0	0.004	0.01
and LDT2	ULEV	0.07	0.055	2.1	0.011	0.01
	LEV	0.07	0.090	4.2	0.018	0.01
	LEV[b]	0.10	0.090	4.2	0.018	0.01
MDV 8,500-	SULEV	0.1	0.100	3.2	0.008	0.06
10,000	ULEV	0.2	0.143	6.4	0.016	0.06
	LEV	0.2	0.195	6.4	0.032	0.12
MDV 10,000-	SULEV	0.2	0.117	3.7	0.010	0.06
14,000	ULEV	0.4	0.167	7.3	0.021	0.06
	LEV	0.4	0.230	7.3	0.040	0.12

[a]PZEV vehicles have the same exhaust standards as SULEV, but have more stringent evaporative and warranty requirements.
[b]Up to 4% of a manufacturers LDT2 may certify to this higher NO_x LEV category.
Source: CARB 1998a.

TABLE 6-4 Tier 2 and Interim Non-tier 2 Full Useful Life (120,000 mi) Exhaust Mass Emission Standards in grams/mile

Bin No.	NO_x	NMOG	CO	HCHO	PM
1	0.00	0.000	0.0	0.000	0.00
2	0.02	0.010	2.1	0.004	0.01
3	0.03	0.055	2.1	0.011	0.01
4	0.04	0.070	2.1	0.011	0.01
5	0.07	0.090	4.2	0.018	0.01
6	0.10	0.090	4.2	0.018	0.01
7	0.15	0.090	4.2	0.018	0.02
8	0.20	0.125/0.156[b,f]	4.2	0.018	0.02
9[a]	0.3	0.090/0.180[b,e]	4.2	0.018	0.06
10[a]	0.6	0.156/0.230[b,d]	4.2/6.4	0.018/0.027	0.08
11[a,c]	0.9	0.280	7.3	0.032	0.12

[a]This bin and its corresponding intermediate life bin are deleted at end of 2006 model year (end of 2008 model year for HLDTs and MDPVs).
[b]Higher NMOG, CO or HCHO values apply for HLDTs and MDPVs only.
[c]This bin is only for MDPVs
[d]Optional NMOG standard of 0.280 g/mi applies for qualifying LDT4s and qualifying MDPVs only.
[e]Optional NMOG standard of 0.130 g/mi applies for qualifying LDT2s only.
[f]Higher NMOG standard deleted at end of 2008 model year.
Source: 65 Fed. Reg. 6855 (2000).

turers will produce larger numbers of near-zero-emission vehicles to satisfy the ACP option to meet their ZEV mandates (CARB 2005e). The figure does not show the difference in future emissions expected in regions with LEV II vehicles compared with regions with Tier 2 vehicles.

PROCESS OF SETTING STANDARDS: LEV II VS TIER 2

The charge to the committee called for an evaluation of the scientific and technical practices used by CARB in setting California emissions standards and for a comparison of its practices with those used by EPA. The comparison of the practices used by CARB and EPA in setting their LEV II and Tier 2 standards, respectively, is an appropriate case study because both processes required extensive regulatory assessments completed relatively recently. LEV II and Tier 2 are the current California and federal LDV and LDT emissions standards and both came into effect with model-year 2004. The committee compared standards for on-road vehicles primarily in terms of two analysis documents: "The Staff Paper—Initial Statement of Reasons" (the staff paper) for LEV II (CARB 1998a) and the "Regulatory Impact Analysis" (RIA) for Tier 2 (EPA 1999A), although a large number of other documents were used for both analyses. Practices used by CARB and EPA to set emissions standards are categorized as four assessments: the need for new standards, technical feasibility, emissions and air quality impacts, and economic impacts.

The Need for New Mobile-Source Emissions Standards

The ultimate objective of regulating any emissions source is the protection of public health and welfare. In the United States, one way that is accomplished is by setting NAAQS for criteria pollutants, which have traditionally been the main focus of mobile-source regulation. California established its own ambient air quality standards (AAQS) for criteria pollutants. As shown in Table 2-1 in Chapter 2, California's AAQS are more stringent than the federal NAAQS although the state does not have AAQS attainment deadlines. Attainment of the NAAQS, for ground-level ozone has in particular been the impetus for regulating mobile-source emissions in recent decades. CARB, as a state agency delegated the responsibility to attain the NAAQS in California, is required by the federal CAA to develop a state implementation plan (SIP) (see discussion in Chapter 3). One element of the SIP is to identify and prescribe source-specific emissions reductions required in each area to attain the NAAQS. EPA sets national emissions standards by authority of the CAA with the goal of progressively attaining the NAAQS nationwide; however, the level of emissions reductions is not tied directly to NAAQS attainment of specific airsheds. The difference is evident in the agencies' assessment of the need for the LEV II and Tier 2 regulations.

CARB submitted a SIP for the South Coast Air Quality Management District (AQMD) to EPA in 1994 with measures to reach attainment for ground-level ozone by 2010. Among other measures, the SIP included a measure called M2 for reducing mobile-source NO_x plus NMHC[4] by 25 tons per day (tpd) in the South Coast (CARB 1994c).

---

[4] Various terms to denote hydrocarbons are used in different regulatory documents. The differences are explained in Appendix A.

Even with M2 reductions and other reductions from all identified sources, CARB determined that more emissions reductions were needed to meet the goals of the SIP. An additional amount, the so-called black-box, required an additional NO_x and NMHC of 75 tpd from to-be-determined sources.[5] The need for the black-box emissions reductions provides evidence of the extent to which control strategies must be pushed to meet air quality goals in the South Coast region. CARB states that "the primary objective of the [LEV II] rulemaking is to implement Measure M2 of the 1994 California SIP for ozone, and to achieve as much additional reactive organic gases (ROG) plus NO_x emissions reductions as are technologically feasible and cost-effective, to be counted against the SIP's additional 75 tpd ROG plus NO_x emission reduction target—the so-called black-box" (CARB 1999).

California's need for stricter emissions standards was based on the emissions reductions needed in the South Coast to attain the 1-hr-average NAAQS for ozone; however, stricter LDV standards were also a means to improve air quality in other parts of the state for ozone, particulate matter (PM), and hazardous air pollutants (HAPs). The assessment of the need for emissions reductions, and thus the need for stricter standards, was performed by the South Coast AQMD as part of the modeling analysis for the SIP. The analysis includes estimation of present and future emissions inventories, including mobile sources. The analysis also includes photochemical grid modeling to determine the expected reductions in ground-level ozone concentrations that result from precursor emissions reductions.

EPA has general authority to prescribe vehicle emissions standards that are not prescribed or limited in CAA section 202(a). As a result of the 1990 amendments, CAA explicitly lists the numerical values of the Tier 1 exhaust emissions standards for LDVs and LDTs, which began with model-year 1994 vehicles. The CAA further directed EPA to submit a report to Congress in 1997 on the need, feasibility, and cost-effectiveness of new vehicle standards for 2004-2006. If the study found

---

[5] The use of unspecified, or black-box, emissions reductions in SIPs was included in Section 182(e)(5) of the federal CAA amendments of 1990. It is only available to areas classified as "extreme" nonattainment for ozone, based on the severity of measured ozone concentrations. To date, the San Joaquin Valley and the South Coast air basins in California are the only areas of the nation to be classified extreme by EPA. It appears that the black-box measures continue to apply in meeting the 8-hr ozone NAAQS in the areas that were classified extreme nonattainment for the 1-hr ozone NAAQS.

that standards were necessary, feasible, and cost-effective, EPA was then directed to set emissions standards via a rule-making. CAA section 202(i) included numerical values for Tier 2 standards as an alternative for EPA to consider if more stringent standards were deemed necessary, feasible, and cost-effective. The study, known as the "Tier 2 Report to Congress," concluded that there would be an air quality need for emissions reductions to aid in meeting and maintaining the NAAQS for both ozone and PM (EPA 1998). The assessment of the need for stricter standards was based on photochemical modeling performed for two other rule-makings: the 8-hr ozone NAAQS and the "Ozone Transport Assessment Group SIP Call"—an EPA rule to promote regional emissions reductions in the eastern United States. Both analyses showed that projected air quality under the existing mobile-source standards would result in several areas nationwide in violation of the 1-hr-average and 8-hr-average NAAQS for ozone and the NAAQS for PM_{10}.[6]

Technical Feasibility Assessment

Projected Technologies

EPA and CARB assess technical feasibility through discussions (some confidential) with auto manufacturers, catalyst and other parts manufacturers, and suppliers. In addition, both agencies rely on bench-scale demonstrations that vehicles can attain the standards. The demonstrations are performed by various groups, including in-house and outside contractors and manufacturers. EPA and CARB present very similar lists and discussions of projected technologies to meet the Tier 2 and LEV II exhaust standards, respectively (Table 6-5). The technologies in the table represent different options that manufacturers could potentially use to meet the standards, although manufacturers can meet the standards with any combination of technologies. The majority of the technologies were already being used to meet the LEV and NLEV standards, and both agencies stated their expectation that those technologies could be improved to achieve even lower emissions. The exception was the advanced catalyst systems that CARB and EPA both projected would exist at the time of implementation. Advanced catalysts were thought to play

[6] Particles with an aerodynamic equivalent diameter ≤ 10 μm.

TABLE 6-5 Ex Ante Estimates of Most Likely Technologies to Meet the CARB LEV II and EPA Tier 2 Standards, Respectively

Technology Type	CARB	EPA
Technologies for improving fuel control	Dual oxygen sensors	Dual oxygen sensors / fast light-off exhaust gas oxygen sensor
	Universal exhaust gas oxygen sensors	Universal exhaust gas oxygen sensors
	Individual cylinder air-fuel control	Individual cylinder air-fuel control
	Adaptive fuel control systems	Adaptive fuel control
	Electronic throttle control systems	Electronic throttle control systems
	Engine calibration techniques	Faster microprocessor
Fuel atomization and delivery	Sequential multipoint fuel injection	Sequential multipoint fuel injection
	Heated fuel injectors	Air-assisted fuel injection
	Air-assisted fuel injectors	Multiple valves and variable valve timing
	Improved induction systems	
Enhanced catalytic converter performance	Heat-optimized exhaust pipes	Heat-optimized exhaust pipe
	Leak-free exhaust systems	Leak-free exhaust system
	Close-coupled catalysts	Close-coupled catalyst
	Electric air injection	Secondary injection of air into exhaust
	Electrically heated catalysts	Manifold with low thermal capacity
	Hydrocarbon adsorber systems	Adsorbers/traps
	Increased catalyst loading	Increased catalyst volume and platinum group metal loading
	Improved high-temperature washcoats	Improved catalyst washcoats
	Abbreviated engine start systems	Retarded spark timing at start-up
Reduced base engine out emissions	Full electronic exhaust gas recirculation	Full electronic exhaust gas recirculation
	Engine designs to reduce oil consumption	Engine modifications
	Reduced combustion chamber crevice volumes	Improved combustion chamber design

Sources: CARB 1998a; EPA 1999A.

an important role in meeting the current California and federal standards (Bertelsen 2001).

The LEV II program also introduced much more stringent evaporative emissions standards for NMHC compared with LEV I. The CARB staff paper presents a discussion of these standards and the technologies expected to achieve them. The standards targeted diurnal and hot-soak fuel emissions and were projected to be met through a combination of better seals and less permeable equipment in the fuel tanks and fuel lines. EPA noted in its Tier 2 rule-making that manufacturers were already certifying vehicles below the level of the Tier 2 evaporative emissions standards; the RIA for the Tier 2 rule-making includes only an abbreviated discussion of evaporative emission-standard technical feasibility. The remaining assessments will focus on the exhaust emissions standards.

Evidence Provided by CARB on Technical Feasibility of LEV II Standards

As one piece of evidence that the LEV II standards were technically feasible, CARB noted that many of the projected LEV II technologies had been in use on vehicles for several years (CARB 1998a). CARB also noted the continuous evolution of emissions-control technologies since the introduction of the original LEV program and noted that the durability and performance of existing components, especially catalysts that heated quickly, had improved greatly.

CARB presented results from two test programs as evidence that vehicles could meet the proposed LEV II emissions limits. In the first program, CARB tested five model-year 1997-1998 passenger cars. The primary modification to all vehicles was the addition of an advanced catalyst system, although other modifications were required, namely, air injection time, oxygen sensor biasing, and ignition retard. The staff paper showed that all vehicles attained emissions below the 50,000-mi LEV II—ULEV level using "green" catalysts. One vehicle was tested after catalysts and oxygen sensors were bench-aged (a technique used at the Southwest Research Institute to simulate 50,000 miles of use) to 50,000 miles, and it met the standards. The test program did not attempt to determine compliance with SULEV standards, although CARB cited some manufacturers' intentions at the time to introduce advanced technology vehicles, such as hybrids, as evidence of meeting SULEV standards.

In the second program, CARB tested two identical SUVs with test weights of 6,000 lb (LDT 2). The new vehicles, as received, showed baseline NO_x and CO emissions below and NMHC emissions slightly above proposed LEV II levels. Oxygen sensors and advanced catalysts were bench-aged. The sensors and catalysts were then added to the two new vehicles, which yielded emissions higher than proposed LEV II levels in many of the tests. CARB reasoned that emissions would be lower with appropriate adjustments to the software that controlled fuel delivery. CARB was not capable of developing such software, but it believed that manufacturers could. CARB modified its test preconditioning methods to simulate the effects of controlling the fuel and adding electronic air injection to one test vehicle in addition to the bench-aged advanced catalysts and oxygen sensors. Subsequent tests attained NHMC, CO, and NO_x emissions just below proposed LEV levels in some tests.

CARB did not perform a feasibility analysis for the heavier trucks (LDT 2 category, 6,000-8,500 lb). The difficulty in achieving the LEV II standards on such vehicles was a primary critique from auto manufacturers during the LEV II rule-making (CARB 1999). CARB (1998a) noted that it had little information about fuel and spark timing strategies that were likely to be part of the manufacturers approach to reducing emissions.

Evidence Provided by EPA on Technical Feasibility of Tier 2 Standards

In its review of model-years 1999 and 2000 certification data, EPA found that several existing engine families already met the new Tier 2 standard NO_x and NMHC fleet averages (EPA 1999a). For model-year 1999, 48 of approximately 400 engines included at least one vehicle configuration, mostly for passenger cars, certified to meet lifetime NO_x emissions below the average of 0.07 g/mi of the proposed emission standard. Approximately half of those vehicles were certified to meet lifetime NO_x levels below 0.04 g/mi.

EPA further cited results from outside testing to support feasibility of Tier 2 standards. Results from fuel sulfur-level testing by the Coordinating Research Council and auto manufacturers showed the potential to go as low as the Tier 2 emissions standards. Seven of 20 LDVs with 100,000-mile-aged catalysts met Tier 2 NO_x and NMHC standards by switching to low-sulfur fuel. Vehicle testing at Southwest Research Institute, sponsored by the Manufacturers of Emissions Controls Association (MECA), demonstrated that two LDVs and one LDT met the Tier 2 NO_x and NMHC standards by replacing the original catalysts with advanced catalysts and by modifying existing secondary air and exhaust gas recirculation. Most vehicles met NO_x and NMOG useful-life design targets, which are generally 50-70% of the standards. Finally, EPA cited the results of CARB's test programs (described above) to demonstrate the feasibility LEV II standards and the evidence that the technology was available to meet Tier 2 standards.

EPA used evidence from external testing but also conducted its own testing program to investigate the feasibility of requiring heavy LDTs (>6,000 lb GVW) to meet the Tier 2 standards at intermediate useful life (50,000 miles). EPA stated that a key element of this test was alteration of engine calibration parameters, including modification of spark timing, exhaust-gas recirculation (EGR), and fuel control. These modifications were made in addition to use of improved catalysts, including increases in volume and precious metal loading and higher cell densities, thermally aged to an equivalent 50,000 miles. Both trucks tested attained NO_x and NMHC emissions below Tier 2 standards in the final tested configurations. EPA stated that considerable tuning of the engine calibration parameters for one vehicle was necessary to reduce NO_x and NMHC emissions, particularly from the initial cold-start phase of the federal test procedure, and to minimize impacts on drivability and fuel economy. Cold-start NO_x emissions were further reduced for this vehicle by using improved low-mass, sealed air-gap exhaust manifolds. EPA also stated that technologies and emissions-control strategies implemented for these heavy LDTs could be used to meet the more stringent standards for MDVs.

Emissions and Air Quality Impacts Assessment

CARB and EPA use similar tools to measure emissions and air quality impacts of proposed regulations, but to different ends. CARB used its own mobile-source emissions model, EMFAC97G, to estimate that the LEV II program would result in a reduction of 6 tpd of ROG (exhaust and evaporative), 51 tpd of NO_x, and 120 tpd of CO in the South Coast basin by 2010 (CARB 1998a). CARB staff determined that these reductions would fulfill the M2 SIP control measure, would pro-

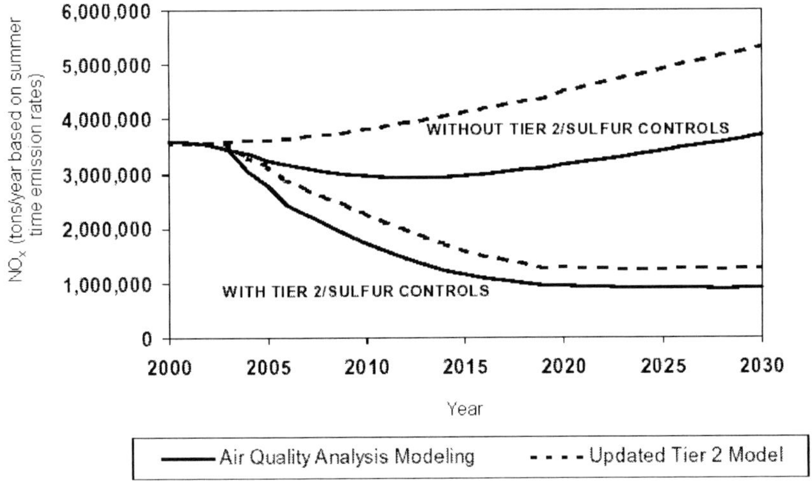

FIGURE 6-2 Forty-seven-state (excluding California) NO_x emissions with the Tier 2/sulfur rule. Vertical axis represents annualized summer tons (tons per year) of NO_x emissions. EPA used two different models of analysis. Source: EPA 1999a.

vide additional reductions toward shortfalls in other programs, and would make progress toward the black-box reductions (M2 and black-box reduction control measures for the South Coast discussed earlier in the chapter). The analysis for the LEV II rule-making did not include any additional photochemical model runs to assess the precise impact on air quality in the South Coast or elsewhere in California. AQMDs in California, including the South Coast, include new LDV emissions reductions when they use modeling to demonstrate attainment of the NAAQS.

EPA used the Tier 2 emissions model, which was similar in function to the planned (at the time) MOBILE 6 emissions model, to project benefits of Tier 2 on nationwide on-road emissions. Figure 6-2 shows a chart from the RIA showing the benefits in NO_x emissions in the lower 47 states (excluding California) with and without the proposed Tier 2 rule ("without" assumed existing NLEV standards for LDV and LDT 1 and Tier 1 standards for LDT 2). In further modeling, EPA used the urban airshed model (UAM), a photochemical grid model, to estimate the impact on ozone pollution with and without the Tier 2 regulations. Two modeling domains were used (East U.S. and West U.S. domains); two episodes were run for the West domain, and three episodes were run for the East domain, where each episode had a different set of meteorological inputs. Each episode was run for a combination of base-year (1996) and future-year (2007 and 2030) emissions. Results of the UAM showed that the number of exceedances of the 1-hr ozone standard was expected to decrease by one-tenth in 2007 and one-third in 2030 in metropolitan areas as a result of the Tier 2 rule. EPA also presented further analysis of the expected Tier-2-related reductions in emissions of PM, diesel PM, and HAPs. The RIA includes estimates of the benefit of Tier 2 regulations on nonattainment of the PM_{10} NAAQS. EPA further quantified and monetized the health and welfare impacts from this range of air quality improvements (discussed below in benefits section).

The analysis of emissions and air quality benefits for each rulemaking highlights some similarities and differences in the agencies' practices in setting emissions standards. CARB and EPA consider technical feasibility by examining how a small subset of vehicles could be redesigned to achieve the proposed standards. CARB sets statewide emissions standards focusing on meeting air quality standards in the worst nonattainment area in the state. An overall emissions budget is developed by AQMDs as part of the SIP planning. The LEV II standards are then developed to achieve as many reductions they perceive to be technically possible to meet the budget. CARB does not examine the air quality benefits of their regulations in isolation from other SIP measures in each AQMD. CARB's LEV II regulations do not include monetary estimates of public health benefits. In EPA's case, standards are set with the primary goal of bringing all of the nation's nonattainment areas closer to attainment, although NAAQS attainment is ultimately a state responsibility. EPA did estimate future emissions, air quality, and health benefits from its proposed Tier 2 standards because it qualified as a major regulation.

Cost Estimates, Cost-Effectiveness, and Benefits Analysis

CARB and EPA looked at the potential costs of implementing the standards and performed some cost-effectiveness analyses that allow comparison of the relative costs of the standards with the costs of other emission-reduction plans. EPA also performed a cost-benefits analysis for the Tier 2 standards. The committee describes the estimates and analyses below.

Cost Estimates

Ideally, costs should be estimated in connection with the expected emissions reductions from setting standards. The costs of achieving a given standard are closely linked to the expected emissions reductions. The level of emissions reduction predicted for the standard should include realistic assumptions about the amount of compliance with the program, the number of vehicles sold, and the effectiveness of controls over the life of the vehicle. If those outcomes or conditions will require additional costs, those costs must be taken into account.

As described in Chapter 5, it is critical when estimating costs to be clear about what constitutes the baseline to which the changed standard is being compared. The incremental costs of changing a vehicle to meet the California SULEV standard will be different if they are compared with the ULEV II standard or with an earlier less stringent standard, such as the LEV. Sometimes, the cost estimates in the regulatory documents are not clear on what standards are being compared.

CARB Cost Estimates

CARB bases its estimates of the cost of compliance with proposed new standards on estimates of the full consumer costs to meet the standards, assuming that such costs are fully passed on. These estimates include variable costs (equipment, assembly, and warranty), support costs, (research and development), investment costs, and dealer costs. (The approach is similar to that detailed in Wang et al. [1993].) In the early years of the California LEV program, there was a great deal of uncertainty about the potential costs of meeting the LEV standards. Table 6-6 summarizes the estimates by CARB and automakers of costs of meeting the LEV standards per vehicle compared with the federal Tier 1 standards and averaged across the California fleet. The ex ante (preproduction) costs estimated by the automakers in 1994 were much higher than the CARB estimates.

In the case of the LEV standards, the actual average cost per vehicle, based on analysis of CARB staff, appear to be much closer to CARB's original estimate than the manufacturer's estimate. CARB reassessed the costs periodically over time as evidence became available about the actual attainment of standards on certain vehicles. The ex post

TABLE 6-6 Estimated Average Cost per Vehicle of California LEV Compared with Tier 1 and Averaged Over the Fleet

	Cost per Vehicle
Preproduction estimates	
CARB (1990)[a]	$170
Sierra Research (1994)	$788
Postproduction estimate	
CARB (1996)[a]	$120

[a]Weighted average of four-, six-, and eight-cylinder vehicles.
Sources: CARB 1990, 1996; Sierra Research 1994.

TABLE 6-7 Estimated Average Cost per Vehicle to Meet California LEV Standard Compared with the Tier 1 Standard (Cost per Vehicle)

Vehicle and Engine[a]	CARB Preproduction Estimate, 1994	CARB Estimate Based on Actual Production, 1998
Honda Civic, four cyl.	$86	$76
Toyota Camry, six cyl.	$137	$79
Ford Crown Victoria, eight cyl.	$139	$152

[a]Costs are to meet the standards for the LEV type, not the other types under the LEV program.
Source: Cackette 1998.

(postproduction) estimate in Table 6-6 shows per vehicle cost averaged across the fleet from the CARB ex post analysis. Average actual costs were even less than the California ex ante estimates to meet the LEV standards. Table 6-7 summarizes the pre- and postpreproduction estimates of costs for the individual vehicles used to obtain the fleetwide averages. The costs of the four- and six-cylinder vehicles were higher ex ante than ex post. This difference was primarily due to lower catalyst cost than anticipated. The eight-cylinder vehicle cost was higher in production than estimated in earlier studies. For that vehicle, costs were slightly higher for equipment and materials, as well as for support and other components.

For the LEV II standards, CARB used a similar approach to determine ex ante average costs per vehicle. Costs were estimated for vehicles of different engine sizes during the time the LEV II standards were being proposed in 1998. An example of the component cost breakdown of vehicles to meet ULEV II over ULEV I standards presented in the staff paper is given in Table 6-8. Engine class costs are then weighted by the

TABLE 6-8 Passenger Car and LDT 1: Incremental Component Costs of a ULEV II Compared with a ULEV I

Emissions-Control Technology	Incremental Cost ($) Four cyl.	Six cyl.	Eight cyl.
UEGO[a]	0	0	0
Air-assist fuel injection[b]	0	0	0
Individual cylinder fuel control[c]	0	0	0
Retarded spark timing at start-up[c]	0	0	0
Low thermal capacity manifold (upgrade)	10	20	20
Greater catalyst loading[d]	14	22	37
Improved double-layer washcoat	2	3	5
Engine modifications[e]	0	10	15
Air injection (electric)[f]	0	33	33
Total incremental component cost	26	88	110

NOTE: Incremental cost represents an average across different models and is based on the estimated component cost multiplied by the additional fraction of new ULEV II models that are projected to use the technology. When an incremental cost is zero, it indicates that the same fraction of ULEV I models used the technology as the fraction of ULEV II models that are projected to use it. Cost estimates are rounded to nearest dollar and retotaled.

[a]Only the front oxygen sensor is a universal exhaust gas oxygen (UEGO) sensor.

[b]Air-assisted fuel injection requires minor redesign of the idle air control valve at no cost and addition of an adaptor to each injector at a cost of $2 each.

[c]Individual cylinder fuel control and retarded spark timing at start up require only software changes at no additional hardware cost.

[d]Catalyst volume on a ULEV II is estimated to be the same as it is on a ULEV I vehicle.

[e]Types of engine modifications may be less uniform throughout the industry and may include such items as an additional spark plug per cylinder, a swirl control valve, or other hardware needed to achieve cold-combustion stability, improved mixing, and better fuel injector targeting.

[f]Cost of air injection includes an electric air pump with integrated filter and relay, wiring, air shutoff valve with integral solenoid, check valve, tubing, and brackets.

Source: Modified from CARB 1998a.

shares of vehicles of different types to find the average costs of meeting the standards. Table 6-9 shows those estimated average costs of meeting the ULEV II and SULEV components of the LEV II program. The costs are generally higher for larger vehicles. In fact, the CARB estimates (CARB 1998a) for SULEV costs are mostly for the four-cylinder engines and a few six-cylinder engines. At the time of the study, CARB engineers believed it likely that most SULEVs would be four-cylinder due to the potential difficulty of meeting the standard for the larger vehicles.

EPA Cost Estimates

EPA, in setting standards, is directed to consider "the need for, and cost-effectiveness of, obtaining further reductions in emissions" (CAA section 202(i)(2)(A)). EPA uses a similar approach for estimating the costs per vehicle similar to the approach of CARB except that EPA develops both short-term and long-term estimates. EPA first assesses the costs of the standards to manufacturers, including variable costs (hardware and assembly), fixed costs (research and development, retooling, and certification), and operating costs (fuel use, fuel cost, and maintenance). The hardware component of the manufacturer's costs are then assumed to be marked up for retail sales to consumers. Cost estimates are made separately for LDVs, the different LDT types, and different engine sizes (four, six, and eight cylinder). EPA attempts to estimate costs (and emissions reductions) for the larger vehicles (for example, 10 cylinder) as compared with CARB analysis. EPA's costs are projected for both variable- and fixed-cost components over the short term, assuming that capital cost recovery will occur in 5 years.

Over the long term, however, cost per vehicle is assumed to fall due to learning, which is the ability to produce lower-emitting vehicles more efficiently.[7] EPA assumes in their Tier 2 analysis that costs will fall to 80% of their earlier value after 2 years in production and will not decrease after that due to learning. Costs per vehicle are estimated to stay the same after 5 years when fixed costs expire. Costs are estimated to be much higher for the larger vehicles. For example, the heavier light-duty trucks (known as LDT 3 and LDT 4) are projected to cost $202 more than an NLEV vehicle, compared with only $70 more for a passenger car and lighter light-duty trucks than an NLEV (both long-run estimates).

[7] For a more complete discussion of the effect of learning on costs, see Chapter 5.

TABLE 6-9 Ex Ante Estimates of the Costs per Vehicle for Meeting the LEV II and Tier 2 Standards (dollars per vehicle)

Source	ULEV-LEV II Compared with ULEV-LEV I	SULEV Compared with ULEV-LEV II	Tier 2 Compared with NLEV[a]: Short term (1-2 yr) per Long term (6+ yr)
CARB 1998a	PC $71	PC $131	
	LDT 1 $46	LDT 1 $105	
	LDT 2 $184	LDT 2 $279	
	MDV 2 $208		
	MDV 3 $209		
	MDV 4 $134		
EPA 1999a			LDV $70/$42
Chapter V			LDT 1 $63/$39
			LDT 2 $107/$80
			LDT 3 $202/$161
			LDT 4 $212/$169

[a]Not specified to which bin the vehicle is certified.
Sources: CARB 1998a; EPA 1999a.

Estimated costs of the federal Tier 2 standards are similar to the California ULEV vehicles, as shown in Table 6-9. In fact, the Tier 2 costs were derived in part from the work that CARB had done on costs for LEV II.

The Union of Concerned Scientists re-analyzed EPA's cost estimate for the average 6-cylinder LDT 2 to assess the costs of meeting the Tier 2 requirements for a Ford Explorer (Mark 1999). They estimated a short-run cost of $138 over a similar NLEV vehicle, which is close to the average LDT 2 estimate by the EPA.

Cost Estimates for Advanced Technology Vehicles

Estimates of the costs for vehicles that will meet the more stringent standards of the California LEV II program are examined here. Table 6-10 shows the estimated incremental costs of PZEVs, AT-PZEVs, and ZEVs taken from different sources. There is a substantial difference in costs between the CARB estimates and other estimates for the PZEVs and AT-PZEVs. A number of PZEVs are already on the road, and these initial models tend to be passenger cars with smaller engines (CARB 2005f, CARB 2006a). AT-PZEV estimates in Table 6-10 refer to gasoline hybrid electric vehicles. A few hybrid models are in production today; however, technology types and costs under future high-volume production scenarios are uncertain, which probably accounts for some of the variability in cost estimates in Table 6-10.

Overall estimates of the costs of these vehicles require assumptions about advancements in technology and distribution of sales across the fleet. There are a range of possible fleet distributions that would result in the standard being met. The actual distribution of types of vehicles actually sold will depend on vehicle prices and consumer preferences. It is difficult to project far into the future what public preferences will be, but some range of plausible fleet distributions might be important to consider. One CARB estimate is that by 2012, the fleet will have to be close to 60% PZEVs to meet the LEV II standards (CARB 2005f). It is not entirely clear how the per-vehicle costs are aggregated across the fleet to meet the standard in the CARB assessments. Small vehicles (four-cylinder) are assumed to be sold in sufficient numbers to meet the 60% PZEV level at the assumed costs. If not, then the costs of having larger vehicles meet the standards would have to be assessed. In several places, the California analysis (CARB 1998a) provides some evidence that CARB and the automakers expect full-sized trucks, SUVs, and vans to

TABLE 6-10 Incremental Manufacturer Cost Estimates of PZEVs, AT-PZEVs, and ZEVs (dollars per vehicle)

			ZEVs		
Source	PZEVs	AT-PZEVs[a]	City EV	Full EV	Fuel Cell
CARB 2003c	$100 compared with SULEVs	$3,300 (2003-2005) $1,500 (2006-2008) $1,200 (2009-2011) $700 (2012 +) compared with SULEVs	$8,000	$17,000	$1,000,000 (2003-2005) $300,000 (2006-2008) $120,000 (2009-2011) $9,300[d] (2012 +)
Dixon et al. 2002[b]	$60-$230 compared with SULEVs	$2,300-$3,300 compared with SULEVs	$5,300-$13,400	$8,000-$26,100	$8,300-$15,200 (2006-2010)
Austin 2005	$740	$6,355[c]			

[a]AT-PZEV in these analyses refer to gasoline hybrid electric vehicle. Estimates do not account for gasoline savings from higher fuel economy. There are likely important differences between studies in the assumed types and costs of technologies that constitute the AT-PZEV.
[b]Estimates for AT-PZEVs and ZEVs for 2003-2007, before high-volume production of these vehicles, except where noted.
[c]Based on variable technology costs for gasoline hybrids found in NESCCAF 2004.
[d]Long-term fuel cell costs based on Browning 2001.

have the greatest difficulty in meeting the standards. An important question that was raises is whether preferences will shift from those vehicles, or whether the standards will be developed for the larger vehicles at higher cost. Given the uncertainty, a range of estimates based on possible outcomes would provide a more thorough assessment.

The estimated costs of the ZEV mandate are notably high for the more distant years, when at least some pure ZEVs must be sold. In all analyses, the assumption is made that over time and as production volumes increase, the costs of producing these vehicles will fall (learning will occur). There is, however, a great deal of uncertainty about what the costs will eventually be and how low they will fall. As described earlier in this chapter, CARB's early estimates of the costs of electric vehicles were low and then the estimates increased over time as implementation of the technology was found to be more difficult. Conversely, by CARB's analysis, LEV costs were lower than expected.

The estimates in Table 6-10 account for the costs of producing ZEVs, but the assumption of how they will be sold to the public is not clear. Fleet distribution issues and how they figure in to the cost estimates are not well explained. If vehicles are not appealing to the public, it will be difficult to sell them without subsidies. The state of California has subsidized the purchase of ZEVs in the past (CARB 2000a), and other financial incentives exist at the local and federal levels to purchase clean vehicles (CARB 2006b, CARB 2005f). In such cases, the vehicle purchasers and taxpayers share the cost of bringing vehicles to the market.

Cost-Effectiveness

EPA and CARB provide estimates of cost-effectiveness of the LEV II and Tier 2 regulations, on basis of the costs and pollutant reductions described above. The reader is referred to CARB (1998a) and EPA (1999a) reports for more details. Table 6-11 shows the cost-effectiveness of the advanced-technology component of the LEV II program. AT-PZEV cost-effectiveness estimates include fuel savings over the life of the vehicle, which necessarily involve assumptions about fuel use and cost. The cost-per-ton-reduced estimates for the ZEV mandate are higher than most other air emissions-control programs. California acknowledges that, but argues that the program has the broader goal of pushing the automakers to develop alternative cleaner technologies.

TABLE 6-11 Estimates of Cost-Effectiveness of Advanced Technology Vehicles (dollars per ton pollutant reduced over the lifetime of the vehicle)

Source	PZEVs	AT-PZEVs[a]
CARB 2004b	\$44,000 per ton ROG + NO_x compared with SULEVs	\$575,000 per ton ROG + NO_x (2005) \$125,000 per ton ROG + NO_x (2006-2008) compared with SULEVs
Dixon et al. 2002[b]	\$18,000-\$71,000 per ton NO_x + NMOG compared with SULEVs	\$650,000-\$1,800,000 per ton NO_x + NMOG compared with PZEVs
Austin 2005	\$65,000 per ton NO_x + NMOG (assumed to be near term)	\$3,000,000 per ton NO_x + NMOG [c] (assumed to be near term)

[a]AT-PZEV in this analysis refers to a gasoline hybrid electric vehicle. In contrast to Table 6-10, these cost-effectiveness estimates include potential fuel savings. Please see corresponding references for methodologies.
[b]Estimates for AT-PZEVs in 2003-2007 before high-volume production of these vehicles.
[c]Based on variable technology costs for a gasoline hybrid in NESCCAF 2004

Benefits Estimates

EPA and other federal agencies are required by Executive Order 12866 and supplements to perform a benefits-cost analysis (BCA) for all major regulations, including the Tier 2 standards. The Tier 2 emissions standards include a BCA described as having four parts: calculation of the emissions impacts of the rule in a future year, estimation of the air quality and environmental change from the rule in the future year, determination of the health and welfare effects in terms of physical effects and monetary value, and calculation of the costs of the standards in the same future year for comparison to monetized benefits (EPA 1999a). EPA estimates benefits occurring in the year 2030 when the entire fleet is expected to be nearly turned over and to consist of mostly Tier 2 vehicles. Because the fleet in this year would represent a nearly complete implementation of the rule, the year would include "the maximum emission reductions (and resultant benefits) and…the lowest costs…on a per mile basis"; and therefore "the resulting benefit-cost ratio will be close to its maximum point" (EPA 1999a).

The emissions and air quality impacts of the rule are calculated using models like those described for environmental impacts; however, it appears that the air quality analyses were repeated specifically for the benefits analysis because the future year was different. Air quality changes included ambient ozone and PM concentrations, airborne nitrogen deposition, and visibility. The health outcomes associated with changes in air quality that EPA considered included both premature mortality and incidences of nonfatal health effects such as asthma. Welfare outcomes considered by EPA included material damage, economic output, visibility, and nutrient loading. The health benefits for the Tier 2 rule are presented in Table 6-12.

The discussion in the RIA notes the difficulties in quantifying and monetizing changes in many of the end points. As a result of these difficulties, some health and welfare outcomes were not quantified and are discussed only qualitatively. In earlier analyses, EPA assessed uncertainty by sensitivity analyses. The RIA includes a qualitative discussion of the uncertainties in estimating benefits and monetizing these benefits. In the Tier 2 rule, information about the possible distribution of benefits was considered for a more statistically accurate estimation of benefits. Uncertainty surrounding cost estimates was not examined. Benefits estimation from air quality regulations was the subject of a separate National Research Council study (NRC 2002b). This NRC study used the benefit estimation from EPA's Tier 2 rule as one of three case studies, and the reader is referred to the NRC (2002b) report for more discussion on EPA's benefits estimation practices.

Two of the most important health benefits of the Tier 2 rule shown in Table 6-12 were PM related: an estimated 4,300 annual avoided cases of premature mortality and 2300 annual avoided cases of chronic bronchitis. EPA presented a preferred estimate of $25 billion (1997 monies) for the benefits from the Tier 2 rule in 2030, not including benefits that were not monetized (EPA 1999b). Twenty-three of the $25 billion account for the health benefits of reduced premature mortality associated with reduced PM concentrations. EPA noted that the estimate is dependent on the method used to value reduced mortality, namely the value of a statistical life (VSL) estimate, which is itself uncertain. EPA also included alternative estimates of the benefits under scenarios where major assumptions were changed. Total costs of the Tier 2 program were developed using the same basis as the benefits analysis (2030 with a full Tier 2 fleet) and methods described earlier in this section. The total costs were estimated to be $5.3 billion (1997 monies).

TABLE 6-12 Annual Health Benefits (Avoided Cases of Mortality and Morbidity and Monetized Value) for Tier 2 Regulation in 2030

Health Outcome	Avoided Cases[a]	Monetized Benefit (1997 $ in millions)[b]
PM-Related Health Outcomes		
Premature mortality (adults, ages 30 and over)	4,300 (2,700-5,900)	23,380
Chronic bronchitis	2,300 (600-4,100)	730
Hospital admissions		
Respiratory causes	1,200 (400-2,100)	10
Cardiovascular causes	500 (100-1,100)	10
Emergency room visits for asthma	900 (400-1,400)	<1
Acute bronchitis (children, ages 8-12)	7,900 (0-16,300)	<1
Lower respiratory symptoms (children, ages 7-14)	87,100 (39,900-131,100)	<5
Upper respiratory symptoms (children with asthma, ages 9-11)	86,500 (25,500-144,600)	<5
Shortness of breath (African Americans with asthma, ages 7-12)	17,400 (4,700-29,500)	<1
Work-loss days (adults, ages 18-65)	682,900 (597,800-771,800)	70
Minor restricted-activity days and acute respiratory symptoms	3,628,500 (3,034,100-4,177,200)	170
Ozone-Related Health Outcomes		
Chronic asthma (adult males, ages 27 and over)	400 (100-800)	10
Hospital admissions		
Respiratory causes	1,000 (200-1,800)	10
Cardiovascular causes	300 (0-500)	<5
Emergency-room visits for asthma	400 (100-600)	<1
Minor restricted-activity days and acute respiratory symptoms	2,226,500 (1,014,400-3,414,800)	100
Decreased worker productivity (adult working population)	Not reported	140

[a]Mean value provided with 5th and 95th percentile values shown in parentheses rounded to the nearest 100.
[b]Mean value of monetized value provided for reference.
Source: NRC 2002b. Adapted from EPA 1999a,b.

Assessment of Approaches Used to Evaluate Stricter Standards

Although the costs for individual vehicles were estimated and the methods documented, it was less clear how costs were aggregated to the fleet level for CARB and EPA. Assumptions about fleet sales and vehicle size mix are important in assessing costs, as discussed above. A related point is that the areas of greatest uncertainty in costs are not addressed. It is clear that there is uncertainty in forecasting costs and emissions reductions, but there are few cases where uncertainty in the costs estimates are assessed. Particularly for future technologies, for which there is a great deal of uncertainty in benefits and costs, some reflection of that uncertainty would be important in assessment of the standards. CARB attempts to reassess costs and emissions reductions as new evidence becomes available about actual methods, costs, and performance. EPA has no requirement to reevaluate or provide ex-post assessments.

There are other costs besides the per vehicle costs that might be important for implementing stricter standards, particularly if those standards are not uniform. The committee found little information about the actual costs of dealing with noncompliant cars, border issues, or other costs of enforcement. However, CARB issues annual reports on enforcement activities that detail the enforcement of mobile-source emissions standards in the state. States adopting the California standards would need to assess whether they would face similar or different compliance issues and costs.

A final important point about the cost analyses is how emissions controls affect the price of the vehicle. This issue is important for assessing not only what the costs may be but also who will pay them. The CARB estimate of the costs of the LEV standards assumes that the costs are paid by buyers of the LEV II vehicles. The estimated average costs per vehicle reflect the costs of the control equipment and a share of the other costs of design and production. However, pricing of vehicles is complex (Sperling et al. 2004a) and does not always reflect the single vehicle cost. There are important marketing and sales considerations. For some vehicles, the emissions costs are likely to be at least partially reflected in the sticker price. If a vehicle type has two versions with different characteristics that consumers can identify, such as a Ford Escape and a Ford Escape Hybrid, the manufacturer can charge different prices. Differences in fuel economy and other features are evident to car buyers and they will pay for different features.[8] For vehicles that are identical, however, except for the emissions equipment, the price charged might have to be the same for both types of vehicles.[9] Therefore, costs are spread across a portion of, or the entire, U.S. vehicle fleet when some states adopt a stricter standard.

ADOPTION OF LEV BY AUTHORITY OF SECTION 177 OF THE CLEAN AIR ACT

In September 1990, New York became the first state to use its authority under section 177 of the federal CAA to adopt California standards by adopting CARB's pre-LEV emissions standards. In October 1991, the 13 Northeast states that were the members of the Ozone Transport Commission (OTC), created by the 1990 CAA amendments, pledged to individually adopt the California LEV standards in their respective states. Massachusetts became the first state to adopt the California LEV standards on January 31, 1992. This action was compelled by a state statute passed in 1990 that required the Massachusetts Department of Environmental Protection to adopt and implement California motor vehicle emissions standards unless the agency can demonstrate, based on substantial evidence, that the California standards will not achieve greater pollution reduction than the federal motor vehicle emissions-control program (M.G.L. c. 111, § 142K). New York followed suit and became the second Northeast state to adopt the California LEV standards in May 1992, followed by Maine in 1993. Most other OTC states, however, did not originally adopt the LEV standards in the face of strong opposition by the auto and oil industries (including a series of federal lawsuits) and skepticism about the benefits of the program by legislators in some states. Some states (for example, Pennsylvania and New Jersey) conditionally adopted the LEV standards to take effect in those states only if a sufficient number of other Northeast states also implemented the program.

---

[8] Fuel economy savings are not always accurately accounted for in consumer vehicle purchase decisions. See Turrentine and Kurani (2005) for a discussion of that point.

[9] One car company told the committee that the same price must be charged for vehicles that look identical except for the emissions-control equipment. Another company representative made the different point that regions accept higher prices when different controls are required in their regions, such as under LEV II.

State-by-state adoption of the LEV program in the entire Northeast region faltered, and on February 1, 1994, the OTC voted 9-4 to recommend to EPA that it mandate adoption of the California LEV standards in the entire region. Virginia, Delaware, New Hampshire, and New Jersey dissented. After extensive proceedings, EPA decided in September 1994 to approve the OTC request and mandated adoption of the LEV standards in all 13 Northeast states on the grounds that such action was necessary to achieve attainment of the NAAQS in the region. EPA determined that states could elect to adopt the California LEV program without the ZEV mandate component, because the ZEV component was a separate requirement, but states would need to comply with the NMOG fleet-average requirement and thus would be required to achieve additional reductions in their LEV programs to replace the expected ZEV contribution to the NMOG fleet average.

Virginia and the vehicle manufacturers went to federal court to challenge EPA's authority to mandate adoption of the LEV standards in the Northeast states. The court held that a state's authority to adopt to the California program under section 177 of the federal CAA was intended to be voluntary and could not be compelled by EPA (Virginia v. EPA, 108 F.3d 1397 [D.C. Cir. 1997]). Moreover, the court held that the EPA decision unlawfully circumvented the provision in the CAA that prohibited EPA from adopting new motor-vehicle emissions standards before the 2004 model year.

By the late nineties, New York, Massachusetts, Maine, and Vermont were the only states to have voluntarily adopted California's LEV program. Implementation began in 1994, 1995, 2001, and 2000, respectively. All four states have since adopted the LEV II regulations that began in model-year 2004. New York and Massachusetts have always included the ZEV mandate as part of their regulations. Maine initially adopted the ZEV mandate but repealed it during adoption of LEV II in 2000. In 2005, however, Maine reinstated the ZEV mandate to begin in model-year 2009. Vermont adopted the ZEV mandate in 2001, but later stated that manufacturers were not required to meet the mandate for model-years 2001-2005. In 2005, Vermont re-affirmed the ZEV mandate to begin in 2007. Each state adopted the ACP option and offered credits toward fulfilling the ZEV requirement. Following California's adoption of LEV II standards, Connecticut, Rhode Island, and New Jersey adopted the California on-road emissions standards, including the ZEV requirement, ACP, and credit system. Implementation of LEV II standards and various components in these states are not scheduled until at least model-year 2008 and have different implementation dates and deadlines.

States' Rationale for Adopting California Emissions Standards

Adoption of the LEV standards in the Northeast has remained a state-by-state issue. States that adopted the LEV program, led by New York and Massachusetts, concluded that the adoption of such standards was a necessary and cost-effective measure to attain the NAAQS. For example, New York said that it was compelled to adopt the California standards because the "emission reductions provided by the adoption of the LEV standards along with a wide array of other mobile, stationary and area source control programs are necessary to attain the NAAQS for ozone in New York as required by federal law" (NYDEC 1992). Other Northeast states came to contrary conclusions. For example, Connecticut decided against implementing the California LEV program because it concluded at the time that the program would be too expensive and would not provide sufficient and timely SIP credits (Dumanoski 1991). The primary reason for states to adopt the California standards has been to obtain additional emissions reductions to meet air quality goals. In each case, adoptions have been based on the conclusion that California emission standards would provide greater mobile-source emissions reductions more quickly than federal standards. That conclusion was almost certain when California's original LEV program began, since EPA was not expected to introduce more stringent Tier 2 standards, if at all, until model-year 2004, as specified in the CAA.

The committee heard about other benefits of California emissions standards that seem to contribute, at least in part, to the rationale behind the adoption decision. One benefit is that the emissions differences between the California program and the federal program will probably become greater as California adopts more stringent phases of the LEV program (beyond LEV II) (NESCAUM 2003), recognizing that no successor to the current Tier 2 program is planned or mandated by the CAA. Regulatory documents from other northeastern states also point out that benefit, stating that the California program is more dynamic than the federal program and will likely be revised before the federal program (MEDEP 2004).

Recent state actions and statements also reflect a growing desire by states to reduce greenhouse gas (GHG) emissions. A statement made to this committee during a public session meeting, for example, reflected Connecticut's position that adoption of the California LEV II standards provides an opportunity to achieve CO_2 reductions to meet its climate plan, because the LEV II program provides incremental GHG emissions reductions over the Tier 2 program and California has added specific

GHG emissions standards to its program (McCarthy 2005). Additional discussion of these GHG standards is provided later in this chapter.

The automobile and oil industries have argued that the LEV II program is not a cost-effective or necessary program compared with the Tier 2 program, which they would otherwise be required to comply with. Manufacturers have also objected that northeastern states were placing the burden of handling costs on out-of-state vehicle manufacturers rather than imposing pollution-reduction costs on in-state industries. Manufacturers could point to such statements as that of Massachusetts Secretary of Environmental Affairs Susan Tierney, who was quoted as saying, "This is a jobs issue for Massachusetts. Every pound we don't take out of automobiles, we have to take out of a sector that has a job at stake" (Dumanoski 1992). Manufacturers also argue that differences in LEV and LEV II implementation dates among the various states complicate vehicle distribution and sales for themselves and their dealers (Babik 2005) and that the additional costs do not outweigh the emissions benefits.

Novel Issues in State Adoption of California Emissions Standards

Beginning with the initial adoption of LEV in the Northeast, automakers and others have objected to states' authority to adopt California emissions standards. Table 6-13 lists several of the objections. Manufacturers raised such objections in public comments to New York and Massachusetts as part of the early state rule-makings to adopt the California LEV standards. The states generally disagreed with the factual or statutory basis of the manufacturers' arguments and proceeded with adoption of the LEV standards as proposed. Trade associations of vehicle manufacturers subsequently filed separate lawsuits in federal district court challenging the LEV standards in New York and Massachusetts.[10] The litigation against the two states proceeded over 7 years and involved extensive motions, briefings, affidavits, discovery, and appeals that were both time-consuming and expensive for manufacturers and states. Twelve separate court decisions were issued by the federal trial and appellate courts for New York and Massachusetts, described in one treatise as "an epic series of court battles" (Wooley and Morss 2004).

---

[10] Manufacturers also filed a lawsuit against Maine's adoption of the LEV standards, but that lawsuit was voluntarily dismissed when Maine agreed to indefinitely delay implementation of the LEV standards.

TABLE 6-13 Manufacturers' Objections to Northeast States' Adoption of the California Emissions Standards

- States adopted California's emissions standards but not the cleaner-burning and lower-sulfur California fuel requirements that were part of the LEV program, resulting in reduced air quality benefits and adversely affecting the advanced emissions-control systems under development to comply with the LEV standards in California.
- The distribution of California-certified vehicles to other states would impose unnecessary burdens on product distribution and availability.
- Auto dealers located near state boundaries would be adversely affected and would be required to stock different versions of the same vehicle model.
- The electric vehicles produced to meet the ZEV mandate would primarily be powered by lead-acid batteries in at least the initial years of the ZEV mandate, and such batteries would produce a much lower range and poorer performance in the colder Northeast states than in California
- Massachusetts and New York adopted the California LEV standards before they had obtained a federal preemption waiver from EPA and without providing the full 2-year lead time required by section 177 of the federal CAA.
- The sales mix of a given manufacturer was different in Northeast states than in California, complicating and jeopardizing compliance with the NMOG fleet-average requirement in each opt-in state.
- The opt-in states failed to provide the same credit allowances provided by California for meeting the NMOG fleet-average and ZEV-mandate requirements.
- States failed to follow California's modifications to its ZEV requirements, resulting in different versions of the LEV program in California and in other states.

The first lawsuit to proceed was in New York, where manufacturers alleged that (1) New York had violated section 177 of the federal CAA by adopting the LEV standards before they had received a preemption waiver and without providing the 2-year lead time required by section 177; (2) New York had failed to adopt the California clean-fuel regulations in conjunction with the LEV standards in violation of the section 177 requirements that state standards be "identical" to California's and not have the effect of creating a "third vehicle"; and (3) New York would violate section 177's prohibitions of a "third vehicle" and restrictions on the sales of any category of California-certified vehicles in its ZEV mandate. The federal district court in New York initially held against manufacturers on some claims but in favor of manufacturers on the ZEV mandate violation and on the failure to adopt California's accompanying fuel standards, requiring manufacturers to build a third vehicle because of the effects of high-sulfur gasoline on the advanced emissions-control equipment being developed to comply with the California LEV standards. In response to New York's motion for reconsideration, the district court later reversed its decision on the fuels and third-vehicle claim and determined that a full trial would be required to resolve that issue.

In two decisions, the Second Circuit Court of Appeals ultimately held in favor of the state on all claims except the lead-time claim (Motor Vehicle Mfrs Ass'n v. NY Dep't of Envtl Conservation, 17 F.3d 52 [2d Cir. 1994]). The Second Circuit construed the third-vehicle prohibition of section 177 narrowly to apply only to major changes in vehicle design directly required by a state and not to changes in vehicle design that a manufacturer chooses to undertake in response to different conditions in a section 177 state, such as different fuels or climate. The Second Circuit described those choices as marketing choices. The Second Circuit also held that the "identicality" requirement of section 177 applied only to vehicle requirements and not to fuel requirements. However, the Second Circuit held that New York had violated the lead-time requirement of section 177 by attempting to apply the standards partway into the 1995 model year and held that New York must apply its standards to an entire model year, thereby postponing the New York standards to the 1996 model year.

In Massachusetts, the vehicle manufacturers filed a motion for a preliminary injunction to temporarily enjoin the Massachusetts regulations until a full trial on the merits could be undertaken. The district court denied the preliminary injunction, and manufacturers dismissed all their claims except the lead-time claim in light of the Second Circuit's decision on the same issues. The decision of the First Circuit Court of Appeals on the lead time differed from that of the Second Circuit, holding that Massachusetts could commence its regulations partway through the 1995 model year (American Auto Mfrs. Ass'n v. Massachusetts Dept. of Envtl. Protection, 31 F.3d 18 [1st online Cir. 1994]).

A second round of litigation under section 177 of the federal CAA occurred after California adopted memorandums of agreement (MOAs) with vehicle manufacturers to provide demonstration programs for advanced-technology ZEVs in place of the rescinded ZEV mandate for model-years 1998-2002. New York and Massachusetts contended that CARB and the vehicle manufacturers had deliberately structured the MOA response to preclude the section 177 states from adopting similar

ZEV programs. Massachusetts decided to adopt regulations with similar requirements as the California MOAs, which had been structured as voluntary agreements rather than regulations in California. New York decided to adhere to the original ZEV mandate for model-years 1998-2002, even though California had rescinded its version of the ZEV mandate for those years.

Vehicle manufacturer trade associations sued New York and Massachusetts in separate lawsuits. In New York, the district court held in favor of the state, holding that the ZEV mandate was not a standard subject to CAA section 209(a) preemption. That decision was then overturned by the Second Circuit Court of Appeals, which held that the New York ZEV mandate was a preempted standard (American Auto. Mfrs. Ass'n v. Cahill, 152 F.3d 196 [2d Cir. 1998]). In Massachusetts, the district court held that the California MOAs were voluntary agreements and not standards preempted under section 209(a), and thus Massachusetts was preempted from adopting regulatory standards with similar substantive requirements. On appeal, the First Circuit initially held that it would defer to EPA's views on whether the California MOAs were standards, but then disagreed with EPA's opinion that the California MOAs were standards under section 209(a) and ultimately held that the Massachusetts ZEV regulations were preempted (Ass'n of Int'l Automobile Mfrs v. Commissioner, Mass. Dep't of Envtl Protection, 208 F.3d 1 [2000]).

In summary, state adoption of the California LEV standards resulted in a series of legal controversies. The federal courts were the only available venue to resolve these issues, and the ensuing litigation took many years and large commitments of resources from both vehicle manufacturers and states. States ultimately prevailed on many of the legal issues, but the manufacturers prevailed on some. Other states that were considering adopting the LEV standards were left in limbo about the legality of adopting the LEV standards, and vehicle manufacturers and dealers subject to the LEV standards in those states that had adopted the LEV standards had to live with uncertainty and confusion about the applicable requirements for many years. A more expedient and less resource-intensive mechanism to resolve these disputes would have benefited all parties.

EPA's Role in Implementation of the LEV Program

The role played by EPA in the adoption of the California LEV program by other states has become complicated and difficult. Section

177 of the CAA does not give EPA an express role in a state's adoption process except for defining the start of a vehicle model year for the 2-year lead-time requirement. The assumption of Congress in enacting section 177 seemed to be that the waiver for California would address all relevant concerns related to economic and technological feasibility, lead time, and other issues; thus, no new issues would be raised when another state adopted California standards. As EPA explained in a letter to Congressmen John Dingell in 1991, before states used section 177 to implement the LEV standards:

> Section 177 does not have an explicit requirement that the state standards adopted under this section's authority be consistent with [the lead time and feasibility criteria of] section 202(a).... Because a section 177 state can adopt only California standards which have received a waiver, and which are therefore consistent with section 202(a), the standards adopted by the section 177 state may also be considered consistent with section 202(a) (Reilly 1991).

EPA was nevertheless called upon to play a central role in the adoption of the LEV program in the northeastern states and to deal with many complex issues that were not anticipated by Congress or EPA. In considering CARB's initial waiver request for the LEV standards, EPA was requested by various parties to consider the impact of adoptions by other states pursuant to section 177. EPA disclaimed any authority to consider issues raised by the attempts of other states to adopt the California standards under section 177 of the CAA (58 Fed. Reg. 4166 [1993]). EPA has, however, taken positions on a number of key issues (discussed in Chapter 3); for example, that the ZEV mandate is segregable from the LEV program, that the fleetwide NMOG average is considered a standard, and that states may adopt but not enforce a standard before the granting of a waiver to California.

Although EPA's position is generally unofficial, it has been influential and even requested by the courts. When vehicle manufacturers sued New York, alleging that its adoption of the LEV standards violated section 177, EPA filed an amicus curiae brief in the Second Circuit Court of Appeals that had a key role in the appellate court's reversal of the district court's decision that the New York ZEV mandate violated section 177 (Brief for the United States as Amicus Curiae, Motor Vehicle Mfrs Ass'n v. N.Y. Dep't of Envtl. Conservation, No. 93-7938 [2d Cir., 1993]).

In the Massachusetts ZEV litigation, the First Circuit Court of Appeals, faced with the issue of whether Massachusetts could adopt ZEV regulations similar to the 1996 MOA between CARB and vehicle manufactures, declined to decide the matter until EPA had issued its view under the doctrine of "primary jurisdiction," which requires courts to defer to an administrative agency on matters within the special competence of that agency (American Automobile Mfrs Assn v. Massachusetts Dep't of Environmental Protection, 163 F.3d 74 [1st Cir. 1988]).

EPA does have one formal opportunity to review section 177 adoptions of California standards when it approves a state SIP revision that incorporates the California standards. This process does not generally provide an effective process for resolving any controversies regarding the state's action. First, EPA has narrow and limited authority to review and question a state action under the SIP review criteria. Second, the SIP review often occurs several years after the state action, limiting EPA's capability to provide a timely response. For example, Massachusetts was the first state to adopt the California LEV standards, effective January 31, 1992, but EPA did not approve the SIP revision incorporating those standards for 3 years (60 Fed. Reg. 6027 [1995]). Finally, some states (for example, Vermont) have never included their adoption of California LEV standards in a SIP submittal, thus precluding any EPA oversight role via this process.

The LEV example provides evidence that, contrary to the original assumption of Congress in enacting section 177, adoptions raise some novel issues (generally of a scientific and technical nature). Most issues involve either conditions that were not considered when EPA issued a waiver to California or consistency with section 177 of the CAA. Although EPA has denied formal authority to address these issues, it has done so informally and the courts have deferred to it as the appropriate agency to do so. This example makes a strong case that EPA could have a role in the administration of section 177, just as it does for virtually every other provision of the CAA. An official EPA decision might also reduce the litigation over states' adoptions of California standards. For example, EPA decisions are challenged in the District of Columbia court of appeals as opposed to district courts where lawsuits against individual states have traditionally been litigated. The lengthy discovery and deposition processes that occur in the latter case would be reduced in a challenge to an EPA decision. Second, the courts traditionally show high deference to EPA determinations, effectively reducing the likelihood of a successful challenge.

State Practices in Adopting California Standards

There are no federal requirements regarding state adoption of California emissions standards beyond the specifications in the CAA section 177. Adopting California mobile-source emissions standards is one of several options available to states to reach the air quality goals embodied in the NAAQS. States generally support their adoption of California standards with a state- or region-specific modeling analysis of emissions benefits. Emissions modeling also serve to quantify the LEV and LEV II program's emissions credits that are included in the state's SIP, and this modeling is usually a required part of the SIP analysis. Air quality modeling—for example, to assess the impact of a new regulation on ozone concentrations—is also a part of the SIP analysis. The committee did not find evidence that states analyze in-state air quality impacts or health impacts of adopting California standards (in isolation from other SIP measures). Furthermore, states typically quote and defer to the technical-feasibility and cost-analysis determination conducted by California when adopting LEV II (NYDEC 2000; MA DEP 1999). A quantification of any improvements in ozone or PM concentrations and health impacts, for example, would be useful support for adopting LEV II because they are the criteria pollutants of most concern in many nonattainment areas. The committee found one example of a nongovernmental organization that included ozone in an assessment of the health impacts of adopting LEV II standards in Connecticut (CFE 2003).

Assessment of emissions benefits is the significant technical practice carried out by states when adopting California standards. The following section discusses these assessments by northeastern states for adoption of California LEV and LEV II emissions standards. Because of the regional nature of air pollution in the Northeast, the states have a history of cooperative analysis of emissions and air quality, such as that by the Ozone Transport Commission (OTC) established by the CAA. A regional air pollution agency, Northeast States for Coordinated Air Use Management (NESCAUM), is directed by the air directors of Maine, Rhode Island, Connecticut, Vermont, New Hampshire, New York, New Jersey, and Massachusetts. NESCAUM promotes cooperation and coordination of technical and policy issues regarding air quality control among the member states. NESCAUM has assessed emissions benefits of adopting California emissions standards in support of its member states.

Comparing the Emissions Benefits of California Versus Federal On-Road Emissions Standards in the Northeast

The northeastern states that have adopted LEV II have supported their rule-making by estimating future emissions using various generations of EPA's MOBILE model. Table 6-14 summarizes the expected emissions benefits found by some of these analyses. The committee did not attempt to analyze the modeling practices and assumptions underlying these analyses. As discussed in Chapter 2, these models are subject to a degree of uncertainty (NRC 2000); however, they are the best tools available to regulatory agencies for assessing emissions benefits. Furthermore, many assumptions are needed to model future mobile-source emissions; for example, sales, fleet mix in future years, types of fuels, and the evolution of vehicle technology and emissions levels. The data presented in Table 6-14 and in the following comparisons serve to show the range of estimates of the emissions benefits of California over federal emissions standards. Most likely, there are significant differences between the models and assumptions used in each study; therefore, caution is needed in comparing these estimates.

The expected emissions benefits of LEV II over Tier 2 in particular have proved to be controversial. In 2003, NESCAUM estimated a 15%-HC mobile-source emission benefit in 2020 in the Northeast based on MOBILE modeling of New York, Massachusetts, and Vermont. The study also included an estimated benefit in mobile-source toxic emissions of 23%. This analysis was used by Connecticut and Rhode Island to support adopting California standards for the first time (CTDEP 2004; RIDEM 2004). Nongovernmental organizations have also used modeling in their analyses in support of state adoption of California emissions standards. The Connecticut Fund for Environment (CFE) estimated a 20%-VOC and a 11%-NO_x benefit in 2025 in that state and a 33%-benefit in HAP emissions (CFE 2003). The Maryland Public Interest Research Group (PIRG) estimated a 13%-VOC benefit, a 11%-NO_x benefit, and a 12-15%-HAP benefit emissions in Maryland in 2025, if the state would adopt LEV II in 2008. The Maryland Department of the Environment estimated a 5%-VOC and a 6%-NO_x benefit of LEV II over Tier 2 in 2020 in Maryland (Snyder 2005).

Manufacturers argue that complying with California standards in a greater number of states comes at a greater cost, and that this cost is not outweighed by the emission benefits of LEV II over Tier 2. The Alliance

TABLE 6-14 Summary of Various Estimates of Emission Reductions Analyses of Adopting California Mobile-Source Emissions Standards in the Northeast[a]

Analysis	Comparison	Area (Year of Benefit)	Emission Reduction Over Federal Standards (%)		
			HC	NO_x	CO
Sierra Research, Inc. 1989	California proposed program and pre-Tier 1 federal program	Typical NESCAUM state (2010)	16.2	27.2	38.7
Pechan/EEA, Inc. 1991	LEV and Tier 1[b]	NESCAUM region (2015)	23-61	26-41	10-33
MA DEP 1999	LEV II (with intermediate CA MDV standards) and Tier 2	Massachusetts (2020)	20	19	17
NY DEC 2000	LEV II and Tier2[c]	New York 2020	8	0	NA
NESCAUM 2003	LEV II and Tier 2[d]	Average between NY, MA, VT (2020)	15.3	NA	NA
NESCAUM 2005; update to 2003 data	LEV II and Tier 2[d]	Average between MA, NY, VT, ME (2020)	7.6	14.7	NA
		Average between CT, RI, NJ (2020)	4.8	10.8	NA

[a]The emissions reduction is the difference in mobile-source emissions between California and federal standards, expressed as a percentage of the federal total.
[b]Only LDV <6,000 lb of GVW; exhaust only.
[c]Unclear whether ZEV mandate or evaporative emissions included.
[d]Only LDV <6,000 lb of GVW.

of Automobile Manufacturers (AAM) presented its own estimates of emissions benefits and a critique of methods used by NESCAUM and other groups (Air, Inc. 2004a,b). In an evaluation of the Connecticut CFE (2003) study, an AAM contractor reported benefits of 7% for VOC, 14% for NO_x and 14% for HAPs in 2025 when modeling Connecticut mobile-source emissions according to EPA guidelines (Air, Inc. 2004b).

EPA has also addressed the modeling methodologies and the benefits estimation of LEV II over Tier 2 in the Northeast. EPA addressed NESCAUM's 2003 modeling estimates in a letter stating that NESCAUM did not follow EPA guidelines for modeling Tier 2 emissions, and this led to overstatements of the benefits of LEV II over Tier 2 (EPA 2004f). EPA stated that emissions benefits of LEV II and Tier 2 should be compared with a non-LEV II and non-Tier 2 baseline, respectively, for a realistic comparison between the two programs. Using these guidelines, EPA estimated that LEV II will provide about a 1% additional reduction in mobile-source HC over Tier 2 in 2020 (EPA 2004f; 70 Fed. Reg. 21959[2005]). NESCAUM (2004) responded to EPA concerns and performed an updated analysis of emissions benefits in the Northeast, including a separate analysis for states that only recently adopted LEV II standards. The revised emissions-benefits estimates are provided in Table 6-14. In responding to EPA, NESCAUM noted that its analysis is intended to support the conclusion that the LEV program provides additional air quality benefits and is not intended to be portrayed as a SIP analysis (NESCAUM 2004). SIP credits, including the numerical value of emissions reductions of the LEV II program, must be evaluated by each state separately in coordination with EPA.

States and NESCAUM have generally documented their analyses well, as evidenced by the ability of outside stakeholders to review and critique their practices. Although the numerical value of the emissions benefits is ultimately important in determining SIP credits, there is no required level of emissions benefits that must be met to adopt the LEV program, and adoption and SIP regulatory processes are usually separate. All of these analyses are presented in terms of additional percentage reductions. Emissions reductions apparently were not translated into estimates of health or environmental benefits. This committee makes no judgment on an appropriate level of emissions benefits (over a similar federal program) to support adoption of a California emission standard. A comparison of benefits and costs for such a decision is in the realm of policy and outside this committee's charge.

Economic Impact Assessment by Opt-in States

Each state is subject to certain requirements for each rule-making. New York, for example, prepared a regulatory impact statement and a job impact statement for adoption of California regulations. The majority of these requirements concern economic impacts. The states that are adopting the California LEV II standards rely for the most part on CARB's ex ante cost estimates as the measure of projected costs of the program. They do not attempt to measure any costs beyond the consumer costs estimated by CARB. For example, states do not appear to measure the cost of program administration, the distribution or dealer costs, or the costs of enforcement of the program. States do not discuss the potential infrastructure costs that might be needed for the ZEV component of the LEV II program. Massachusetts, for example, stated that there will be no significant economic effects on dealers. It suggested, however, that dealers will need to incur more training costs for mechanics because of additional types of vehicles, but these costs are assumed to be recouped through increased sales of vehicles, such as the PZEVs and the AT-PZEVs (MADEP 1999).

In an analysis by New York, the state argued that the ZEV requirements are not expected to have any effect on dealers and that any change in required paperwork will not be substantial. Using the same argument as Massachusetts, New York stated that there must be additional mechanic training but that this will generate new jobs for the region. However, fewer vehicles of other types will be sold, and the mix of services needed will change for mechanics. The costs of retraining represents costs to dealers or car buyers, depending on how the costs are distributed. On the ZEV mandate, Maryland, one of the states that decided against adopting the LEV II standards, stated that it had identified substantial infrastructure costs associated with the ZEV component of LEV II (Snyder 2005).

There is little discussion from the states or in the NESCAUM analysis of the LEV II program about the potential distribution issues or pricing policies for LEVs. The auto companies argue that there are sales issues, especially at the border of the LEV II states. Often, dealers near state borders swap vehicles over state lines, potentially complicating compliance with the standards for the automakers. For that reason, at least one automaker (J. German, Honda, personal commun., May 18, 2005) argued that it designs vehicles to meet both Tier 2 and LEV II standards when possible. As noted in Chapter 5, however, cost data are often difficult to obtain, thus limiting the scope of any state economic impact analysis.

An additional problem is that the prices of an entire fleet in a region might require adjustment to sell a vehicle fleet that meets the emissions standards. Pricing this fleet at the cost of the emissions controls might not result in sales. In that case, subsidies of certain vehicles might be necessary, and the costs of bringing the fleet to market might change. There is little discussion of any of these issues in economic impact analyses of LEV standards by California and opt-in states. The committee heard some evidence of a proposed Northeast-wide NMOG fleet average that manufacturers could use to certify vehicles. This proposal might provide more flexibility to the manufacturers in selling vehicles that meet the standard. Such flexibility on sales and distribution might ease the economic burden for manufacturers and dealers and decrease the limitations of consumer choice. The committee encourages such flexibility, provided air quality is not compromised.

CALIFORNIA MOTOR VEHICLE GREENHOUSE GAS STANDARDS

Legislative Mandate

The California legislature in 2002 concluded that the effects of global warming in California would be substantial. The Pavley bill (California AB 1493, Feb. 23, 2001) in particular found that global-warming impacts on the state included: potential reductions in the state's water supply, adverse health effects from increases in air pollution caused by higher temperatures, adverse impacts on agriculture caused by impacts on water supplies and increases in pestilence outbreaks, increases in wildfires, and potential damage to the state's coastlines and ocean ecosystems. In response, the legislature passed and Governor Davis signed into law the Pavely bill. This legislation directed CARB to develop and issue greenhouse gas (GHG) emissions standards for LDVs, using its authority to set its own mobile-source emissions standards (CAA section 209). The legislature directed CARB to "develop and adopt regulations that achieved the maximum feasible and cost-effective reduction of greenhouse gases from motor vehicles." The Pavey bill focuses on LDVs because approximately 40% of GHG emissions in the state are estimated to come from this source.

TABLE 6-15 CO_2 Equivalent Emissions Standards for Model-Years 2009 through 2016

		CO_2 Equivalent Emissions Standard by Vehicle Category (g/mi)	
	Year	Passenger Cars and Small Trucks/SUVs	Large Trucks/SUVs
Near-term	2009	323	439
	2010	301	420
	2011	267	390
	2012	233	361
Mid-term	2013	227	355
	2014	222	350
	2015	213	341
	2016	205	332

Source: CARB 2004d.

Proposed Regulation

In response, CARB (2004c) proposed a set of standards in its staff report; these standards were approved and adopted by CARB on September 23, 2004. Table 6-15 displays the proposed standards, which are given in grams per mile of CO_2 equivalent.[11] The proposal recognized four sources of motor vehicle GHG emissions: tailpipe emissions of carbon dioxide (CO_2), methane (CH_4), and nitrous oxide (N_2O) emissions; CO_2 emissions due to operating the vehicle air conditioning system; hydrofluorocarbon (HFC) emissions from the air conditioning system; and upstream GHG emissions associated with the production of fuels. The CO_2 equivalent standards are incorporated into the existing LEV program and, like existing LEV II standards, there is a separate fleet average for passenger cars and small trucks/SUVs and large trucks/SUVs.

[11] Emissions of other GHGs are translated into estimates of the equivalent amount of CO_2 by use of global warming potential (GWP). Because the atmospheric lifetimes of other GHGs are different from CO_2, a GWP depends on the time horizon. If a 100-year time horizon is used, CH4 has a GWP of 23, N_2O has a GWP of 296, HFC-134a has a GWP of 1,300, HFC-152a (a possible replacement for HFC-134a) has a GWP of 1,300. The 100-year time horizon is recommended by the Intergovernmental Panel of Climate Change and CARB.

Scientific and Technical Analysis

The CARB staff report (2004c) includes an assessment of the technology and fuels that could help meet the standards, the level of reductions that could be achieved, and the costs. The CARB assessment was based on vehicle simulation and cost analysis of GHG emissions reductions contained in a report done for the Northeast States Center for a Clean Air Future (NESCCAF 2004). NESCCAF (2004) predicted the GHG emissions impacts of various technology combinations for five types of vehicles (small car, large car, minivan, small truck, and large truck) by using a model that estimates operating characteristics and emissions for different motor vehicle designs. NESCCAF (2004) also provided cost estimates for the various technology combinations, including available technologies and technologies that had been demonstrated in prototype form.

CARB (2004d) estimated that the near-term (2009-2012) standards would result in a 22% reduction in GHG emissions compared with the 2002 fleet, and the mid-term (2013-2016) standards would decrease fleet emissions by 30%. These estimates translate into an 88,000 ton/day reduction in CO_2-equivalent emissions by 2020 and a 155,000 ton/day reduction by 2030 (CARB 2004d). Table 6-16 shows the estimated fleet-wide incremental costs of controls to meet these standards developed by CARB. Some portion of the cost of reductions in CO_2 can be recouped in

TABLE 6-16 Average Cost of Control

		Average Cost of Control	
	Year	Passenger Cars, Small Trucks, and SUVs	Large Trucks and SUVs
Near-term	2009	17	36
	2010	58	85
	2011	230	176
	2012	367	277
Mid-term	2013	504	434
	2014	609	581
	2015	836	804
	2016	1,064	1,029

Source: CARB 2004d.

fuel savings over time. CARB asserts that nearly all technology combinations will result in reductions in lifetime operating costs that exceed retail price of the technology, assuming a fuel cost of $1.74 per gallon; cost recovery time would presumably be smaller at higher gasoline prices. For example, CARB (2004d) looked at the potential increase in monthly payments over a 5-year vehicle loan versus the monthly decrease in operating costs and concluded that the proposed standards would result in a monthly savings of $3.50 to $7.00.

Manufacturers and others have submitted comments critical of CARB's analysis (CARB 2005f). CARB's cost estimates are disputed by an analysis performed by Sierra Research for the Alliance of Automobile Manufacturers (Sierra Research 2004), which concluded that average compliance costs would be approximately $3,000 per vehicle, assuming nationwide compliance with the CARB standards, and higher than $3,000 if separate vehicles are produced for states that enforce the California standards. Sierra Research also argued that the increased vehicle prices would decrease the sales of new vehicles, slowing the introduction of cleaner vehicles into the fleet, and that lower operating costs would result in vehicle owners driving more, thus increasing emissions.

Another objection to the standards is that California LDV emissions are a small fraction of the global totals and will have no discernable effect on global climate change (CARB 2004c). CARB pointed out, however, that its GHG standards represent a "no regrets" policy (reducing emissions while providing a net cost savings to vehicle owners) and that the state's action will prompt other states and countries to follow course (CARB 2004c).

Opt-in States

Recent state actions and statements also reflect a growing desire by states to reduce GHG emissions. Massachusetts, New York, New Jersey, Connecticut, Rhode Island, Vermont, and Maine have amended their prior adoption of the LEV II program to include the California GHG emissions standards. In addition, Washington state recently adopted the LEV II program contingent on Oregon's adoption. Oregon has adopted temporary rules on the LEV II program and is scheduled to propose permanent rules in the summer of 2006. Neither state has areas in nonattainment of the 1-hr or 8-hr ozone standard. Both states are focusing on the GHG emission-reduction benefits of the California program (Office of the Governor of Washington 2004; State of Oregon 2004).

Legal and Administrative Status of Greenhouse Gas Initiative

CARB approved the GHG standards in September 2004, and the final rule-making package was approved by California's Office of Administrative Law on September 15, 2005. The standards face major challenges in terms of obtaining a waiver from EPA and from lawsuits filed by automakers. A primary objection raised by automakers is that, because most reductions in GHG emissions are from decreases in CO_2 emissions due to improved fuel economy, the GHG standards are fuel economy standards, which are preempted by federal laws and regulations mandating uniform, nationwide standards for fuel economy. That objection will be one of the primary arguments made by plaintiffs in the lawsuit, *Central Valley Chrysler et al. v. Witherspoon,* which is scheduled to begin trial in the Federal District Court in late 2006.

CONCLUSIONS

The history of LDV regulation provides insights into how EPA and CARB set emissions standards and how other states adopt standards. EPA and CARB's present-day LDV emissions standards integrate fuel and emissions limits, certification flexibility, and, in CARB's case, a mandate for promoting advanced-technology vehicles. The following conclusions are drawn from the information presented in this chapter on the practices and issues related to LDV emissions standards.

- CARB and EPA analyses of their emissions standards are found in CARB's staff paper and EPA's regulatory impact analysis, respectively. Both documents include a technical feasibility assessment, emissions impacts analysis, engineering cost assessments, and cost-effectiveness estimates. For major rules, such as the Tier 2 LDV regulations, EPA is required to conduct a health benefits assessment to compare with total costs. CARB does not consider public health benefits directly in its regulatory analysis of emissions standards because it uses its proposed standards to attain health-based NAAQS, which EPA has already assessed for public health benefits.
- The majority of available cost estimates of emissions standards are for LDVs, but these estimates vary substantially and are uncertain. It is difficult to determine what parties bear what fraction of the costs of emissions standards. Manufacturers closely guard cost and pricing data to avoid placing themselves at a competitive disadvantage.

- With the LEV program, California has used its authority as Congress envisioned: to implement more aggressive measures than the rest of the country and to serve as a laboratory for technological innovation. There have been successes, such as CARB's early recognition of the need to couple fuel composition with emissions control, and failures, such as the promotion of widespread use of electric vehicles.
- As a rationale for adopting California LDV standards, some states expect that California will continue to reduce standards earlier than the federal program. Some states have also adopted or expressed interest in adopting the California GHG emissions standards.
- To date, CAA section 177 authority has been used primarily by various northeastern states to adopt California LDV standards. Manufacturers of mobile sources have raised objections to the adoption of California standards by other states. Up to this point, adopting states and manufacturers have turned to the courts to resolve their technical and legal disputes when direct negotiations have failed. Although EPA is an appropriate entity to comment on some of these disputes, it has no authority over states' adoption decisions.

7

Other Case Studies

Light-duty vehicles (LDVs) have had the longest emissions-regulation history of any mobile sources and have raised many of the key policy and technical issues related to mobile-source emissions control. Regulation of non-LDV vehicles and engines, however, has raised several new issues that are important to state emissions standards. This chapter presents four case studies of emissions standards for non-LDV vehicles and engines:

- *Spark-ignition marine outboard and personal watercraft engines.* This case study of a newly regulated source is also one of the few instances where a state (New York) has exercised its opt-in authority to adopt a California emissions standard for a nonroad engine. (Texas has also adopted California's standards for large spark-ignition engine.)
- *On-road heavy-duty engines.* EPA's 2007 standards, which include fuel sulfur limits, are technology forcing and projected to result in widespread use of exhaust after-treatment.[1] California has adopted nearly identical standards after 2007, although there are differences for 2005 and 2006. Some other states have chosen to adopt California's program over federal regulations.
- *Small nonroad spark-ignition (SI) engines.* Small SI engines, primarily in lawn and garden equipment, are important sources of non-

[1] After-treatment is used here and throughout this chapter to mean removal of pollutants in the exhaust gases after they have exited the engine. This term refers to only one part of typical emissions control on present day mobile sources, which integrates fuel properties, engine modifications, and exhaust treatment.

road emissions nationwide. The most recent California emissions standards regulate evaporative emissions on such engines for the first time and are expected to require catalyst after-treatment on more types of engines than in previous regulations. These California standards drew increased attention because of economic and safety issues related to controlling emissions from small SI engine equipment.

- *Voluntary programs.* A number of incentive-based nonregulatory programs have provided for significant cost-effective emissions reductions in nonattainment areas.

SPARK-IGNITION MARINE OUTBOARD AND PERSONAL WATERCRAFT ENGINES

The 1990 Clean Air Act (CAA) amendments focused regulatory attention on reducing emissions from nonroad vehicles and engines. In the 1991 "Nonroad Vehicle study" to Congress, EPA found that recreational marine engines contributed approximately 30% of the total hydrocarbon (HC) emissions from nonroad sources, second only to small SI lawn and garden engines (EPA 1991). Data presented in the figures in Chapter 2 confirm that percentage in 2002. Most recreational marine emissions come from SI engines, which can be divided into two groups: outboard and inboard engines. Outboard engines typically hang on the hull of a boat and are traditionally light two-stroke engines. Personal watercraft (PWC) commonly have two-stroke jet drives and have been regulated with marine outboard engines as discussed below. Inboard engines are within the hull of the boat and are mostly derivations of four-stroke automobile engines.

Separate standards apply to SI outboard and PWC engines, and this case study focuses on those sources. Early rule-makings note that EPA was most concerned about outboard and PWC engines since they use two-stroke engine technology with much higher rates of HC emissions then inboard or stern-drive engines (61 Fed. Reg. 52087 [1996]).

EPA Standards for Outboard and Personal Watercraft Engines

Stringency

The U.S. Environmental Protection Agency (EPA) finalized first-time HC and nitrogen-oxide (NO_x) emissions standards in 1996 for out-

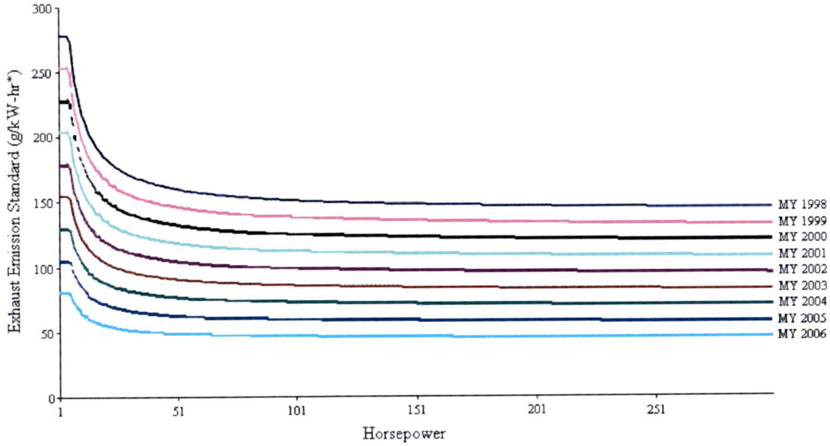

FIGURE 7-1 EPA model-year 1998-2006 exhaust (HC + NO_x) emissions standards curves for recreational marine outboard and personal watercraft engines. Note that the emissions standards at the smallest levels, approximately less than 5 horsepower, reach an upper maximum and do not correspond to the equation given in the text. *Grams per kilowatt-hour. Source: CARB 1998b.

boards and PWC. Because it is typically more difficult to reduce emissions on a given type of engine as its rated power decreases, the EPA emissions standards are a function of the rated power of the engine. Furthermore, the standards become progressively more stringent from the beginning of the regulations in model-year 1998 to 2006. The formula (EPA 1996a) for calculating the emissions standards as a function of rated power of the engine is

$$HC + NO_X = A^* \left(151 + \frac{557}{P^{0.9}}\right) + B,$$

where HC + NO_x is the level of the emissions standards in grams per kilowatt hour, P is the rated power of the engine, and A^* and B are coefficients that decrease each year between 1998 and 2006.

Manufacturers are allowed to meet the standard as a corporate average. A graphic depiction of the formula is presented in Figure 7-1.

Scientific and Technical Analysis

As noted above, EPA identified a need for emissions standards for the SI sources based on their contribution to emissions. A regulatory impact analysis (RIA) accompanying the regulations included technical feasibility, economic impacts, and environmental impact analyses (EPA 1996a). These analyses are summarized below.

HCs are the pollutant of concern from outboards and PWC because large quantities are emitted from standard two-stroke engines. Candidate technologies listed in the RIA to meet EPA standards were four-stroke engines, direct-injection two-stroke technology, and catalyst additions. Switching to four-stroke engines or direct-injection technology would result in both significant reductions in HC emissions and improvements in fuel economy. Four-stroke engines were especially promising because of the growing availability of these products. Manufacturer and in-house test data demonstrated a 75-95% reduction in HC emissions by switching from traditional two-stroke to four-stroke engines, a 75-90% reduction by switching from traditional two-stroke to two-stroke engines with direct injection, and a 65-75% reduction by adding a catalytic converter (EPA 1996a). EPA also found that switching to four-stroke or direct-injection increased NO_x emissions, although the increases were small compared with the HC benefits. EPA states that manufacturers could use such technologies as exhaust gas recirculation (EGR) and better air and fuel control to counter the NO_x increases.

Projected engine changes to meet the EPA standards were expected to have several other impacts. For example, switching from traditional two-stroke engines would mean using approximately 30% less fuel, which previously was exhausted unburned (EPA 1996a). EPA stated that engines would be easier to start and have improved performance, faster accelerations, and less smoke, fumes, and noise (61 Fed. Reg. 52087 [1996]). EPA stated that, in its view, the regulations did not violate or conflict with safety mandates. Although EPA acknowledged that the Coast Guard was concerned about fuel-injection systems on marine vessels, EPA believed that manufacturers would work with the Coast Guard to ensure the safety of fuel-injection systems, which were already in use on some outboard engines (EPA 1996a).

EPA's estimates of technology costs were based on confidential data provided by manufacturers. Cost and cost effectiveness estimates were not presented for specific technologies because of the possibility of associating specific technologies with specific manufacturers based on those data. Rather, EPA presented its cost method for a fictitious engine family and presented the range of marginal manufacturing costs and cost-effectiveness across potential combinations of technologies and engines without identifying which technologies were used for which engine family (EPA 1996a). In assessing economic impacts, EPA noted that a small number of engine manufacturers dominated the SI marine industry. As part of the RIA, EPA contracted with two independent companies to gather information on engine and vessel markets. EPA estimated that total annualized costs due to the rule-making would reach a near-term peak of $370 million in 2006, roughly 8% of projected retail expenditures in that year. The estimated average per-engine-cost increase to the consumer would be $700 in 2006, which takes into account $480 in fuel savings.

EPA estimated a 75% reduction in HC emissions, approximately 550,000 tons nationwide, from the sources under consideration by 2025 as a result of its rule. A qualitative discussion of air quality and of health and welfare benefits was also included for ozone; no modeling was completed to estimate the effects of the rule on ambient concentrations. Estimated emission-reduction benefits for benzene, 1,3-butadiene, formaldehyde, acetaldehyde, and carbon monoxide were presented along with an assessment of the expected health and welfare impacts of lower concentrations of each pollutant. However, these health and welfare impacts were not quantified or monetized (EPA 1996a).

The California Air Resources Board Standards for Outboard and Personal Watercraft Engines

Stringency

In 1998, the California Air Resources Board (CARB) adopted new exhaust emissions standards for outboard and PWC marine engines to take effect in model-year 2001. In adopting the rule, CARB stated that the regulations were designed to harmonize as closely as possible with federal rules. The regulations included corporate averaging and used EPA test procedures and test cycles for certification and testing. CARB's standards took EPA's model-year 2006 exhaust standard for HC and NO_x and applied it to model-year 2001. CARB also set two lower tiers of standards at 80% and 35% of the 2006 value to begin in 2004 and 2008, respectively. The emissions standards calculated as a function of power rating are shown graphically in Figure 7-2. Water quality was also ex-

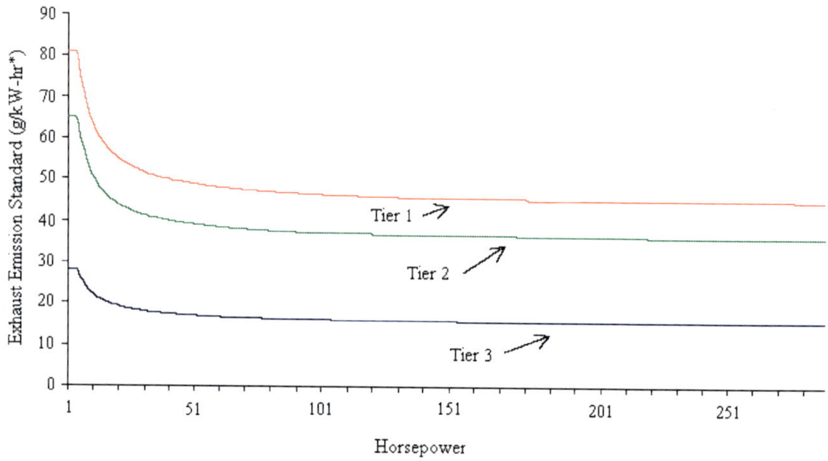

FIGURE 7-2 HC and NO_x Tier 1 will begin in 2001 (EPA's 2006 model year standard); Tier 2, which is 80% of Tier 1, will begin in 2004; and Tier 3, which is 35% of Tier 1, will begin in 2008 Note that the emissions standards at the smallest levels, approximately less than 5 horsepower, reach an upper maximum and do not correspond to the equation given in the text. *Grams per kilowatt-hour. Source: CARB 1998b.

pected to benefit from reduced unburned-fuel emissions, and CARB accounted that as an important benefit of the rule.

Scientific and Technical Analysis

CARB's technical feasibility assessment for their proposed standards included an evaluation of commercially available two-stroke carbureted and direct-injection engines and four-stroke engines (CARB 1998b). Data from federally certified direct- injection two-stroke engines showed emissions to be about 85% lower than emissions from carbureted two-stroke outboard engines. Compared with the emissions from conventional carbureted two-stroke engines, the emissions from four-stroke engines were typically 75-90% lower. Direct-injection two-stroke engines could meet the first and second tiers of the proposed standards, but compliance with the third tier would probably require addition of a catalyst (CARB 1998b). Emissions data collected by EPA demonstrated that existing four-stroke engines could easily comply with the proposed California Tier 1 and Tier 2 standards, and many already complied with the proposed Tier 3 standards. Engine manufacturers had expressed concern about 4-stroke engines, including their larger size, heavier weight, and increased cost. However, four-stroke engines were found to offer similar power-to-weight ratios and consume less fuel and oil, thereby offsetting increases in purchase costs. For exhaust after-treatment, engine modification and use of catalytic converters were considered technically feasible (CARB 1998b).

Cost analysis for compliance with Tier 1 and Tier 2 standards for outboard engines took into account horsepower rating, annual engine sales, emissions-control requirements under the national and California standards, carbureted two-stroke and controlled engine emissions levels, incremental engine prices, and fuel-economy improvements and associated savings. For different horsepower engines, CARB estimated the number of additional emissions-controlled engines that would have to be sold, the associated retail prices, and the lifetime emissions benefits (CARB 1998b). Estimating cost of compliance with Tier 3 standards required a different approach because outboard engines and PWC were not manufactured with catalysts at the time of the assessment. Therefore, CARB conducted a cost assessment similar to that for adding a catalyst to four-stoke engines.

CARB estimated additional statewide emissions reductions for outboard engines and PWC by 2010 and 2020 that would be achieved through the accelerated implementation of the EPA emissions standards (CARB 1998b). Estimates of reductions were obtained using the OFF-ROAD inventory computer model (CARB 2004e), which showed substantial additional reductions for reactive organic gas (ROG) and smaller additional reductions for NO_x. One benefit of accelerated implementation of the EPA standards was the faster elimination of carbureted two-stroke engines. These engines, as noted above, eject as much as 30% of their fuel into the air and water uncombusted and are a significant source of HCs (some of which are ozone precursors) and hazardous air pollutants (HAPs) (CARB 1998b).

Section 11346.3 of the California Government Code requires state agencies to assess the potential for adverse economic impacts on California business enterprises and individuals when proposing to adopt or amend any administrative regulation. The assessment must include a consideration of the impact of the proposed regulation on California jobs; business expansion, elimination, or creation; and the ability of California business to compete. CARB expected that the proposed regulations would not impose a significant cost burden on marine engine manufacturers within the state because most manufacturers are large and located outside California. Annual costs of the proposed regulations were estimated to be around $33 million in 2001, $20 million in 2004, and $21 million in 2008. Those costs were expected to be passed on by manufacturers to marine engine buyers, resulting in an increase of about 14% in average retail prices of a marine engine. The impact on retail sales of these additional costs was anticipated to be minimal, as the most important factors in the purchase of a marine vehicle include the cost of maintenance, which should be reduced in the newer engines, and fuel efficiency, which should improve substantially with the availability of direct-injection two-stroke engines and improved four-stroke engines (CARB 1998b).

New York Adoption of CARB Standards

During the summer of 1999, New York was out of attainment of the National Ambient Air Quality Standards (NAAQS) for the 8-hr ozone standard for 38 days, and the New York City metropolitan area was out of attainment for the 8-hr standard for 26 days. To help bring nonattainment areas into compliance with the NAAQS, the state sought additional ways to reduce emissions of compounds that result in ozone formation. Emissions from PWC were identified as one such source; PWC were responsible for 8,850 tons of volatile organic compounds (VOCs) and 39 tons of NO_x emissions in 1999 (NYDEC 2003).

In 2000, the New York state legislature amended its Environmental Conservation Law by adding language that would allow for the adoption of regulations consistent with the California emissions reductions and labeling regulations for new SI marine engines used in PWC. The legislative objective of these changes was to "reduce emission of hydrocarbon and oxides of nitrogen into the air as well as exhausted into the water, from SI engines, specifically personal watercraft engines." The state cited reductions in particulate matter (PM) emissions as an additional benefit of the proposed regulations. The adoption of the PWC regulations would be incorporated into the state implementation plan (SIP).

The proposed PWC regulations were identical to California PWC regulations. The New York program included the implementation of increasingly more stringent PWC emissions standards between 2001 and 2008, and the application of new test procedures for new and in-use engines. New York found that the CARB standards resulted in PWC emissions 59% lower than the federal program for the average horsepower engine for 2001 and 49% lower for 2004. The total estimated cost of implementation for New York was determined from the incremental cost estimates generated by CARB and the PWC fleet estimated for the year 2004. The New York Department of Environmental Conservation (DEC) agreed with CARB that cost increases would be passed on to the consumer but found that the impact on sales would be reduced by the resulting increase in fuel efficiency, lower maintenance costs, and a demand for new technologies offering performance advantages. No substantial impact on employment was expected, as the marine engine manufacturing industry accounted for only 0.9% of manufacturing jobs in the state (NYDEC 2003).

HEAVY-DUTY-VEHICLE ENGINE STANDARDS

Current Standards

Emissions standards for highway heavy-duty trucks and buses have traditionally been less stringent than those for light-duty passenger cars and trucks. Emissions standards apply to the engines and not the vehicles, and early regulations have been met primarily by engine modifications rather than exhaust after-treatment. More details on early standards for these engines are given in Chapter 4.

In 1995, EPA, CARB, and engine manufacturers signed a statement of principles (SOP) that recognized the need for significant controls on highway HDV engines, particularly for NO_x and PM, and ensured regulatory certainty for the industry (EPA 1995). The SOP included agreements to achieve the following:

- Reduce NO_x emissions standards to roughly 50% of 1998 standards beginning in model-year 2004.
- Harmonize certain California and federal standards for HDV engines.
- Evaluate the role of fuel in achieving even lower future emissions.
- Research achieving NO_x and PM emissions as low as 1.0 grams per brake-horsepower-hour (g/bhp-hr) and 0.05 g/bhp-hr, respectively. The latter PM emissions limit already applied to urban buses.

Model-Year 2004 Rule

EPA adopted rules in 1997 (62 Fed. Reg. 54694 [1996]) that included a 2.4 g/bhp-hr (NO_x plus nonmethane hydrocarbon [NMHC])[2] emission standard to take affect in model year 2004. Emission standards for carbon monoxide (CO) and PM continued at their 1998 levels of 15.5 and 0.10 g/bhp-hr, respectively. In 1999, California adopted those same emissions standards for model-year 2004 HDVs. Since most HDV engines are in trucks that conduct interstate transport, many of these sources in California are not registered there; thus, California has an interest in a stringent national standard. In 1999, EPA also reaffirmed their standards after determining that the technology would be available to achieve those levels of emissions by 2004.

Consent Decree

In 1998, the U.S. Department of Justice, EPA, CARB, and seven major engine manufacturers reached a settlement over the manufacturers' use of software programs that allowed better fuel economy at cruise speeds but resulted in excess NO_x emissions. The software had been used on most model-year engines in the 1990s, and the NO_x emission increases were not detected with standard federal test procedure (FTP) certification. Although manufacturers declared that existing regulations allowed this fuel-saving strategy, EPA declared this practice illegal. The result of the settlement included civil penalties and a consent decree in which manufacturers agreed to perform the following:

- Reprogram the software controlling the engine if the engine was rebuilt, also referred to as chip reflash.
- Meet the model-year 2004 emissions standard for NO_x and NHMC in 2002 nationwide.
- Certify engines using a supplemental steady-state test procedure based on European certification tests (the EURO III European stationary cycle [ESC test]) and require that engine emissions not exceed 1.25 times the FTP emissions limits under certain operating conditions (the not-to-exceed [NTE] rule). The ESC and NTE test procedures would

---

[2] Manufacturers had the option of certifying at 2.4 g/bhp-hr the NMHC and NO_x standard or at 2.5 g/bhp-hr with a limit on NMHC of 0.5 g/bhp-hr.

be required in addition to the FTP and applied up to model-year 2004 engines when the consent decree expired. CARB later adopted a rule requiring the additional ESC and NTE test procedures to apply to model-years 2005 and 2006 engines as well.

Model-Year 2007 Rule

In 2001, EPA finalized the most stringent emissions standards to date for HDV engine emissions standards for model year 2007, referred to hereafter as 2007 HDV engine standards. A PM emission standard of 0.01 g/bhp-hr is to begin with model-year 2007. NO_x and NMHC standards of 0.20 g/bhp-hr and 0.14 g/bhp-hr, respectively, will be phased in for model-years 2007-2010 diesel engines; the standards will apply to 50% of sales in model-years 2007-2009 and 100% of sales in model year 2010.[3] A major part of this control effort is the coincidental low-sulfur fuel regulations, which limit diesel-fuel sulfur content to 15 parts per million (ppm) beginning phase-in in 2006. Present day diesel-fuel sulfur limits are 500 ppm; however EPA estimated in 2000 that diesel fuel in the United States on average contained sulfur at 340 ppm (EPA 2000c). Manufacturers are required to certify engines by using the supplemental steady-state test procedure and NTE rule beginning with model-year 2007 as part of these standards. California will adopt the same emissions standards and test procedures for HDV diesel engines beginning with model-year 2007. The emissions standards require a reduction of more than 90% in NO_x and PM from new engines by 2010.

The 2007 HDV engine standards are technology-forcing and are expected to require widespread use of new types of exhaust after-treatment devices. EPA conducted two biennial assessments of the feasibility of meeting these standards. The last assessment in March 2004 concluded that manufacturers were on track to achieve the required emissions reductions.

Despite EPA, CARB, and manufacturers having worked cooperatively to achieve a uniform standard for HDV engines, a number of legal disputes arose over the standard. Engine manufacturers and their trade association (EMA) challenged the NTE rule in courts, arguing that NTE

---

[3] Although not discussed here, gasoline engines for HDVs are also subject to the same standards beginning with 2008 and 2009. Separate standards that apply to the whole vehicle, as opposed to the engine, for HDV gasoline engines also apply during this period.

limits are illegal and technically infeasible. Manufacturers argued that imposing NTE limits is effectively imposing an emissions standard rather than a certification test. Further, they argued that laboratory tests cannot ensure compliance with a NTE cap under all real-world driving conditions. These suits led to a settlement among manufacturers, CARB, and EPA in which manufacturers agreed to a manufacturer-run, in-use compliance testing program to coincide with the 2007 emissions standards (EPA 2003d). Manufacturers have also opposed a CARB rule to require installation of modified engine-control software (chip reflash) when in-use engines are rebuilt. CARB found that engines were not being rebuilt as quickly as anticipated and started a voluntary reflash program in 2004. CARB subsequently found that the voluntary program was not meeting its goals, and adopted a rule in 2004 to make the reflash mandatory. Finally, the South Coast Air Quality Management District (SCAQMD) attempted to achieve further emissions reduction in its airshed by mandating that HDV fleets include advanced technology engines. As discussed in Chapter 3, the Supreme Court ruled that this regulation was an emissions standard and that the SCAQMD had no authority to establish it.

Scientific and Technical Practices in Setting 2007 HDV Engine Standards

The HDV engine standards for 2007 have much in common with the LDV standards of the 1970s: The 2007 standards are highly technology-forcing, providing substantial emissions reductions and requiring exhaust after-treatment. The amount of analysis that supported the 2007 HDV standards, however, is much greater than the amount that supported the standard setting for LDVs in the 1970s. EPA took the lead in setting the HDV standards, and this section focuses mostly on EPA's practices as described in the RIA (EPA 2000c). As in the LDV case study, standard-setting practices were divided into health and environmental impacts, technical feasibility, and economic impact analyses. As mentioned above, a major part of this rule-making was the simultaneous regulation to reduce sulfur content in fuel. Because the engine and fuel are viewed as an integrated system and the 2007 standards can be met only by using low-sulfur fuel, the RIA includes a similar analysis for the technology and costs required to produce low-sulfur diesel fuel, including the costs for production and distribution. The fuel production costs and the engine technology costs were used to compute the cost-effectiveness and cost-benefit analysis that is presented in the RIA.

Health and Welfare Effects

The RIA for the 2007 HDV engine standards described the health and welfare effects of reducing ozone concentrations and emissions of its precursors. EPA performed photochemical grid modeling analysis for the eastern United States to project the effects of the proposed rule on ozone in 2020 and 2030 relative to a baseline of no proposed rule. EPA also considered the effect of the rule based on analysis of a number of local SIPs. The HDV engine standard was projected to reduce ozone concentrations in nonattainment areas and reduce the number of nonattainment areas for ozone. The RIA also discussed the health and welfare effects of the proposed rule on PM concentrations (and PM nonattainment areas), diesel exhaust emissions, HAP concentrations, visibility and regional haze, acid deposition, eutrophication and nitrification, polycyclic organic matter deposition, and CO concentrations.

Emissions benefits were estimated by EPA using MOBILE5b (NO_x, VOC, and CO) and PART5 (PM and sulfur dioxide [SO_2]). Some adjustments were made to these models to account for updated data in MOBILE6 (which had not been released at the time of the analysis). HAP emissions were estimated by an EPA contractor using MOBTOX5b, a model based on MOBILE5b that includes HC toxic fractions by vehicle type and driving conditions. The RIA presented estimated emissions reductions of benzene, formaldehyde, acetaldehyde, and 1,3 butadiene. Emissions were estimated from diesel engines, including crankcase emissions, and from HDV gasoline engines, including evaporative HC emissions. EPA also included estimates of the emissions benefits of the low-sulfur requirement for non-HDVs. Figure 7-3 shows projected NO_x emissions resulting from the 2007 HDV engine standards (controlled) relative to the 2004 HDV engine standards (baseline). The 2007 standards were estimated to reduce HDV engine on-road NO_x emissions by 90% in 2030. The RIA also estimated a 90% reduction in PM_{10} emissions in 2030 and a 30% reduction in NMHC emissions in 2030 (EPA 2000c).

Health benefits from the 2007 HDV rule were estimated for many of the same respiratory health end points considered in the Tier 2 rule (see Table 6-12 in Chapter 6). The rule was expected to result in 8,300

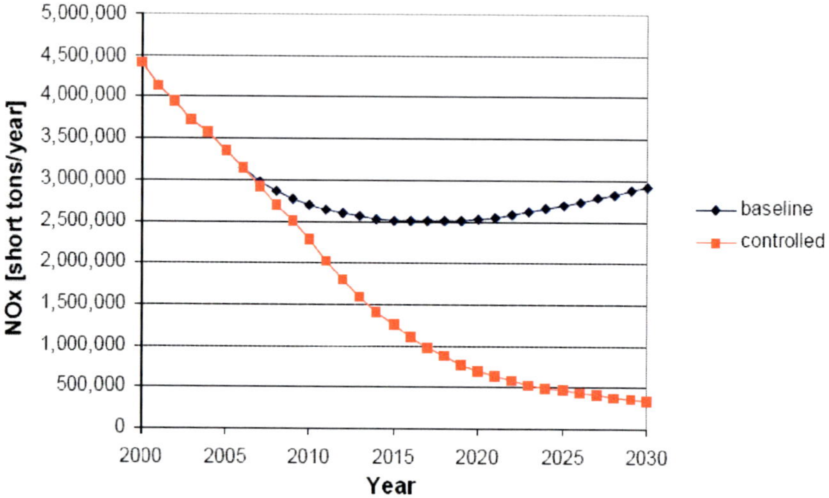

FIGURE 7-3 Projected nationwide NO_x emissions with and without the 2007 HDV engines and fuel sulfur standards. Source: EPA 2000c.

avoided cases of premature mortality, 5,500 avoided cases of chronic bronchitis, and 361,400 avoided asthma attacks annually in 2030. The reader is referred to a 2002 NRC report on estimating public health benefits of proposed air pollution regulation, which includes the 2007 HDV rule as a case study and lists the estimated types, numbers and monetized value of the health benefits (NRC 2002b).

Technical Feasibility

The 2007 rule includes standards for HDV diesel engines and for HDV gasoline engines. Because the standards are most important for diesel engines, only these technologies are summarized here. Candidate technologies target PM and NO_x—the pollutants controlled to the greatest degree by the standards. Earlier standards, including the 2004 HDV engine standards, were met largely by reducing the formation of pollutants in the engine during combustion (engine-out emissions), for example, by introducing electronic fuel systems. EPA stated in its RIA that cooled EGR would be used widely to meet the 2004 and 2007 standards. Although improvements to EGR technology were expected and it would be more effective with low-sulfur fuel, this strategy alone was not thought to be sufficient to meet the 2007 HDV engine standards.

Catalyzed diesel particulate filters (CDPFs), in conjunction with low-sulfur fuel, were identified as the only control devices capable of meeting the 2007 standards of 0.01 g/bhp for PM. Particulate filters work by filtering PM on a ceramic or metal filter, which is later burned off (filter regeneration). Catalyzed filters contain precious metals and base metal catalyst components that allow regeneration to occur at the temperatures of normal engine operation (passive regeneration). Because CDPFs constantly regenerate, they maintain low back pressure and thus minimize fuel economy penalties. CDPFs are able to achieve 90% reduction in exhaust PM emissions by using 15-ppm sulfur fuel. EPA states that some manufacturers had already developed and field-tested these filters and determined their feasibility and durability. Other research and field evidence, such as CDPF use on European trucks, also supports the feasibility of this technology for control of PM mass and ultrafine PM number concentrations (EPA 2000c).

NO_x emission control for light-duty vehicles is provided in part by EGR, when used, and by efficient three-way catalyst technology that achieves more than 95% NO_x reduction under stoichiometric combustion conditions where oxygen content is less than 1% before the catalyst. Diesel exhaust under typical driving conditions has excess oxygen (6% to 18%), which reduces PM formation but makes it more difficult to remove NO_x. EPA (2000c) discussed three main technologies, described in Table 7-1, that reduce NO_x in lean-burn conditions and that have the potential to be used to meet the 2007 standards.

Field testing and research evidence for NO_x control are presented in the RIA only for NO_x adsorbers, which EPA considered, at the time, the leading candidate technology to meet the most stringent 2010 NO_x emissions limits. EPA found later that technologies such as selective catalytic reduction (SCR) will probably be used by some manufacturers to meet the standards (EPA 2004g). EPA noted in the RIA that HDV engine companies had already introduced NO_x adsorber technology on products in Japan, Europe, and the United States, and all were achieving significant NO_x reductions. EPA noted that research was conducted on NO_x adsorber technology by the U.S. Department of Energy and the Oak Ridge National Laboratory. Finally, EPA staff conducted bench-scale testing of four NO_x adsorber systems provided by the Manufacturers of Emissions Control Association (MECA) at the National Vehicle and Fuel Emissions Laboratory (NVFEL) and showed that the systems could

TABLE 7-1 NO_x After-Treatment Devices for HDV Diesel Engines

1. *Lean NO_x catalysts.* A zeolite coating on a catalyst adsorbs hydrocarbons that oxidize and create an oxygen poor environment conducive to NO_x reduction. Even when HCs are actively introduced into the exhaust stream, these systems are expected to achieve only a 30% NO_x reduction.

2. *NO_x adsorber.* An adsorber is an advancement of the three-way catalyst in that it contains a mechanism that stores NO_x under lean-burn conditions and then releases it for conversion by the three-way catalyst under periodic, short, fuel-rich, and stoichiometric exhaust conditions. Fuel sulfur content is critical to the performance and durability of the adsorber. NO_x adsorbers have been shown to achieve greater than 90% NO_x control efficiency. The majority of the technology description in this RIA covers NO_x adsorbers.

3. *Selective catalytic reduction.* Reduces NO_x over a vanadium-titanium catalyst by introducing ammonia upstream of the catalyst in the form of a urea ($[NH2]_2CO$) solution. A diesel oxidation catalyst (DOC) may be introduced up- and downstream of the vanadium-titanium catalysts to convert nitrogen oxide to the desirable nitrogen dioxide form and to oxidize any unreacted ammonia. EPA cites evidence that 70-80% of NO_x reduction could be achieved in the temperature conditions typical of diesel exhaust. EPA identified two difficulties with this technology: the necessity for a urea infrastructure and the lack of incentives for users to replenish their urea supply.

Source: EPA 2000c.

achieve greater than 90% control over a broad range of operating conditions. A detailed description of performance and durability testing and results is presented in the RIA (EPA 2000c). EPA (2004g) identified four improvements necessary for commercial application of NO_x adsorbers: (1) broaden the temperature range over which the adsorbers were effective, (2) improve thermal durability, (3) improve methods for desulfation (desulfation is the process by which sulfur is cleansed from the NO_x absorbers), and (4) improve the integration of NO_x adsorbers with engines to achieve regeneration and maximize fuel economy (EPA 2004g).

The RIA included cost estimates for meeting the 2007 HDV engine standards, which included variable costs (incremental hardware costs, assembly costs, and associated markups) and fixed costs (tooling, research and development, and certification) (EPA 2000c). EPA also accounted for a "learning curve," which reduces manufacturer costs over time as they innovate and lower the operating costs needed to meet the standards. Costs were estimated for four classes: light HDVs (8,500-19,500 lb gross vehicle weight [GVW]), medium HDVs (19,501-33,000 lb GVW), heavy HDVs (33,001+ lb GVW), and urban buses. EPA used contractor estimates of some of the variable costs of projected control equipment as well as maintenance savings associated with the use of low-sulfur diesel fuel. Cost estimates for the heavy HDVs are presented in Table 7-2. EPA also presented a cost-effectiveness analysis comparing these regulations with others, and a cost-benefit analysis as required under Executive Order 12866 (discussed in Chapter 3). EPA estimated the total monetized benefits in 2030 of the 2007 HDV rule to be $70.4 billion (in 1999 dollars), 90% of which is due to the avoided cases of premature mortality due to reduced PM concentrations. Annual costs of the rule were estimated to be $4.2 billion (in 1999 dollars). To characterize uncertainty, EPA provided alternative calculations of the benefits using different assumptions and values for the key parameters used in the calculations. The reader is referred to the RIA (EPA 2000c) and the NRC's report on health benefits (NRC 2002b) for details on these analyses.

Scientific and Technical Practices in CARB's Adoption of HDV 2007 Standards

CARB (2002) adopted slightly modified versions of the EPA 2007 standards and test procedures for HDV diesel engines. Despite the differences, the two sets of standards are widely considered to be identical. CARB's proposed rule was accompanied by a staff paper ("Initial Statement of Reasons") that assessed emissions benefits, technical feasibility, and economic impacts (CARB 2001b). CARB noted that the emissions reductions proposed in this rule were in addition to those in the 1994 SIP but that additional reductions from HDV engines would be necessary for some areas in California to attain the NAAQS. CARB briefly reviewed technologies with a reference to information presented in EPA's RIA and reviewed ongoing research and demonstration projects. Cost estimates were also taken from EPA's RIA. CARB calculated emissions benefits statewide and for various basins by calculating a ratio of the 2007and 2004 emissions standards and applying it to the projected emissions resulting from the 2004 standards. No separate emissions modeling or air quality modeling was done for this rule. CARB presented cost-effectiveness estimates of the rule and compared them with other California mobile-source regulations.

TABLE 7-2 Per-Vehicle Incremental Cost Estimates of 2007 Heavy (33,001+ lb gross vehicle weight) HDV Diesel-Engine Standards Compliance

Item	Fixed Cost, $	Variable Cost, $	Operating Cost, $ (net present value in year of sale of lifetime operating costs)
Near-Term (2007) (1999 dollars per engine)			
NO_X adsorber catalyst	191	1,456	0
Catalyzed diesel particulate filter	89	1,103	208
HC and H_2S cleanup catalyst	0	338	0
Closed crankcase system	0	49	218
Low-sulfur diesel fuel	0	0	3,969
Maintenance savings	0	0	(610)
Total	280	2,946	3,785
Long-Term (2012–) (1999 dollars per engine)			
NO_X adsorber catalyst	0	932	0
Catalyzed diesel particulate filter	0	686	208
HC and H_2S cleanup catalyst	0	216	0
Closed crankcase system	0	32	172
Low-sulfur diesel fuel	0	0	4,209
Maintenance savings	0	0	(610)
Total	0	1,866	3,979

Abbreviation: H_2S, hydrogen sulfide.
Source: EPA 2000c.

Biennial Reviews and Status of 2007 Rule Implementation

In the preamble to the 2007 HDV engine rule, EPA committed itself to biennial assessments of heavy-duty NO_x adsorber technology (66 Fed. Reg. 5002 [2001]); two of them have been released (EPA 2002b, 2004g). Although the initial intent of the assessments was public periodic review of the status of the NO_x adsorber technologies, the assessments have reported more broadly on industry progress toward all aspects of the 2007 standards. Some main points from the 2004 review are the following:

- CDPFs will be used by all manufacturers to meet the 0.01-g/bhp-hr standard in 2007 and are already being successfully used where fuel meets the 15-ppm sulfur.
- Because of flexibility in the phase-in requirements in 2007-2009, most manufacturers appear to have chosen to meet a fleet-average standard of 1.2 g/bhp-hr for NO_x in those years (as opposed to 50% of the fleet meeting the 0.20-g/bhp-hr standard and 50% meeting the existing 2.5-g/bhp-hr standard), followed by a jump to 100% compliance to meet the 0.20-g/bhp-hr standard in 2010. Improvements to EGR systems could also allow manufacturers to meet the interim 1.2-g/bhp-hr standard without NO_x adsorbers. Some manufacturers may also use selective catalytic reduction.
- Despite the lack of widespread use of NO_x adsorbers, they continue to improve and are used in light heavy-duty trucks in Japan.
- Prototype vehicles with NO_x adsorbers will be introduced in 2005 in the United States for early customer fleet testing.
- Discussions with manufacturers led EPA to conclude that manufacturers have solutions to meet the standards. Manufacturers provided evidence that compliance with the standards was integrated into their business plans, and resources toward compliance were allocated.

California Standards for HDVs and State Adoption of California Standards

Differences between California and Federal Standards

California's standards are widely considered to harmonize with EPA's standards for HDV diesel engines for model-year 2007 and beyond. Nevertheless, several states, using their authority under CAA section 177, adopted California's standards for those engines. As discussed above, manufacturers were required to use the ESC and NTE tests in addition to the FTP test for engine certification for model-years 2002-2004 and will be required to certify engines using all three tests for model-year 2007. EMA challenged these requirements in the courts. California adopted a regulation requiring that NTE and ESC tests be used for California-certified engines for model-years 2005 and 2006 to ensure that engines would continue to achieve low emissions for those interim model years (the "NTE rule").

State Adoption of NTE Testing

The State and Territorial Air Pollution Program Administrators and Association of Local Air Pollution Control Officials (STAPPA/ ALAPCO 2001) published a model rule for states to adopt NTE test procedures for model-years 2005 and 2006 engines (STAPPA/ ALAPCO 2001). As of March 2003, California's NTE rule had been adopted by 12 states: North Carolina, New Jersey, Maryland, Delaware, Georgia, Massachusetts, Maine, Texas, Rhode Island, New York, Pennsylvania, and Connecticut as well as the District of Columbia. STAPPA/ALAPCO states that these adoptions would affect one-third of national truck sales and produce emissions benefits equivalent to removing 30 million cars from the road (Becker 2004).

State Adoption of California 2007 Standards

STAPPA/ALAPCO (2004) published a model rule for states to adopt California's HDV diesel-engine standards, although the standards and test procedures for model year 2007 and beyond are the same as EPA's. The rationale behind this model rule is to ensure that states keep the level of emissions control achieved by the 2007 standards even if EPA changes the rule or otherwise relaxes the standards (Becker 2004). The perception that EPA could change the rule stems from strong opposition to the 2007 standards from the trucking industry, engine manufacturers, and the fuel industry concerning technical feasibility, the ability to distribute low-sulfur fuel, and the cost to industry of new engines and fuel. The American Trucking Association (ATA), the American Petroleum Industry (API), and Mack Trucks/Volvo Powertrain all filed petitions asking EPA to reconsider the 2007 standards; all petitions were rejected by EPA (2001a,b,c). In 2004, the trucking industry brought their concerns to Congress, who in turn requested a Government Accountability Office (GAO) study on the issue. The study found that refiners and engine makers were on track to meet the 2007 standards but urged that an independent review panel be appointed to assess the technical feasibility issues and to consider financial incentives to ensure industry compliance (GAO 2004). Executives from the trucking industry also publicly called for delay of the rules (Clean Air Trust 2003). ATA has since endorsed the 2007 standards (Heilprin 2004); however, it has expressed doubts that the expected emissions benefits will be realized, especially if truck buyers stock up on less-expensive higher-emitting trucks before the first model year of regulations (Barber and Sween 2005). The majority of engine manufacturers are confident they will meet the 2007 standards (EPA 2004g).

As a result of concerns that pressure on EPA could cause a delay or change in the 2007 standards, nine states adopted California's program for HDV engines: North Carolina, New Jersey, Maryland, Delaware, Georgia, Maine, New York, Pennsylvania, and Connecticut (STAPPA/ALAPCO 2005). States view California's program as having a higher certainty of delivering the HDV diesel emissions reductions that many states have included in their SIP and air quality plans. Similar to the adoption of the NTE testing, industry strongly opposes state adoption of California's 2007 standards (Clean Air Report 2004).

OFF-ROAD EQUIPMENT WITH SMALL SPARK-IGNITION ENGINES

Off-road equipment with small spark-ignition (SI) engines includes lawn mowers, lawn tractors, generator sets, pumps, augers, leaf blowers, brush cutters, and string trimmers. The majority of small SI engines are in lawn and garden equipment; others are in commercial-turf equipment, golf-course related equipment, and handheld equipment such as chain saws. For regulation purposes, small engines are defined by CARB and EPA as engines rated below 19 kilowatts (kw), or 25 hp. Nonroad emissions are an appreciable part of the anthropogenic emissions inventory in the United States and, as discussed in Chapter 2, represent important controllable emissions for states to meet the NAAQS.

Federal Standards

The CAA amendments of 1990 designated for the first time that EPA set emissions standards for nonroad mobile sources. As discussed in Chapter 3, CAA section 209(e), also added in 1990, extended California's authority to set their own standards to most nonroad sources, including small SI engines. EPA first approved a waiver for California for small-engine emissions standards in 1994. In legislation passed by Congress in 2003, states other than California were explicitly preempted from setting or adopting standards separate from EPA's for small SI engines (P.L. 108-199 Division G Section 428). (This law is sometimes referred to as the Bond Amendment of 2003.) The Bond Amendment effectively creates a new category of regulated mobile sources. Previously, mobile sources, such as locomotives, aircraft, and some agricultural equipment, could have either one standard in all states or a second standard in California and in any state that chose to adopt that standard (CAA section 177). Small SI engines are now separate in how they may be regulated: they may be subject to a second (California) standard but, unlike other sources with California standards, the second standard may not be adopted by other states under section 177 of the CAA.

EPA finalized their Phase I rule-making for most new small SI engines on handheld and nonhandheld equipment in 1995 for model-year 1997. Phase I emissions standards were expected to reduce HC emissions by 32% and CO emissions by 7% in 2020 when complete fleet turnover was expected (60 Fed. Reg. 34582 [1995]). Phase I standards promoted the conversion from two-stroke to four-stroke engine design.

In 1996, EPA signed two statement of principles (SOPs) with manufactures to reduce emissions from handheld and nonhandheld small SI engines (EPA 1996b,c). The SOPs, among other things, outlined technologies and laid the framework for achieving Phase II emissions standards that would further reduce HC and NO_x emissions. EPA finalized their Phase II rule-making for nonhandheld engines in 1999 and for handheld engines in 2000 for exhaust emissions of HC and NO_x. Phase II standards for nonhandheld engines were predicted to reduce emissions of HC and NO_x by 59% by 2027 (EPA 1999c) and for handheld engines were predicted to reduce emissions of HC and NO_x by 70% by 2027 (EPA 2000d) nationwide. Phase II standards also included some efforts to harmonize EPA and CARB's certification procedures. The Phase II standards is expected to result primarily in engine redesign to reduce engine-out emissions and to increase durability, although after-treatment (catalysts) was expected for some handheld equipment. Nonhandheld engines were expected to switch from side-valve to cleaner over-head-valve technology to comply. EPA is expected to introduce new national standards for small engines in 2006 per the Bond Amendment. Table 7-3 summarizes EPA's existing small-engine exhaust emissions standards. The table shows emissions standards for various engine displacement classes defined in the RIA for nonhandheld (EPA 1999c) and handheld engines (EPA 2000d).

California Standards

CARB first adopted rules for small engines in 1990 that consisted of exhaust emissions standards, test procedures, warranty provisions, and compliance programs. The first Tier 1—standards became effective with model-year 1995 engines. Later Tier 2 standards became effective with model-year 2000 engines and included exhaust emissions standards for HC and NO_x, CO, and PM (two-stroke engines only). Although CARB stopped separating engines into handheld and nonhandheld equipment, engines less than 65 cubic centimeters (cc) were assumed to be used primarily in handheld equipment and were allowed higher emissions than engines greater than 65 cc. In 2004, CARB amended its small-engine standards (known as Tier 3) by setting stricter emissions limits for HC and NO_x exhaust and by introducing evaporative emissions standards for the first time for these engines. CARB harmonized with EPA's HC and NO_x exhaust standard for engines less than 50 cc and expanded its "small" handheld range to include engines less than 80 cc. CARB's Tier 3 exhaust emissions limits for larger engines (more than 80 cc) are the same as EPA's for model-year 2005, but they become more stringent than EPA's in future model years and are then expected to require widespread use of exhaust after-treatment in the form of low- to medium-efficiency catalysts. Table 7-4 summarizes CARB's existing small-engine exhaust emissions standards.

CARB also adopted HC evaporative emissions standards for small engines in 2004. The standards are aimed at reducing leaks in the fuel storage and delivery systems, and the rules combine performance and design standards. Performance standards apply to emissions from an entire piece of equipment or vehicle. Certification for performance standards involves measuring evaporative emissions from equipment in an

TABLE 7-3 Existing EPA Phase II Exhaust Emissions Standards for Small SI Engines in Handheld and Nonhandheld Equipment

		Year, HC + NO_x, g/kW-hr					
EPA Category	Engine Displacement	2002	2003	2004	2005	2006	2007+
Handheld	<20 cc (Class III) [a]	238	175	113	50	50	50
	20-50 cc (Class IV) [a]	196	148	99	50	50	50
	>50 cc (Class V) [a]	Ph 1	Ph 1	143	119	96	72
Nonhandheld	<66 cc (Class IA)	Ph 1	16.1[b]	16.1[b]	16.1[b]	16.1[b]	16.1
	66-100 cc (Class IB)	Ph 1	16.1[b]	16.1[b]	16.1[b]	16.1[b]	All
	100-225 cc (Class I)	Ph 1	16.1[b]	16.1[b]	16.1[b]	16.1[b]	engines
	>225 cc (Class II)	16.6	15.0	13.6	12.1	12.1	12.1

[a]The finalized standards are based on a 25%, 50%, 75%, and 100 % phase-in of 50 g/kW-h for Classes III and IV and 72 g/kW-h for Class V.
[b]For any new engine family produced after Aug 1, 2003.
Source: EPA 1999c, 2000d.

TABLE 7-4 Existing CARB Small-Engine Exhaust Emissions Standards, Adopted in 2004 (also referred to as Tier 3)

Engine	HC + NO_x, g/kW-h				CO, g/kW-h
Displacement	2005	2006	2007	2008+	2005+
<50 cc	50	50	50	50	536
50-80 cc	72	72	72	72	536[a]
80-225 cc	16.1	16.1	**10.0**	**10.0**	549
>225 cc	12.1	12.1	12.1	**8.0**	549

[a]Slightly lower CO standard for vertical shaft engines in this category for model-year 2005. Bold numbers show where the standards deviate significantly from EPA Phase II.
Source: CARB 2004f.

enclosure. Design standards, in contrast, are defined as specifications (for example, for permeability) of equipment components. Certification of each component does not involve measuring emissions. Design standards were adopted at manufacturers' request so that the numerous small-engine equipment makers could avoid the expense of measuring emissions for certification. Manufacturers have the option to certify small engines according to either design or performance standards for some larger engines. The reader is referred to CARB's rule-making web page for the exact limits, which vary by engine displacement and model year (CARB 2005g).

CARB's Tier 3 exhaust and evaporative emissions standards were met with opposition from some engine manufacturers, their trade organizations, and other stakeholders. Three general concerns were raised. The first was that compliance with two emissions standards in different areas of the country would be particularly difficult in a nonintegrated industry where engine manufacturers have little control over product distribution. The second was that the costs of complying with regulations that require catalyst after-treatment would be particularly burdensome for the small-engine manufacturing industry because the costs of after-treatment devices are high compared with the prices of some engines. The third was that the use of heat-generating catalysts on equipment close to the operator and in close contact with grass and other flammable matter is an important fire hazard. The remainder of this case study focuses on the three concerns listed above.

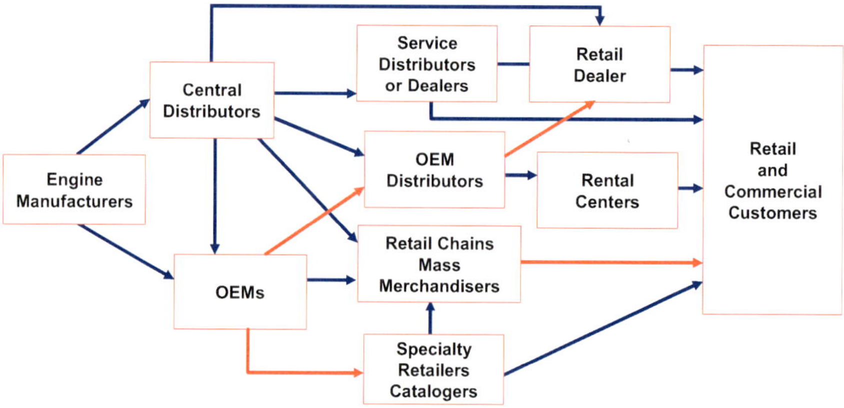

FIGURE 7-4 Flow chart of small SI engine distribution process. Source: Guerry 2004.

Distribution of Engines Certified to Different Emissions Standards

A major concern in meeting California's emissions standards was that distributing two separate products, a federal- and a California-certified engine, is problematic because of the nature of the small-engine equipment market. This market is largely nonintegrated and dominated by hundreds of small manufacturers and component suppliers. Many engine manufacturers sell engines to original equipment manufacturers (OEM), who then produce and distribute equipment though multistep national supply chains, as shown in Figure 7-4 (Guerry 2004). A more detailed 1992 description of the industry can be found in the Appendix of EPA's Phase II RIA (EPA 2000d). Some companies are more integrated than others because some manufacture both engines and original equipment, such as riding lawn-tractors. Another difference is that some engine and equipment makers sell products directly to retailers, and some sell them to wholesalers and distributors who then sell to retailers ("two-tier" distribution). The committee heard arguments that distributing a new product to small geographic areas through a multistep national supply chain is difficult and expensive (Guerry 2004). A multistep distribution chain also increases the chances of federal-certified engines being sold in California, raising compliance issues for industry. California's economic impact assessment when adopting a rule is focused on in-state businesses[4] and does not include an assessment of distribution and inventory costs associated with compliance with a separate standard, either in California or elsewhere. The committee did not come across any quantitative analysis, regulatory or otherwise, of the impacts of a second California standard on distribution of small-engine equipment. A qualitative discussion of distribution costs under uniform versus multiple emissions-standard scenarios is presented in Chapter 5; however, a detailed assessment of product distribution for the small-engine and equipment industry is beyond the scope of this study and the publicly available information.

Increased Manufacturing Costs for Small Spark-Ignition Engines

A second concern raised by manufacturers was the economic burden imposed on the small-engine equipment industry by the additional costs of emissions controls. Although catalysts were used in the past on some small-engine handheld equipment, CARB's Tier 3 regulations seemed to necessitate use of after-treatment devices on many more engines of varying size, horsepower, and price. Catalyst use had not been widespread on nonhandheld equipment, because standards had been met by reducing engine-out emissions alone. Furthermore, CARB's evaporative emissions standards would require more expensive, less permeable parts. CARB estimated that the retail cost increase of adding a catalyst and complying with Tier 3 evaporative standards on a Tier-2 nonhandheld engine would be \$37-\$52 for engines 80 to 225 cc and \$71-\$179 for engines more than 225 cc (CARB 2003d).[5] Some manufacturers and their trade associations argued that these costs are large fractions of the retail prices of small-engine equipment, and imposing those cost increases on consumers would reduce demand for equipment (Guerry 2004). For example, some equipment such as walk-behind lawnmowers may retail for prices on the order of \$100, only a fraction of which is the cost to manufacture the engine. The committee heard testimony from the Outdoor Power Equipment Institute (OPEI) that CARB's Tier 3 proposal was expected to increase the costs of certain lower-priced products by one-third (Guerry 2004). Briggs and Stratton also testified that the average walk-

---

[4] California estimates engineering costs of compliance, but in this discussion that is considered separate from economic impacts.

[5] These estimates were based on CARB's initial proposed exhaust emissions limits, which were later raised.

behind mower could increase in price by 20% as a result of the standards (Hanz and Hotz 2005).

Briggs and Stratton, whose manufacturing operations are located outside California, stated that economic consequences of complying with CARB's Tier 3 regulations could force them to shift manufacturing offshore (Briggs and Stratton 2003). A contractor to Briggs and Stratton reported that this relocation could result in the loss of up to 22,000 jobs at parts suppliers and manufacturers in multiple states (NERA 2004). These economic impacts were used in support of the Bond Amendment, which restricted states' ability to adopt California's emissions standards for small engines (*Congressional Record* S14469, Nov. 12, 2003). The committee did not have any economic information or analysis in support of Briggs and Stratton's statements that widespread adoption of California standards would shift manufacturing offshore. As discussed in previous chapters, economic data for such an analysis is often considered confidential and is guarded by companies. Because neither EPA nor CARB assesses the national economic impacts of a second standard, the committee did not have sufficient information available to assess practices or develop findings on the economic impacts of a second small-engine emission standard on specific companies, the small-engine industry, or the national economy.

Safety of Emissions-Control Equipment on Small Spark-Ignition Engine Equipment

A third concern that received attention during CARB's Tier 3 rulemaking was safety. Some manufacturers, trade associations, fire-safety officials, and the Consumer Product Safety Commission (CPSC) raised concerns about the use of heat-generating catalysts that could burn equipment operators or ignite fires in dried vegetation or other flammable material. Forest and vegetation fire hazards were of particular concern because of the potential use of catalysts on lawn and garden equipment. Debris build up from grass clippings, leaves, and other vegetative materials can become dry and combustible during equipment storage, posing a danger when the equipment is next used. Some equipment might include shielding devices or insulation to protect against high surface temperatures; however, insulation can prolong cool-down time, resulting in the equipment being at higher temperatures when placed in storage. Refueling of heated equipment might also pose a fire risk, for example, fumes igniting from a small spark from carbon or from smoldering debris. CARB's initially proposed evaporative regulations requiring pressurized fuel tanks also raised concerns as a potential explosive hazard.

The issue of burn and fire hazards from emissions-control equipment was not a new concern. LDV and motorcycle manufacturers voiced similar concerns when the widespread use of catalytic converters was first promoted by emissions standards in the 1970s. In addition, catalytic converters had been used on handheld equipment preceding EPA's Phase II standard for this equipment (EPA 2000d). EPA and CARB have given consideration to safety in their rule-makings. EPA is required by the CAA to "take into account safety factors associated with application of emission control technologies to non-road engines and vehicles" (CAA section 213(a)(4)). CARB is required to "determine the technological feasibility of the adoption or amendment of the standard or regulation. That determination shall include, but is not limited to, the availability; effectiveness, reliability, and safety expected of the proposed technology in an application that is representative of the proposed use" (California Health and Safety Code section 43013(e)(2)).

Safety concerns associated with catalyst use were discussed qualitatively in the Phase II RIAs for handheld and nonhandheld small-engine standards. EPA stated that low-efficiency catalysts had been in use on some handheld products and met existing standards. To comply with new emissions standards and meet Forest Service equipment temperature requirements, EPA acknowledged the limitations of using high-efficiency (high-temperature) catalysts and discussed how catalyst use would need to be balanced with reductions in engine-out emissions. EPA (2000d) concluded that "the engine and equipment manufacturer must carefully consider the cooling and safety implications of catalyst installation and reflect this in its design strategy for the engine and equipment."

CARB addressed safety to some extent in its Tier 3 emissions standards. In a proof-of-concept study for CARB performed by Southwest Research Institute (SwRI), catalysts were attached to stock engines and equipment to demonstrate the ability to meet proposed emissions limits. As part of this study, SwRI also measured temperatures of the mufflers with and without catalysts for an estimate of how surface temperatures change with the addition of a catalyst (see Figure 7-5). CARB (2003d) suggested that a catalyst combined with a heat shield resulted in equipment surface temperatures in the range of those measured near mufflers from existing noncatalyzed engines (see Figure 7-5). CARB's staff paper

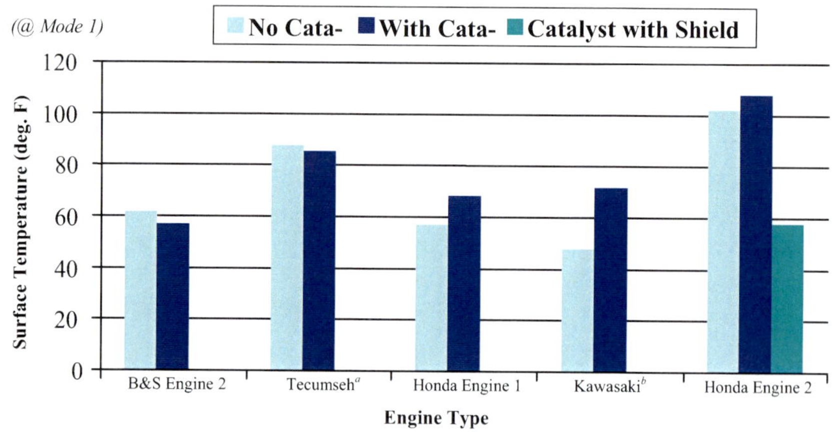

FIGURE 7-5 Results of research at Southwest Research Institute under contract to CARB to test surface-temperature effects of adding catalysts to small SI engines. a, at 250 hr; b, at 125 hr. Abbreviation: B&S, Briggs and Stratton. Source: CARB 2004g.

includes a discussion of manufacturers concerns and points out that heat shields and other insulating material were being used by many manufacturers to reduce temperature around mufflers. Manufacturers, however, critiqued SwRI methods in comments submitted to CARB, raising concerns that they did not measure the areas of highest temperatures, which were observed by manufacturers to reach up to 1100°F in some muffler locations without the use of a heat shield (CARB 2004g). In response to some early manufacturer safety concerns, CARB relaxed its proposed standards, thus reducing the required catalyst efficiency and resulting in lower exhaust and equipment temperatures (CARB 2004g). With respect to practical catalyst use, CARB states that "much of the effort needed to make the catalyst system operate on equipment is the responsibility of product engineers working for either the engine or equipment manufacturers" (CARB 2004g).

Manufacturers remained concerned about safety after CARB's regulations were adopted, and voiced concerns to this committee on the issue (Guerry 2004). Furthermore, the National Association of State Fire Marshals (NASFM), the California Fire Chiefs Association (CFCA), and the CPSC also expressed concerns to CARB of the potential burn and fire hazard associated with their proposed and final regulations (inferred from comments in CARB 2004g). The CFCA and NASFM later sent letters to members of Congress and EPA voicing their concern that CARB had moved forward with catalyst-forcing standards while ignoring safety concerns (*Congressional Record* S14469, Nov. 12, 2003). CARB agreed to participate in a pre-implementation cooperative safety study with manufacturers and fire officials, although its regulations remained unchanged. Congress later passed the Bond Amendment, which specifically charged EPA to "give appropriate consideration to safety factors (including the potential increased risk of burn and fire) associated with compliance with the California standards" when considering a waiver request from California for small SI engine standards (P.L. 108-199 Division G Section 428). Furthermore, Congress passed legislation in 2005 requiring EPA to perform a technical study on safety issues before EPA proposes new small-engine emissions standards (P.L. 109-54 Title II Section 205). The results of the CARB/manufacturer study and the EPA study were not available to the committee before finalizing this report. Although the focus of the safety discussion for small SI engines was on fire and burn hazards, the committee notes that stricter emissions standards reduce inhalation exposure of hazardous fumes for operators of small SI engine equipment. (Inhalation exposure is discussed, for example, in CARB [2000e] and Dost [2003].) It might be appropriate to consider reduced inhalation exposure in combination with the increased burn and fire hazard when considering safety of small-engine equipment.

VOLUNTARY PROGRAMS

Incentive Programs

A number of NAAQS nonattainment areas faced with the need to reduce emissions dramatically have used voluntary programs or programs that provide financial incentives for reducing emissions from federally preempted sources. These programs are important in that they provide an alternative to the development of state emissions standards. Because the programs typically track what types of sources are being reduced and the cost-effectiveness of emissions reductions, they provide an opportunity to compare the costs of emissions controls for preempted sources with the costs of controls on non-preempted sources. Many of the voluntary or incentive-based programs that have been or are being implemented at the federal, state, and local levels in the United States focus on either on-road HDV diesel-engine fleets or nonroad equipment.

These programs include the following:

- **Federal Programs**
  — U.S. Department of Transportation (DOT) Transportation Equity Act for the Twenty-First Century (TEA-21)
    - Congestion Mitigation and Air Quality Improvement Program (CMAQ)

  — U.S. Department of Energy (DOE) Energy Policy Act of 1992 (EPAct)
    - State and Alternative Fuel Provider Program
    - Federal Fleet Program
    - Private and Local Government Fleet Program
    - Alternative Fuel Petition Program
    - Clean Cities Program

  — EPA Office of Air and Radiation Programs
    - Voluntary Diesel Retrofit Program
    - Clean School Bus USA
- **California Programs**
  — California Air Resources Board
    - Diesel Risk Reduction Plan
    - Carl Moyer Program

  — California Department of Transportation (Caltrans) Greening the Fleet Program
  — South Coast Air Quality Management District's Mobile Source Air Pollution Reduction Review Committee Funding Program
  — Sacramento Emergency Clean Air and Transportation Program
  — San Joaquin Valley Emergency Clean Air Attainment Program
  — Gateway Cities Clean Air Program
  — Port of Los Angeles and Port of Oakland Clean Air Programs
  — Bay Area Air Quality Management District's Transportation Fund for Clean Air Program
- **Programs in Other States**
  — Texas Commission on Environmental Quality's Texas Emissions Reduction Plan
  — Houston-Galveston Area Council Clean Cities/Clean Vehicles Program
  — New York State Department of Environment Conservation's Clean Water/Clean Air Bond Act Program
  — Puget Sound Clean Air Agency's Diesel Solutions Program

Two of the largest incentive-based programs are the California Carl Moyer Program and the Texas Emission Reduction Program.

Carl Moyer Program

The State of California implements a voluntary incentive-based program called the Carl Moyer Program, which provides grant funds for the incremental cost of reducing emissions of NO_x from federally preempted sources. The incentives are available for on-road and off-road sources, including heavy-duty trucks, marine engines, locomotive engines, stationary agricultural pump engines, forklifts, airport ground-support equipment, and auxiliary power units. The Carl Moyer Program is administrated by CARB, and approximately $150 million in grants were made between 1998 and 2002. Funds are provided by the state and by local air quality management districts. Recent legislation has expanded funding for the program to an anticipated $140 million per year. Both private companies and public agencies operating HDV engines in California are eligible to apply for the grants. CARB is currently revising the guidelines to address PM as well as HC emissions reductions.

Participation in the Carl Moyer Program is summarized in Table 7-5, and the types of projects funded are shown in Table 7-6. The data in Tables 7-5 and 7-6 reveal a number of key features of the Carl Moyer Program. First, the estimated cost-effectiveness of the emission reductions is generally well below the cost of many of the control measures implemented in California. CARB cost-effectiveness estimates of proposed SIP measures average approximately $8,300 per ton of NO_x reduced, the range being $1,000-$22,000 per ton of NO_x reduced (CARB 2004h). Second, the emission reductions achieved through the program (4-16 tons per day of NO_x statewide) are a small fraction of the emission inventory and the required emission reductions. Total NO_x emission in the South Coast Air Quality Management District in 1997, for example, was estimated to be 1,024 tons per day (annual average day) (SCAQMD 2003). Finally, the program has grown because of its ability to continue to provide relatively low-cost emission reductions.

Texas Emissions Reduction Plan

The Texas Emissions Reduction Plan (TERP) was established in fiscal year (FY) 2002 by the Texas state legislature. The goal of TERP is

TABLE 7-5 Program Summary by Fiscal Year of the Carl Moyer Program (through mid-2003)

Fiscal Year	Number of Participating Districts	Carl Moyer Funding ($ millions)	Matching Funds from Districts ($ millions)	NO_x Reduction (tons per day)	Average Cost-Effectiveness of All Projects Statewide ($ per ton)
1998-1999	16	24.5	12.25	4	3,000
1999-2000	20	19	9.31	7	<5,000
2000-2001	21	45	12.00	14	4,000
2001-2002	NA	16	~8	16	NA
2002-2003	NA	19.68	~9.8	NA	NA

Source: Chan et al. 2005. Reprinted with permission; copyright 2005, ENVIRON.

TABLE 7-6 Project Summary for the First Three Fiscal Years of the Carl Moyer Program (through mid-2003)

Source Category and Equipment Type	NO_x (tons per year)	Cost-Effectiveness ($ per ton)
Heavy-duty line haul	41	2,570
Refuse haulers	432	6,563
Urban transit buses	413	4,715
School buses	4	10,039
Other on-road sources	116	5,756
Farm equipment	36	4,179
Construction	54	3,627
Other nonroad sources	52	3,587
Locomotives	22	1,160
Marine vessels	698	3,044

Agricultural irrigation pumps	1,767	2,353
Forklifts (electric)	163	5,057
Total	3,798	

Source: Chan et al. 2005. Reprinted with permission; copyright 2005, ENVIRON.

to reduce emissions of NO_x from federally preempted sources in the nonattainment and near-nonattainment areas of Texas through voluntary incentive programs. TERP was initially funded at a level of approximately \$20 million per year, and that funding has been increased to a level of approximately \$130 million per fiscal year in 2004 and 2005. The current level of funding is projected to continue through 2008. The Texas Commission on Environmental Quality (TCEQ) administers TERP grants and other TERP financial incentives.

TERP provides funding for cleaner on- and off-road engines, energy efficiency programs, cleaner fuel and other infrastructure programs, and research and development of new technologies. Emission reductions to be achieved through TERP have been incorporated into the ozone SIPs for the Houston-Galveston-Brazoria area (HGB) and the Dallas-Fort Worth area (DFW). Tables 7-7 and 7-8 summarize data on TERP-funded projects for FYs 2002, 2003, and part of 2004 in the HGB and DFW areas. To put the numbers in Table 7-8 in context, the 2007 NO_x SIP budget for HGB is approximately 525 tons per day, 33 tons per day of which are TERP projects. Total NO_x emissions in the HGB area in 1996 were projected to be approximately 1250 tons per day.

Comparison of Tables 7-4 through 7-8 reveals many similarities between the TERP and Carl Moyer programs. In both states, the estimated cost-effectiveness of the emissions reductions is generally below the cost of many of the control measures that are implemented in the states. Second, the emissions reductions achieved through the program are a small fraction of the emissions inventory and the required emissions reductions. In TERP, specific goals were set for emissions reductions in the DFW and HGB areas. Although substantial resources have been expended on the program, and the program provides cost-effective emissions reductions, the HGB and DFW areas still have a shortfall of needed emission reductions, as shown in Figure 7-6 for HGB. (A similar shortfall is found in Dallas-Fort Worth.)

CONCLUSIONS

As discussed in Chapter 2, mobile sources other than LDVs are important contributors to emissions inventories. EPA's and CARB's regulations for these other mobile sources raised a variety of issues, as discussed in Chapter 7. The chapter also described incentive-based mobile-source emissions reductions programs an alternative to emissions stan-

TABLE 7-7 Summary of TERP-Funded and -Recommended Projects to Date in the HGB and DFW Areas (as of November 2, 2004)

	HGB	DFW	HGB and DFW
NO_x Emission Reduction (tons per day)			
FY 02	0.16	1.17	1.33
FY 03	0.62	0.33	0.95
FY 04	4.73	2.97	7.71
Total to Date	5.51	4.47	9.98
2007 SIP Goal	32.9	16.3	49.2
TERP Funding ($ in millions)			
FY 02	3.0	8.8	11.9
FY 03	12.8	1.7	14.5
FY 04	44.6	23.2	67.8
Total to Date	60.4	33.7	94.2
TERP Projects			
FY 02	24	11	35
FY 03	21	11	32
FY 04	76	97	173
Total to Date	121	119	240
TERP Average Cost-Effectiveness ($ per ton)			
FY 02	10,005	4,367	5,101
FY 03	8,104	3,009	6,792
FY 04	6,218	5,210	5,832
Average to Date	6,675	4,795	5,853

Source: Chan et al. 2005. Reprinted with permission; copyright 2005, ENVIRON.

dards. The following conclusions are drawn from the case studies presented in this chapter:

- The biennial reviews of the 2007 HDV rule are similar to the biennial reviews of the zero-emission-vehicle mandate discussed in Chapter 6. These reviews of technological progress are beneficial to stakeholders and the public.
- Some states have indicated that they consider the adoption of California's 2007 HDV standards, which are nearly identical to EPA's, to be a safety net in case EPA delays federal standards.
- Small-engine emissions control poses special challenges to the design, production, and distribution of small-engine equipment. CARB has shown some flexibility in setting emissions standards for small engines to deal with some of the difficulties inherent in the nonintegrate

TABLE 7-8 Summary of TERP-Funded and -Recommended Projects to Date in the HGB and DFW Areas by Emission Sources (as of November 2, 2004)

Project Types	Approved Amount ($)	Total Project NO_x Reduction (tons)	Total NO_x Emission Reduction (tons per day)	Cost-Effectiveness ($ per ton)
FYs 2002, 2003 and 2004 Projects by Type: HGB/DFW Combined				
Railroad projects	27,800,000	5,400	3.57	5,100
On-road vehicles	28,900,000	4,600	2.27	6,300
Nonroad vehicles and equipment	32,000,000	5,100	3.61	6,200
Commercial marine vessels	5,400,000	940	0.53	5,700
Total	94,200,000	16,000	9.98	5,900

Source: Adapted from Chan et al. 2005. Reprinted with permission; copyright 2005, ENVIRON.

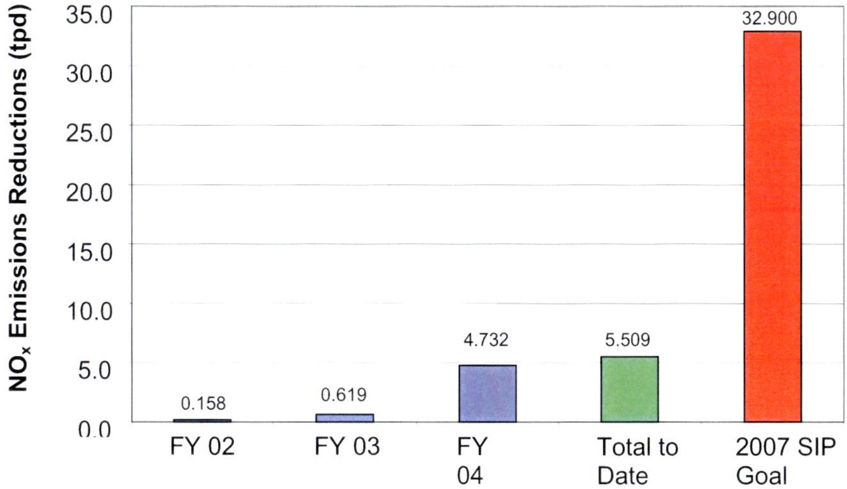

FIGURE 7-6 TERP NO_x emission reductions to date versus the 2007 SIP TERP goal in the HGB area. Source: Chan et al. 2005. Reprinted with permission; copyright 2005, ENVIRON.

industry. Manufacturers and fire safety officials have raised safety concerns about widespread use of heat-generating catalysts on lawn and garden equipment. Safety studies beyond normal EPA and CARB practices are ongoing; the committee did not have enough information to assess safety concerns fully.

- As a result of a 2003 law passed by Congress, small SI engines form a new regulatory category different from other mobile sources. California may set a second standard for those sources; however, the California standard for small SI engines is the only California standard that other states may not adopt under CAA section 177.
- Some states have devised nonregulatory, incentive-based emissions-control programs for mobile sources. These programs to date have provided emissions control from mobile sources that would not otherwise have been provided by regulation.

8

Recommendations

The committee's recommendations are summarized below by topic. Preceding each set of recommendations is a brief summation of the important findings and conclusions leading to the recommendations. Chapter numbers are listed to show where information can be found that supports the committee's conclusions.

CALIFORNIA'S ROLE IN MOBILE-SOURCE EMISSIONS REGULATION

The original reasons for which Congress authorized California to have a separate set of standards remain valid. California's authority to set its own mobile-source emissions standards inevitably imposes additional risks and costs, such as design, production, and distribution costs, though the costs and benefits are difficult to quantify. However, experience to date indicates that the California program has been beneficial overall for air quality by improving mobile-source emissions control (Chapters 2, 3, 6 and 7).

Recommendation

California should continue its pioneering role in setting mobile-source emissions standards. The role will aid the state's efforts to achieve air quality goals and will allow it to continue to be a proving ground for new emissions-control technologies that benefit California and the rest of the nation.

EPA AND CARB TECHNICAL AND SCIENTIFIC PRACTICES IN SETTING STANDARDS

CARB and EPA have essentially the same starting point and motivation for setting new or stricter standards: attainment of the National Ambient Air Quality Standards (NAAQS). Each agency follows a series of procedural steps leading to a finalized regulation. These steps include identification of the need for new emissions standards, evaluation of potential control strategies, publication of proposed regulations, and solicitation of public comments on proposals before promulgating the regulations (Chapters 6 and 7). Some differences exist in the scope of CARB and EPA regulatory assessments as a result of the different procedures that the agencies must follow (Chapters 3, 6 and 7).

Recommendations

Consistent with a 2000 NRC report on modeling mobile-source emissions, CARB and EPA should work in tandem to improve mobile-source emissions models. In particular, consistent with this earlier NRC report, CARB and EPA should complete long-range plans that address improvements or new approaches to mobile-source emissions models. Such plans will improve estimations of emissions reductions. The estimations are a major part of assessing the impacts of emissions standards. The committee also recommends that CARB and EPA include, to the extent possible, air quality impact assessments as part of each rule-making, because the effect of reducing mobile-source emissions on ambient pollutant concentrations will vary from region to region.

Although the committee did not have sufficient information to evaluate the safety issues associated with past regulations, it recommends that safety issues continue to be given careful consideration by EPA and CARB when setting mobile-source emissions standards.

Given that CARB and EPA emissions standards tend to require new technological developments, the committee also recommends that periodic assessments of technological feasibility be continued by the agencies for some of the more important standards. Examples of such standards include CARB's biennial review of the zero-emission-vehicle mandate and EPA's biennial review of on-road diesel-engine standards. Periodic assessments will allow the standards to be based on the most current understanding of the science and technology.

THE WAIVER PROCESS

The waiver review process usually takes several years to complete, and waivers are often granted shortly before the vehicles and engines that meet the standards are in the market. In some cases, waivers have been approved after vehicles and engines that meet the standards are already in the market. Although many California waiver requests are relatively straightforward and uncontroversial, EPA must nevertheless provide the opportunity for full public participation and subsequent technical analyses. This time-consuming process creates uncertainty for California, other states considering adopting those California standards, and manufacturers (Chapter 3).

Recommendations

California, other states, and manufacturers all have a strong interest in obtaining EPA waiver decisions well before the applicable standards take effect. The committee recommends establishment of a two-track system for waiver requests. Many California waiver requests have not been controversial, and EPA has not received any significant comments. EPA could expedite waiver requests that it considers noncontroversial, approving the waiver with a minimal analysis in a direct final decision without a full notice-and-comment process. The final decision would be published in the *Federal Register*, and if any interested party raised a substantive objection to the decision, it would be withdrawn and subjected to the full waiver process. This expedited process would allow EPA to process quickly and efficiently those waiver requests that are noncontroversial, freeing up resources to focus on those that require more time and discussion.

The committee also recommends consideration of a mandatory time limit for EPA to review and issue a waiver decision for controversial waiver requests. The time limit could be based on existing timetables for the EPA waiver process. California is required to provide adequate lead time between adoption of state regulations and their implementation: usually at least 2 years for on-road sources and at least 2 years for non-road sources. A time limit of 2 years or less for EPA review would place the review process between the adoption of the standards by California and the time that the standards take effect. Given the importance of the EPA waiver review and the need to conclude such reviews more quickly, EPA should ensure that sufficient resources are devoted to the waiver review process so that the quality of the review is not sacrificed to comply with new time limits.

ADOPTION OF CALIFORNIA EMISSION STANDARDS BY OTHER STATES

Manufacturers of mobile sources have raised objections to the adoption of California standards by other states. Up to this point, adopting states and manufacturers have resorted to the courts to resolve their technical and legal disputes when direct negotiations have failed. Although EPA is an appropriate entity to comment on some of these disputes, it has no authority over states' adoption decisions (Chapter 6).

Recommendations

The process by which a state adopts California emissions standards should be improved to aid in the resolution of the legal and technical disputes that often arise. As the agency that has the overall authority for implementing the CAA, including the mobile-source provisions, EPA should consistently participate in the process of the adoption of California standards by another state. EPA's current role in the state adoption process includes the authority to approve or disapprove the state SIP claims for emissions benefits from California emissions standards. The committee discussed additional roles for EPA to improve the state adoption process and considered two possible alternatives.

1. Each time a state intends to adopt a California emission standard, EPA would provide formal guidance to aid the state's adoption decision. EPA would determine whether any new issues have arisen that were not considered in the California waiver for the same standard (for example, issues related to technological feasibility, lead time, identicality, and cost) and whether these issues provide cause for states to reject the stan-

dard. EPA would further determine whether the state action is consistent with the requirements specified in section 177 of the CAA. EPA's determinations would be developed with the aim of deterring litigation over potential disputes. However, EPA's determinations would not be binding, and states would retain their ability to adopt California standards at their discretion.

2. EPA is given the authority to review and, under limited circumstances, deny a state adoption decision using a truncated waiver determination process. In its review, EPA would consider whether the state's adoption of the California standard raises any issues not considered in the original California waiver and whether the state action is consistent with section 177 of the CAA. In this scenario, it is important that EPA's waiver determination not delay or otherwise impede adoption of a California standard. EPA would be required to approve automatically any state's adoption request that had not been denied after 18 months of submittal. It is also important that EPA give the same deference to section 177 state findings as it does to California's findings when making a waiver determination. Under this alternative, EPA's determinations would be binding in the same manner as EPA's determination of a California waiver application.

The committee also discussed whether EPA's review under alternative 2 should include an assessment of the necessity or usefulness of the adoption for states to attain their air quality goals. Such an assessment would have to balance the benefits of additional emissions reductions, increased flexibility for states to develop air quality management plans, and wider distribution of new technologies against the costs to industry and consumers.

What role EPA is to have in the state adoption process is a policy decision that goes beyond scientific and technical considerations. The committee disagreed as to which of the two approaches described above would be most effective. However, even if there is no change in the adoption process, non-California states should continue their efforts to work with manufacturers to minimize compliance burdens. As an example, the committee encourages northeastern states that have adopted California light-duty-vehicle emissions standards to implement a regionwide fleet-average emission standard rather than having each state meet a separate fleet-average standard.

SMALL-ENGINE EMISSIONS STANDARDS

Compared with light-duty vehicle emissions control, small-engine emissions control poses special design, production, and distribution challenges (Chapters 5 and 7). CARB has demonstrated some flexibility in setting emissions standards for small engines to deal with some of the difficulties inherent in the non-integrated industry (Chapter 7).

Recommendations

California should continue its pioneering role when setting emissions standards for small engines to aid its efforts to improve air quality and be a proving ground for new emissions-control technologies. The committee encourages CARB to use the flexibility it has shown in revising standards based on new scientific and technical information for regulating small engines. The committee also recommends that the suggested alternatives for improving the state adoption process be used if a decision is made in the future to allow states to adopt California small-engine standards.

COST ANALYSES

One element of relying on technology-forcing regulations and, in California's case, of serving as a laboratory for mobile-source emissions-control technologies is the considerable uncertainty in estimating the cost of complying with emissions standards. In addition, the committee finds that it is difficult to determine what parties bear what fraction of the costs of emissions standards (Chapters 5, 6 and 7).

Recommendations

To address the uncertainty inherent in prospectively estimating costs to comply with mobile-source emissions standards, the committee recommends that agencies and stakeholders attempt to improve communication about the uncertainty by providing a range of costs rather than a single point estimate, especially for new technologies. In addition, because costs are such an important element for understanding the impacts of state emissions standards, the committee finds a need for a comprehensive study of the costs of state standards. This study should include the difference in costs for the states that adopt California standards compared with costs for California, the distribution of those costs, and their cost-effectiveness. Costs should be viewed broadly and include the costs to manufacturers and distributors to develop and distribute products certified under two emissions standards and the costs to states to implement, enforce, and maintain the program.

HARMONIZATION OF STANDARDS AND PROCEDURES

Recognizing the needs of some states to adopt more stringent mobile-source emissions standards to help meet air quality goals, a desirable objective is harmonization of CARB's and EPA's certification procedures. Although harmonization is a worthy pursuit when the interests of the federal government and the states coincide, there are areas where their priorities diverge (Chapters 5 and 6).

Recommendations

Regulators should make a determined effort to harmonize the procedures for testing and certification and look for opportunities to harmonize the emissions standards. Domestically, CARB and EPA should conduct a biennial assessment, either through a written report or public meeting, of where emissions testing and certification procedures can be harmonized and what emissions standards can be harmonized. The committee recognizes that EPA is leading the U.S. participation in international efforts to harmonize emissions standards and testing procedures. EPA should continue these efforts and encourage international participation in the biennial harmonization assessments. The committee recognizes that many countries will lag in the adoption of mobile-source emissions standards; therefore, global efforts to harmonize may need to focus initially on emissions testing and certification procedures.

References

Air, Inc. (Air Improvement Resources, Inc.). 2004a. Review of Cambridge Systematics LEV II Study. Memorandum to Greg Dana, Alliance of Automobile Manufacturers, from Jeremy G. Heiken, Air Improvement Resources, Inc., Novi, MI. January 5, 2004.

Air, Inc. (Air Improvement Resources, Inc.). 2004b. Review of Connecticut Fund for the Environment LEV II Studies. Memorandum to Greg Dana, Alliance of Automobile Manufacturers, from Jeremy G. Heiken, Air Improvement Resources, Inc., Novi, MI. January 28, 2004.

Allred, E.N., E.R. Bleecker, B.R. Chaitman, T.E. Dahms, S.O. Gottlieb, J.D. Hackney, D. Hayes, M. Pagano, R.H. Selvester, S.M. Walden, et al. 1989. Acute Effects of Carbon Monoxide Exposure on Individuals with Coronary Artery Disease. Health Research Report 25. Cambridge, MA: Health Effects Institute.

Anderman, M., F.R. Kalhammer, and D. MacArthur. 2000. Advanced Batteries for Electric Vehicles: An Assessment of Performance, Cost and Availability. Prepared for California Air Resources Board, Sacramento, CA, by The Year 2000 Battery Technology Advisory Panel [online]. Available: http://ntl.bts.gov/card_view.cfm?docid=22893 [accessed July 19, 2005].

Anderson, J.F., and T. Sherwood. 2002. Comparison of EPA and Other Estimates of Mobile Source Rule Costs to Actual Price Changes. SAE Technical Paper 2002-01-1980. Warrendale, PA: Society of Automotive Engineers.

Ando, A.W., V. McConnell, and W. Harrington. 2000. Costs, emissions reductions and vehicle repair: Evidence from Arizona. J. Air Waste Manage. Assoc. 50(4):509-521.

Aplet, J.H., and G. Meade. 1997. Mobile source emissions regulation in California. Pp. 39-70 in Air Pollution and Regional Economic Performance:

A Case Study, J.V. Hall, ed. Advances in the Economics of Environmental Resources, Vol. 2. Greenwich, CT: JAI Press.

Argote, L., and D. Epple. 1990. Learning curves in manufacturing. Science 247(4945):920-924.

Ashbaugh, L.L., and D.R. Lawson. 1991. A Comparison of Emissions from Mobile Sources Using Random Roadside Surveys Conducted in 1985, 1987, and 1989. AWMA Reprint No. 91-180.58. Presented at 84th Annual AWMA Meeting, Vancouver, BC. June.

ATSDR (Agency for Toxic Substances and Disease Registry). 1993. Toxicological Profile for 1,3-Butadiene. U.S. Department of Health and Human Services, Agency for Toxic Substances and Disease Registry, Atlanta, GA [online]. Available: http://www.atsdr.cdc.gov/toxprofiles/tp28.html#bookmark03 [accessed Feb. 1, 2006].

ATSDR (Agency for Toxic Substances and Disease Registry). 2004. Medical Management Guidelines (MMGs) for Formaldehyde (HCHO) [online]. Available: http://www.atsdr.cdc.gov/MHMI/mmg111.html [accessed May 24, 2005].

Austin, T. 2005. Incremental Costs and Cost-Effectiveness of California Emissions Standards. Presentation at the Fourth Meeting on State Practices in Setting Mobile Source Emissions Standards, April 15th, 2005, Boston, MA.

Babik, R. 2005. State Practices in Setting Mobile Source Emissions Standards: Perspectives of General Motors. Presentation at the Third Meeting on State Practices in Setting Mobile Source Emissions Standards, January 19th, 2005, Dearborn, MI.

Barber, M., and D. Sween. 2005. Truckers doubt they'll be ale to meet new standards designed to cut diesel engine emissions starting in 2007. Enterprise 34(36):5. February 28, 2005.

Becker, S.W. 2004. State and Local Mobile Source and Fuel Standards A Briefing for the National Research Council. Presentation by STAPPA/ALAPCO at the First Meeting of the Committee on State Practices in Setting Mobile Source Emissions Standards, June 28th, 2004, Washington D.C.

Bell, M.L., A.D. McDermott, S.L. Zeger, J.M. Samet, and F.Dominici. 2004. Ozone and short-term mortality in 95 US Urban Communities, 1987-2000. JAMA 292(19):2372-2378.

Benson, J.D., V. Burns, R.A. Gorse, A.M. Hochhauser, W.J. Koehl, L.J. Painter, and R.M. Reuter. 1991. Effects of Gasoline Sulfur Level on Mass Exhaust Emissions - Auto/Oil Air Quality Improvement Research Program. SAE 912323. Warrendale, PA: Society of Automotive Engineers.

Bertelsen, B.I. 2001. Future U.S. motor vehicle emission standards and the role of advanced emission control technology in meeting those standards. Top. Catal. 16/17(1-4):15-22.

Bishop, G.A., and D.H. Stedman. 1990. On-road carbon monoxide emission measurement comparisons for the 1988-1989 Colorado oxy-fuels program Environ. Sci. Technol. 24(6):843-847.

Bishop, G.A., S.S. Pokharel, and D.H. Stedman. 1999. On-Road Remote Sensing of Automobile Emissions in the Phoenix Area: Year 1. Contract No. E-23-4. Prepared for Coordinating Research Council, Inc., Atlanta, GA, by Department of Chemistry and Biochemistry, University of Denver, Denver, CO [online]. Available: http://www.crcao.com/ [accessed July 12, 2005].

Bishop, G.A., S.S. Pokharel, and D.H. Stedman. 2000. On-Road Remote Sensing of Automobile Emissions in the Los Angeles Area: Year 1. Prepared for Coordinating Research Council, Inc., Alpharetta, GA, by Department of Chemistry and Biochemistry, University of Denver, Denver, CO [online]. Available: http://www.crcao.com/ [accessed July 12, 2005].

Bishop, G.A., M.J. Williams, D.A. Burgard, and D.H. Stedman. 2003. On-Road Remote Sensing of Automobile Emissions in the Chicago Area: Year 5, September 2002. Final Report. Contract No. E-23-4. Prepared for Coordinating Research Council, Inc., Alpharetta, GA, by Department of Chemistry and Biochemistry, University of Denver, Denver, CO [online]. Available: http://www.feat.biochem.du.edu/assets/databases/Illinois/Arlhghts/Chicago_Year_5_CRC2002.pdf [accessed Jan. 21, 2005].

Blanchard, C.L., and S.J. Tanenbaum. 2003. Differences between weekday and weekend air pollutant levels in southern California. J. Air Waste Manage. Assoc. 53(7):816-828.

Blanchard, C.L., and S.J. Tanenbaum. 2005. Weekday/Weekend Differences in Ambient Concentrations of Primary and Secondary Air Pollutants in Atlanta, Baltimore, Chicago, Dallas-Fort Worth, Denver, Houston, New York, Phoenix, Washington and Surrounding Areas. NERL Project ES04-1. NERL Subcontract No. LDX-4-44213-01. Envair, Albany, CA. May 16, 2005.

Boyle, M. 2002. Dirty little engines get cleaner. Fortune 145(10):146B.

Bresnahan, T., and D. Yao. 1985. The nonpecuniary costs of automobile emissions standards. Rand J. Econ. 16(4):437-455.

Briggs and Stratton. 2003. Extreme California Emissions Proposal Would Cost the Heartland 22000 Manufacturing Jobs. News Releases: September 16th, 2003 [online]. Available: http://www.briggsandstratton.com/display/router.asp?DocID=64034 [accessed Oct. 20, 2005].

Browning, L. 2001. Projected Automotive Fuel Cell Use in California. Consultant Report. P600-01-022F. Prepared for California Energy Commission, by Arthur D. Little Inc., Cupertino, CA. October 2001 [online]. Available: http://www.energy.ca.gov/reports/2002-02-06_600-01-022F.PDF [accessed Jan. 25, 2006].

Burke, A.F., K.S. Kurani, and E.J. Kenney. 2000. Study of the Secondary Benefits of the ZEV Mandate. UCD-ITS-RR-00-7. Prepared for California Air Resources Board, by Institute of Transportation Studies, University of California-Davis. August 2000 [online]. Available: http://its.ucdavis.edu/publications/2000/RR-00-07.pdf [accessed July 14, 2005].

Cackette, T. 1998. The Cost of Emission Controls Motor Vehicles and Fuels: Two Case Studies. Massachusetts Institute of Technology, Cambridge, MA. Presentation at the 1998 Summer Symposium of the EPA Center on Airborne Organics, July 9-10, Dedham, MA.

Canale, R.P., S.R. Winegarden, C.R. Carlson, and D.L. Miles. 1978. General Motors Phase II Catalyst System. SAE Trans. 87:843-852.

CARB (California Air Resources Board). 1990. Proposed Regulations for Low-Emission Vehicles and Clean Fuels. Technical Support Document. Sacramento, CA: California Environmental Protection Air Resources Board. August 13, 1990.

CARB (California Air Resources Board). 1991. Proposed Regulation for Low-Emission Vehicles and Clean Fuels: Final Statement of Reasons. Sacramento, CA: California Environmental Protection Air Resources Board. July 1991.

CARB (California Air Resources Board). 1994a. Zero-Emission Vehicle Update. Draft Technical Document for the Low-Emission Vehicle and Zero-Emission Vehicle Workshop on March 25, 1994. CARB Mobile Source Division, El Monte, CA. March 2, 1994.

CARB (California Air Resources Board). 1994b. 1994 Low Emission Vehicle and Zero-Emission Vehicle Program Review: Staff Report. Sacramento, CA: California Environmental Protection Agency Air Resources Board. April 1994.

CARB (California Air Resources Board). 1994c. The California State Implementation Plan for Ozone. Adopted Nov. 15, 1994. California Environmental Protection Agency Air Resources Board [online]. Available: http://www.arb.ca.gov/planning/sip/94sip/94sip.htm [accessed July 14, 2005]

CARB (California Air Resources Board). 1996. Pp. 5-6 in Resolution 96-12, March 28, 1996. Zero Emission Vehicle Program. California Air Resources Board [online]. Available: http://www.arb.ca.gov/regact/zev/zevresol.pdf [accessed July 19, 2005].

CARB (California Air Resources Board). 1998a. Staff Report: Initial Statement of Reasons Proposed Amendments to California Exhaust and Evaporative Emission Standards and Test Procedures for Passenger Cars, Light-Duty Trucks and Medium-Duty Vehicles "LEV II" and Proposed Amendments to California Motor Vehicle Certification, Assembly-Line and In-use Test Requirements " CAP 2000". Mobile Source Control Division, California Environmental Protection Agency Air Resources Board, El Monte, CA [online]. Available: http://www.arb.ca.gov/regact/levii/LEVii/isor.pdf [accessed July 14, 2005].

CARB (California Air Resources Board). 1998b. Public Hearing to Consider Adaptation of Emission Standards and Test Procedures for New 2001 and Later Model Year Spark-Ignition Marine Engines. Staff Report. California Air Resources Board, Sacramento, CA. October 23, 1998 [online]. Available: http://www.arb.ca.gov/regact/marine/marine.htm [accessed August 25, 2005].

CARB (California Air Resources Board). 1999. Final Statement of Reasons. "LEV II" and "CAP 2000" Amendments to California Exhaust and Evaporative Emission Standards and Test Procedures for Passenger Cars, Light-Duty Trucks and Medium-Duty Vehicles, and the Evaporative Emission Requirements for Heavy-Duty Vehicles. California Environmental Protection Agency Air Resources Board. September 1999 [online]. Available: http://www.arb.ca.gov/regact/levii/to_oal/leviifso.pdf [accessed July 14, 2005].

CARB (California Air Resources Board). 2000a. Staff Report: 2000 Zero Emission Vehicle Program Biennial Review. California Environmental Protection Agency Air Resources Board. August 7, 2000 [online]. Available: http://www.arb.ca.gov/msprog/zevprog/2000review/staffreportfinal.pdf [accessed July 14, 2005].

CARB (California Air Resources Board). 2000b. Preliminary Draft Staff Report, May 31 Workshop: ZEV 2000 Biennial Review. California Environmental Protection Agency Air Resources Board. May 22, 2000 [online]. Available: http://www.arb.ca.gov/msprog/zevprog/2000review/reportv7.pdf [accessed July 14, 2005].

CARB (California Air Resources Board). 2000c. Resolution 00-29, September 7, 2000. California Air Resources Board [online]. Available: http://www.acpropulsion.com/CARB/Res%2000-29%20Fin3.pdf [accessed July 19, 2005].

CARB (California Air Resources Board). 2000d. Public Workshop to Discuss Issues Related to the Zero Emission Vehicle Regulation. Mail Out #MSC-00-23. California Air Resources Board. October 16, 2000 [online] Available: http://www.arb.ca.gov/msprog/mailouts/mouts_00.htm [acces-sed July 19, 2005].

CARB (California Air Resources Board). 2000e. A Report to the California Legislature on the Potential Health and Environmental Impacts of Leaf Blowers. Mobile Source Control Division, California Air Resources Board, Sacramento, CA. February 2000 [online]. Available: http://www.energy.ca.gov/reports/2002-02-06_600-01-022F.PDF [accessed Jan. 25, 2006].

CARB (California Air Resources Board). 2001a. ARB Staff Review of Report Entitled "Impacts of Alternative ZEV Sales Mandates on California Motor Vehicle Emissions: A Comprehensive Study". California Air Resources

Board. October 31, 2001 [online]. Available: http://www.arb.ca.gov/regact/zev2001/StaffResp.PDF. [accessed July 28, 2005].

CARB (California Air Resources Board). 2001b. Public Hearing to Consider Amendments Adopting More Stringent Emissions Standards for 2007 and Subsequent Model Year New Heavy-Duty Diesel Engines. Staff Report: Initial Statement of Reasons. California Air Resources Board, Sacramento, CA. September 7, 2001 [online]. Available: http://www.arb.ca.gov/regact/hdd2007/isor.pdf. [accessed August 25, 2005].

CARB (California Air Resources Board). 2002. California Exhaust Emission Standards and Test Procedures for 2004 and Subsequent Model Heavy-Duty Diesel-Engines and Vehicles. Adopted: December 12, 2002. California Air Resources Board, Sacramento, CA [online]. Available: http://www.arb.ca.gov/regact/levhdg02/ftp6.pdf [accessed August 25, 2005].

CARB (California Air Resources Board). 2003a. Ambient Air Quality Standards. California Air Resources Board. July 9, 2003 [online]. Available: http://www.arb.ca.gov/aqs/aaqs2.pdf [accessed May 20, 2005].

CARB (California Air Resources Board). 2003b. ARB Modifies Zero Emission Vehicle Regulation. California Air Resources Board News Release, April 24, 2003 [online]. Available: http://www.arb.ca.gov/newsrel/nr042403.htm [accessed July 19, 2005].

CARB (California Air Resources Board). 2003c. Staff Report: Initial Statement of Reasons. 2003 Proposed Amendments to the California Zero Emission Vehicle Program Regulations. California Environmental Protection Agency Air Resources Board. January 10, 2003 [online]. Available: http://www.arb.ca.gov/regact/zev2003/isor.pdf [accessed Jan. 6, 2006].

CARB (California Air Resources Board). 2003d. Initial Statement of Reasons for Proposed Rulemaking Public Hearing to Consider the Adoption of Exhaust and Evaporative Emission Control Requirements for Small Off-Road Equipment and Engines Less Than or Equal to 19 Kilowatts. Staff Report. California Air Resources Board, Sacramento, CA. August 8, 2003 [online]. Available: http://www.arb.ca.gov/regact/sore03/isor.pdf [accessed August 25, 2005].

CARB (California Air Resources Board). 2004a. Report of Enforcement Activities for January 1 – December 31, 2003. California Air Resources Board [online]. Available: http://www.arb.ca.gov/enf/reports/03enfrpt.pdf [accessed January 5, 2006].

CARB (California Air Resources Board). 2004b. Final Statement of Reasons: The 2003 Amendments to the California Zero Emission Vehicle Program Regulations. California Environmental Protection Agency Air Resources Board. January 2004 [online]. Available: http://www.arb.ca.gov/regact/zev2003/fsor.pdf [accessed July 14, 2005].

CARB (California Air Resources Board). 2004c. Staff Report: Initial Statement of Reasons for Proposed Rulemaking, Public Hearing to Consider

Adoption of Regulations to Control GHG Emissions from Motor Vehicles. California Environmental Protection Agency Air Resources Board [online]. Available: http://www.arb.ca.gov/regact/hrnhsgas/isor.pdf [accessed July 14, 2005].

CARB (California Air Resources Board). 2004d. Report to the Legislature and the Governor on Regulations to Control GHG Emissions from Motor Vehicles. California Environmental Protection Agency Air Resources Board. December 2004 [online]. Available: http://www.arb.ca.gov/cc/reports/ab1493_legreport.pdf [accessed July 14, 2005].

CARB (California Air Resources Board). 2004e. Updates to the Off-Road Emissions Model [online]. Available: http://www.arb.ca.gov/msei/off-road/updates.htm [accessed August 30, 2005].

CARB (California Air Resources Board). 2004f. Updated Informative Digest. Adaptation of Exhaust and Evaporative Emission Control Requirements for Small Off-Road Engines Less Than or Equal to 19 Kilowatts and Equipment That Use Such Engines. California Air Resources Board, Sacramento, CA. August 10, 2004 [online]. Available: http://arb.ca.gov/regact/sore03/uip.pdf [accessed August 26, 2005].

CARB (California Air Resources Board). 2004g. Public Hearing to Consider the Adaptation of Exhaust and Evaporative Emission Control Requirements for Small Off-Road Equipment and Engines Less Than or Equal to 19 Kilowatts. "Revised" Final Statement of Reasons. California Air Resources Board, Sacramento, CA August 2004 [online]. Available: http://arb.ca.gov/regact/sore03/rfsor.pdf [accessed August 26, 2005].

CARB (California Air Resources Board). 2004h. The Carl Moyer Program Annual Status Report: The Carl Moyer Memorial Air Quality Standards Attainment Program: Incentives for Lower-Emission Heavy-Duty Engines. California Air Resources Board, Sacramento, CA. February 2004 [online]. Available: http://www.arb.ca.gov/msprog/moyer/moyer_2004_report.pdf [accessed July 20, 2005].

CARB (California Air Resources Board). 2005a. Air Quality Data Statistics. Aerometric Data Analysis and Management System, California Air Resources Board [online]. Available: http://www.arb.ca.gov/adam/welcome.html [accessed May 20, 2005].

CARB (California Air Resources Board). 2005b. Emission Data. Emission Inventory, California Air Resources Board [online]. Available: http://www.arb.ca.gov/ei/emissiondata.htm [accessed Feb. 2005].

CARB (California Air Resources Board). 2005c. The California Almanac of Emissions and Air Quality – 2005 Edition. Planning and Technical Support Division, California Air Resources Board [online]. Available: http://www.arb.ca.gov/aqd/almanac/almanac05/almanac05.htm [accessed May 16, 2005].

CARB (California Air Resources Board). 2005d. Updated Informative Digest. Amendments to Adopt More Stringent Emission Standards for 2007 and

Subsequent Model Year New Heavy-Duty Diesel Engines. California Air Resources Board [online]. Available: http://www.arb.ca.gov/regact/hdde2007/uid.pdf [accessed July 29, 2005].

CARB (California Air Resources Board). 2005e. Response to Testimony Presented at the Committee's Fourth Meeting in Boston. Letter to John Holmes, National Research Council, Washington, DC, from Tom Cackette, Chief Deputy Executive Officer, Air Resources Board, California Environmental Protection Agency, Sacramento, CA. April 28, 2005.

CARB (California Air Resources Board). 2005f. Final Statement of Reasons: Regulations to Control Greenhouse Gas Emissions From Motor Vehicles. California Environmental Protection Agency Air Resources Board. August 4, 2005 [online]. Available: http://www.arb.ca.gov/regact/grnhsgas/fsor.pdf [accessed Jan. 24, 2006].

CARB (California Air Resources Board). 2005g. Small Off-Road Engines and Equipment [online]. Available: http://www.arb.ca.gov/msprog/offroad/sore/sore.htm [accessed August 26, 2005].

CARB (California Air Resources Board). 2006a. Cleaner Gas Cars. A Zero & Near Zero Emission Vehicle Guide. California Air Resources Board [online]. Available: http://www.driveclean.ca.gov/en/gv/driveclean/vtype_cleaner.asp [accessed Jan. 27, 2006].

CARB (California Air Resources Board). 2006b. Incentives. A Zero & Near Zero Emission Vehicle Guide. California Air Resources Board [online]. Available: http://www.driveclean.ca.gov/en/gv/incentives/index.asp [accessed Jan. 27, 2006].

CDC (Centers for Disease Control and Prevention). 1997. Outdoor Carbon Monoxide Attributed to Tractor Exhaust-Kentucky, 1997. MMWR 46(51):1224-1227 [online]. Available: http://www.cdc.gov/Publications/mmwr/wk/mm4651.pdf [accessed May 23, 2005].

CDC (Centers for Disease Control and Prevention). 2004. Carbon Monoxide Poisonings Resulting from Open Air Exposures to Operating Motorboats-Lake Havasu City, Arizona, 2003. MMWR 53(15):314-318 [online]. Available: http://www.cdc.gov/mmwr/preview/mmwrhtml/mm5315a3.htm [accessed May 23, 2005].

CFE (Connecticut Fund for the Environment). 2003. The Drive for Cleaner Air in Connecticut: The Benefits of Adopting the California Low-Emission Vehicle Standard for Cars and Light Duty Trucks. Connecticut Fund for the Environment, New Haven, CT [online]. Available: http://www.cfenv.org/air/cars/Drive-for-Cleaner-Air-report.pdf [accessed July 21, 2005].

Chameides, W., D. Greenbaum, R. Wassel, K.J. Holmes, A. Staudt, and K. Gustavson. 2005. Air quality management in the United States. EM(July):22-29.

Chan, L.M., A. Pollack, C. Lindhjem, P. Chandraker, S. Rao, and C. Tran. 2005. On-Road and Non-Road Emission Reductions from TERP in 2007 and 2010 in the Houston, Galveston, and Dallas Fort-Worth Areas. Final report. Prepared for Houston Advanced Research Center, The Woodlands, TX, by ENVIRON, Novato, CA [online]. Available: http://files.harc.edu/harc/Projects/AirQuality/ Projects/H042.T1/H42FinalReport.pdf [accessed Oct. 20, 2005].

Chinkin, L.R., D.L. Coe, T.H. Funk, H.R. Hafner, P.T. Roberts, P.A. Ryan, and D.R. Lawson. 2003. Weekday versus weekend activity patterns for ozone precursor emissions in California's South Coast Air Basin. J. Air Waste Manage. Assoc. 53(7):829-843.

Clean Air Report. 2004. Industry Blasts State Diesel Efforts, Says Federal Rule Delay Possible. Inside EPA Clean Air Report 15(21):14-15.

Clean Air Trust. 2003. Trucking Company Execs Named Clean Air "Villains" of the Month. Clean Air Villains. June 2003 [online]. Available: http://www.cleanairtrust.org/villain.0603.html [accessed August 26, 2005].

Cobb, N., and R.A. Etzel. 1991. Unintentional carbon monoxide-related deaths in the United States, 1979 through 1988. JAMA. 266(5):659-663.

Coulter-Burke, S., D. Calkins, and G. Yarwood. 2002. Analysis of Emission Reduction Control Measures for Attaining the 8-Hour Ozone NAAQS in the Tyler/Longview/Marshall Near Nonattainment Area. Draft Final Report. Prepared for East Texas Council of Governments, Kilgore, TX, By ENVIRON International Corporation, Novato, CA. September 9, 2002 [online]. Available: http://www.tceq.state.tx.us/assets/public/implementation/air/sip/sipdocs/2004-06-NET/04077sipapg_pro.pdf [accessed Jan. 21, 2006].

Crandall, R.W., H.K. Gruenspecht, T.E. Keeler, and L.B. Lave. 1986. Regulating the Automobile. Washington, DC: The Brookings Institution.

Croes, B.E., L.J. Dolislager, L.C. Larsen, and J.N. Pitts. 2003. The O_3 "weekend effect" and NO_x control strategies: Scientific and public health findings and their regulatory implications. EM (July):27-37.

CT DEP (Connecticut Department of Environmental Protection). 2004. Hearing Report Regarding Regulations of the Abatement of Air Pollution: Proposed Adoption of Section 22a-174-36b and Proposed Amendment of Section 22a-174-36(i) of the Regulations of Connecticut State Agencies. May 7, 2004 [online]. Available: http://dep.state.ct.us/air2/regs/callev.pdf [accessed July 14, 2005].

Currie, D.P. 1970. Motor vehicle air pollution: State authority and federal preemption. Mich. L. Rev. 68:1083, 1089.

Dana, G. 2004. Presentation at the First Meeting on State Practices in Setting Mobile Source Emissions Standards, June 28th, 2004, Washington, DC.

Davis, S.C., and S.W. Diegel. 2004. Transportation Energy Data Book: Edition 24. ORNL-6973. Prepared by Oak Ridge National Laboratory, Oak Ridge,

TN, for Office of Planning, Budget Formulation and Analysis, Energy Efficiency and Renewable Energy, U.S. Department of Energy [online]. Available: http://cta.ornl.gov/data/download24.shtml [accessed May 16, 2005].

Devlin, R.B., A.J. Ghio, H. Kehrl, G. Sanders, and W. Cascio. 2003. Elderly humans exposed to concentrated air pollution particles have decreased heart rate variability. Eur. Respir. J. 40(suppl.):76s-80s.

Dixon, L., I. Porche, and J. Kulick. 2002. Driving Emissions to Zero: Are the Benefits of California's Zero Emission Vehicle Program Worth the Costs? MR-1578-RC/S&T. Santa Monica, CA: RAND [online]. Available: http://www.rand.org/publications/MR/MR1578/ [accessed July 21, 2005].

Dockery, D.W., C.A. Pope, III, X. Xu, J.D. Spengler, J.H. Ware, M.E. Fay, B.G. Ferris, Jr., and F.E. Speizer. 1993. An association between air pollution and mortality in six U.S. cities. N. Engl. J. Med. 329(24):1753-1759.

Dost, F. N. 2003. Toxicology and potential Health Risk of Chemicals that May be Encountered by Workers Using Forest Vegetation Management Options, Part 1: Risk to Workers Associated with Exposure to Emissions from Power Saws. Title No. 3. Forest Practices Branch, BC Ministry of Forests [online]. Available: http://www.for.gov.bc.ca/hfp/publications/00012/3-Dost-PowersawEmissions.pdf [accessed Jan. 25, 2006].

Dumanoski, D. 1991. Mass. charges Conn. hurting antismog plan. Boston Globe, Nov. 13, 1991, P. 29.

Dumanoski, D. 1992. Emissions law hit by oil, auto lobbyists. Boston Globe, March 17, 1992, P.20.

Durbin, D.A. 2005. Hybrid Car Sales Soar in U.S. in 2004. Associated Press, April 25, 2005 [online]. Available: http://www.commondreams.org/headlines05/0425-03.htm [accessed July 12, 2005].

Ehlmann, J., and G. Wolff. 2005. Automobile emissions: The road toward zero. EM (Jan.): 33-37.

EMA (Engine Manufacturers Association). 2004. Supreme Court Ruling-EMA vs. SCAQMD. Engine Manufacturers Association [online]. Available: http://www.enginemanufacturers.org/news/whatsnew.asp [accessed Oct. 13, 2005].

ENVIRON. 2000. Evaluation of Attainment Control Strategies for the Dallas-Fort Worth State Implementation Plan. Final Report. Prepared for North Central Texas Council of Governments, Arlington, TX, by ENVIRON International Corp, E.H. Pechan & Associates, Springfield and Earth Matters, Jerome, AZ. October 2000.

EPA (U.S. Environmental Protection Agency). 1991. Nonroad Vehicle and Engine Emissions Study-Report. EPA460/3-91-02. Office of Air and Radiation, U.S. Environmental Protection Agency, Washington, DC. November 1991 [online]. Available: http://www.epa.gov/nonroad/nrstudy.pdf [accessed August 26, 2005].

EPA (U.S. Environmental Protection Agency). 1995. EPA, California Air Resources Board, and Manufacturers of Heavy-Duty Engines Sign "Statement of Principles". Environmental Fact Sheet. EPA420-F-95-010a. Office of Air and Radiation, Office of Mobile Sources, U.S. Environmental Protection Agency. July 1995 [online]. Available: http://www.epa.gov/otaq/regs/hd-hwy/noxsop2.txt [accessed August 26, 2005].

EPA (U.S. Environmental Protection Agency). 1996a. Regulatory Impact Analysis: Control of Air Pollution Emission Standards for New Nonroad Spark-Ignition Marine Engines. ANR-443. Office of Mobile Sources, Office of Air and Radiation, U.S. Environmental protection Agency, Washington, DC. June 1996 [online]. Available: http://www.epa.gov/otaq/regs/nonroad/marine/marnfria.pdf [accessed August 29, 2005].

EPA (U.S. Environmental Protection Agency). 1996b. Statement of Principles for the Regulation of Exhaust Emissions from Handheld Spark-Ignited Engines at or Below 19 Kilowatts. Office of Transportation and Air Quality, U.S. Environmental Protection Agency. May 9, 1996 [online]. Available: http://www.epa.gov/otaq/regs/nonroad/equip-ld/hhsop.txt. [accessed August 26, 2005].

EPA (U.S. Environmental Protection Agency). 1996c. Small Nonhandheld Spark-Ignited Nonroad Engine Statement of Principles. Office of Transportation and Air Quality, U.S. Environmental Protection Agency. December 1996 [online]. Available: http://www.epa.gov/otaq/regs/nonroad/equip-ld/nhh-sop.pdf [accessed August 26, 2005].

EPA (U.S. Environmental Protection Agency). 1997. Benefits and Costs of the Clean Air Act, 1970 to 1990. Final Report to U.S. Congress. EPA 410/R-97-002. Office of Air Quality Planning and Standards, U.S. Environmental Protection Agency, Research Triangle Park, NC [online]. Available: http://www.epa.gov/oar/sect812/contsetc.pdf [accessed Jan. 20, 2006].

EPA (U.S. Environmental Protection Agency). 1998. Tier 2 Report to Congress. EPA420-R-98-008. Office of Air and Radiation, U.S. Environmental Protection Agency [online]. Available: http://www.epa.gov/oms/tr2study.htm#Documents [accessed July 20, 2005].

EPA (U.S. Environmental Protection Agency). 1999a. Regulatory Impact Analysis: Control of Air Pollution from New Motor Vehicles: Tier 2 Motor Vehicle Emissions Standards and Gasoline Sulfur Control Requirements. EPA-420/D-99-001. Office of Air and Radiation and Office of Mobile Sources. U.S. Environmental Protection Agency, Ann Arbor, MI [online]. Available: http://www.epa.gov/otaq/regs/ld-hwy/tier-2/frm/ria/r99023.pdf [accessed July 25, 2005].

EPA (U.S. Environmental Protection Agency). 1999b. Final Tier 2 Rule: Air Quality Estimation, Selected Health and Welfare Benefits Methods, and Benefit Analysis Results. EPA420-R-99-032. Office of Air Quality Planning and Standards, U.S. Environmental Protection Agency, Research

Triangle Park, NC. December 1999 [online]. Available: http://www.epa.gov/otaq/regs/ld-hwy/tier-2/frm/tsd/r99032.pdf [accessed Jan. 27, 2006].

EPA (U.S. Environmental Protection Agency). 1999c. Final Regulatory Impact Analysis Phase 2: Emission Standards for New Nonroad Nonhandheld Spark-Ignition Engines At or Below 19 Kilowatts. EPA420-R-99-003. Office of Air and Radiation, U.S. Environmental Protection Agency. March 1999[online]. Available: http://www.epa.gov/otaq/regs/ nonroad/equip-ld/ph2frm/r99003.pdf [accessed August 26, 2005].

EPA (U.S. Environmental Protection Agency). 2000a. Draft Technical Support Document: Control of Emissions of Hazardous Air Pollutants from Motor Vehicles and Motor Vehicle Fuels. EPA-420/D-00-003. Assessment and Standards Division, Office of Transportation and Air Quality, U.S. Environmental Protection Agency [online]. Available: http://www.epa.gov/otaq/regs/toxics/d00003.pdf [accessed Jan. 20, 2006].

EPA (U.S. Environmental Protection Agency). 2000b. Guidelines for Preparing Economic Analysis. EPA240-R-00-003. National Center for Environmental Economics, U.S. Environmental Protection Agency [online]. Available: http://yosemite.epa.gov/ee/epa/eed.nsf/webpages/Guidelines.html [accessed January 6, 2006].

EPA (U.S. Environmental Protection Agency). 2000c. Regulatory Impact Analysis: Heavy-Duty Engine and Vehicle Standards and Highway Diesel Fuel Sulfur Control Requirements. EPA420-R-00-026. Office of Air and Radiation, U.S. Environmental Protection Agency. December 2000 [online]. Available: http://www.epa.gov/oms/diesel.htm [accessed August 29, 2005].

EPA (U.S. Environmental Protection Agency). 2000d. Final Regulatory Impact Analysis Phase 2 Final Rule: Emission Standards for New Nonroad Handheld Spark-Ignition Engines At or Below 19 Kilowatts. EPA420-R-00-004. Office of Transportation and Air Quality, Office of Air and Radiation, U.S. Environmental Protection Agency. March 2000 [online]. Available: http://www.epa.gov/otaq/regs/nonroad/equip-ld/hhsfrm/r00004.pdf [accessed August 29, 2005].

EPA (U.S. Environmental Protection Agency). 2001a. Response to API's Request for Reconsideration of the Phase 2 Final Rule (Heavy-Duty Engines and Trucks). Letter to Red Cavaney, President and CEO, American Petroleum Institute, Washington, DC, from Christine Todd Whitman, Office of Transportation and Air Quality, U.S. Environmental Protection Agency, Washington, DC. August 28, 2001 [online]. Available: http://www.epa.gov/otaq/regs/hd2007/frm/apirecon.pdf [accessed August 29, 2005].

EPA (U.S. Environmental Protection Agency). 2001b. Response to American Trucking Associations' Request for Reconsideration of the Phase 2 Final Rule (Heavy-Duty Engines and Trucks). Letter to Thomas Richichi , Beveridge & Diamond, PC, Washington, DC, from Christine Todd

Whitman, Office of Transportation and Air Quality, U.S. Environmental Protection Agency, Washington, DC. August 28, 2001 [online]. Available: http://www.epa.gov/otaq/regs/hd2007/frm/atarecon.pdf [accessed August 29, 2005].

EPA (U.S. Environmental Protection Agency). 2001c. Response to Mack Trucks'/Volvo Powertrain's Request for Reconsideration of the Phase 2 Final Rule (Heavy-Duty Engines and Trucks). Letter to Julie Domike, Wallace, King, Marraro & Branson, PLLC, Washington, DC, from Christine Todd Whitman, Office of Transportation and Air Quality, U.S. Environmental Protection Agency, Washington, DC. August 28, 2001 [online]. Available: http://www.epa.gov/otaq/regs/hd2007/frm/mack.pdf [accessed August 29, 2005].

EPA (U.S. Environmental Protection Agency). 2002a. 1,3-Butadiene (CASRN 106-99-0). Integrated Risk Information System, U.S. Environmental Protection Agency [online]. Available: http://www.epa.gov/iris/subst/0139.htm [accessed May 23, 2005].

EPA (U.S. Environmental Protection Agency). 2002b. Highway Diesel Progress Review. EPA420-R-02-016. Office of Transportation and Air Quality, Office of Air and Radiation, U.S. Environmental Protection Agency. June 2002 [online]. Available: http://www.epa.gov/otaq/regs/hd2007/420r02016.pdf [accessed August 29, 2005].

EPA (U.S. Environmental Protection Agency). 2003a. National Air Quality and Emissions Trends Report, 2003 Special Studies Edition. EPA 454/R-03-005. Air Quality Trends Analysis Group, Emissions Monitoring and Analysis Division, Office of Air Quality Planning and Standards, U.S. Environmental Protection Agency, Research Triangle Park, NC. September 2003 [online]. Available: http://www.epa.gov/air/airtrends/aqtrnd03/ [accessed May 20, 2005].

EPA (U.S. Environmental Protection Agency). 2003b. User's Guide to MOBILE 6.1 and 6.2 – Mobile Source Emissions Factor Model. EPA420-R-03-010. Office of Air and Radiation, U.S. Environmental Protection Agency. August 2003 [online]. Available: http://www.epa.gov/otaq/models/mobile6/420r03010.pdf [accessed May 20, 2005].

EPA (U.S. Environmental Protection Agency). 2003c. Diesel Engine Exhaust (CASRN N.A.). Integrated Risk Information System, U.S. Environmental Protection Agency [online]. Available: http://www.epa.gov/iris/subst/0642.htm [accessed May 24, 2005].

EPA (U.S. Environmental Protection Agency). 2003d. Notice of a Settlement Agreement with the Engine Manufacturers Association to Propose a Manufacturer-Run, In-Use Emissions Testing Program for Heavy-Duty Diesel Trucks and To Provide Engine Certification Guidance. EPA420-F-03-019. Office of Transportation and Air Quality, Office of Air and Radiation, U.S. Environmental Protection Agency. June 2003 [online].

Available: http://www.epa.gov/otaq/regs/hd-hwy/ncp/f03019.pdf [accessed August 29, 2005].

EPA (U.S. Environmental Protection Agency). 2004a. AirData: Generating Reports and Maps. Office of Air and Radiation, U.S. Environmental Protection Agency [online]. Available: http://www.epa.gov/air/data/geosel.html [accessed May 20, 2005].

EPA (U.S. Environmental Protection Agency). 2004b. The Particle Pollution Report: Current Understanding of Air Quality and Emissions through 2003. EPA454-R-04-002. Office of Air and Radiation, U.S. Environmental Protection Agency [online]. Available: http://www.epa.gov/airtrends/pm.html [accessed Jan. 20, 2006].

EPA (U.S. Environmental Protection Agency). 2004c. Nonroad Engine Population Estimates. EPA420-P-04-006. Office of Air and Radiation, U.S. Environmental Protection Agency [online]. Available: http://www.epa.gov/otaq/models/nonrdmdl/nonrdmdl2004/420p04006.pdf [accessed May 20, 2005].

EPA (U.S. Environmental Protection Agency). 2004d. Nitrogen Dioxide. Air Trends. Office of Air and Radiation, U.S. Environmental Protection Agency [online]. Available: http://www.epa.gov/airtrends/nitrogen.html [accessed May 23, 2005].

EPA (U.S. Environmental Protection Agency). 2004e. Final Regulatory Analysis: Control of Emissions from Nonroad Diesel Engines. EPA420-R-04-007. Office of Transportation and Air Quality, U.S. Environmental Protection Agency [online]. Available: http://www.epa.gov/nonroad-diesel/2004fr.htm#ria [accessed May 23, 2005].

EPA (U.S. Environmental Protection Agency). 2004f. EPA Office of Transportation and Air Quality, Comments on NESCAUM White Paper, March 25, 2004. Letter to Kenneth A. Colburn, Executive Director, NESCAUM, Boston, MA, from Chester France, Director, Assessment and Standards Division, National Vehicle and Fuel Emission Laboratory, U.S. Environmental Protection Agency, Ann Arbor, MI. March 26, 2004.

EPA (U.S. Environmental Protection Agency). 2004g. Highway Diesel Progress Review Report 2. EPA420-R-04-004. Office of Transportation and Air Quality, Office of Air and Radiation, U.S. Environmental Protection Agency. March 2004 [online]. Available: http://www.epa.gov/otaq/regs/hd2007/420r04004.pdf [accessed August 29, 2005].

EPA (U.S. Environmental Protection Agency). 2005a. Nonattainment Area Map. Office of Air Quality Planning and Standards, U.S. Environmental Protection Agency [online]. Available: http://www.epa.gov/oar/oaqps/greenbk/mapnpoll.html [accessed Oct. 20, 2005].

EPA (U.S. Environmental Protection Agency). 2005b. Particulate Matter Nonattainment Area Map. Office of Air Quality Planning and Standards, Office of Air and Radiation, U.S. Environmental Protection Agency [online]. Available: http://www.epa.gov/oar/oaqps/greenbk/mappm 25o3.html [accessed Nov 9, 2005].

EPA (U.S. Environmental Protection Agency). 2005c. 1970-2002 Average Annual Emission, All Criteria Pollutants, January 2005. Air Pollutant Emission Trends Data and Estimation Procedures, Clearinghouse for Inventories & Emission Factors, Technology Transfer Network, U.S. Environmental Protection Agency [online]. Available: http://www.epa.gov/ttn/chief/trends/index.html#tables [accessed May 24, 2005].

EPA (U.S. Environmental Protection Agency). 2005d. MOVES (Motor Vehicle Emission Simulator). Office of Transportation and Air Quality, U.S. Environmental Protection Agency [online]. Available: http://www.epa.gov/otaq/ngm.htm [accessed July 18, 2005].

EPA (U.S. Environmental Protection Agency). 2005e. Benzene. Air Toxics Website, Technology Transfer Network, U.S. Environmental Protection Agency [online]. Available: http://www.epa.gov/ttn/atw/hlthef/benzene.html [accessed May 23, 2005].

EPA (U.S. Environmental Protection Agency). 2005f. FY 2005 Annual Performance Plan. Budget Documents. Office of Chief Financial Officer, U.S. Environmental Protection Agency [online]. Available: http://www.epa.gov/ocfopage/budget/2005/2005ap/2005ap.htm [accessed Oct. 14, 2005].

EPA (U.S. Environmental Protection Agency). 2006. NONROAD Model (Nonroad Engines, Equipment, and Vehicles). Office of Transportation and Air Quality, U.S. Environmental Protection Agency [online]. Available: http://www.epa.gov/oms/nonrdmdl.htm [accessed Jan. 15, 2006].

Faiz, A., C.S. Weaver, and M.P. Walsh. 1996. Air Pollution from Motor Vehicles: Standards and Technologies for Controlling Emissions. Washington, DC: World Bank.

FHWA (Federal Highway Administration). 2005. Highway Statistics 2003. Federal Highway Administration, U.S. Department of Transportation [online]. Available: http://www.fhwa.dot.gov/policy/ohim/hs03/index.htm [accessed May 18, 2005].

Finlayson-Pitts, B.J., and J.N. Pitts. 1999. Chemistry of the Upper and Lower Atmosphere: Theory, Experiments, and Applications. San Diego: Academic Press.

Fletcher, G. 2004. A Taste of the engines to come: New Chrysler motors more powerful, efficient, while costing less. National Post, Dec. 31, 2004, DT:13.

Frampton, M.W., W. Zareba, C.C. Daigle, G. Oberdörster, and M.J. Utell. 2002. Inhalation of ultrafine particles alters myocardial repolarization in humans [abstract]. Am. J. Respir. Crit. Care Med. 165:B16.

France, C. 2005. EPA's Process for Developing Mobile Source Emissions Standards Regulations. Presentation at the Third Meeting on State

Practices in Setting Mobile Source Emissions Standards, January 19, 2005, Dearborn, MI.

French, T. 2004. The Critical Role of Federal Preemption. Presentation by Engine Manufacturers Association at the Second Meeting on State Practices in Setting Mobile Source Emissions Standards, October 4, 2004, Irvine, CA.

Fujita, E.M., B.E. Croes, C.L. Bennett, D.R. Lawson, F.W. Lurmann, H.H. Main. 1992. Comparison of emission inventory and ambient concentration ratios of CO, NMOG, and NOx in California's South Coast Air Basin. J. Air Waste Manage. Assoc. 42(3):264-276.

Fujita, E.M., J.G. Watson, J.C. Chow, N.F. Robinson, L.W. Richards, and N. Kumar. 1998. Northern Front Range Air Quality Study, Volume C: Source Apportionment and Simulation Methods and Evaluation. Final Report. Colorado State University, Fort Collins, CO.

Fujita, E.M., W.R. Stockwell, D.E. Campbell, R.E. Keislar, and D.R. Lawson. 2003. Evolution of the magnitude and spatial extent of the weekend ozone effect in California's South Coast Air Basin, 1981-2000. J. Air Waste Manage. Assoc. 53(7):802-815.

GAO (Government Accountability Office). 2004. EPA Could Take Additional Steps to Help Maximize Benefits form the 2007 Diesel Emissions Standards. GAO 04-313. Washington, DC: GAO. March 2004 [online]. Available: http://www.gao.gov/new.items/d04313.pdf [accessed August 29, 2005].

Gerard, D., and L.B. Lave. 2002. Technology-Forcing Automobile Emissions Provisions of the 1970 Clean Air Act: Forcing Invention or Innovation? Presentation at the Center for Integrated Study of the Human Dimensions of Global Change, November 6, 2002, Carnegie Mellon University, Pittsburgh, PA.

Gerard, D. and L.B. Lave. 2005. Implementing technology-forcing policies: The 1970 Clean Air Act Amendments and the introduction of advanced automotive emissions controls in the United States. Technol. Forecast. Soc. Change 72(7):761-778.

Ghio, A.J., C. Kim, and R.B. Devlin. 2000. Concentrated ambient air particles induce mild pulmonary inflammation healthy human volunteers. Am. J. Respir. Crit. Care Med. 162(3Pt.1):981-988.

Gillies, J.A., and A.W. Gertler. 2000. Comparison and evaluation of chemically speciated mobile source $PM_{2.5}$ particulate matter profiles. J. Air Waste Manage. Assoc. 50(8):1459-1480.

GM (General Motors Corporation). 2005. EV1: Lessons Learned. Hybrids, EV1 Program [online]. Available: http://www.gm.com/company/gmability/adv_tech/300_hybrids/hyb_ev1.html [accessed July 19, 2005].

Goodstein, E., and H. Hodges. 1997. Behind the numbers: Polluted data. The American Prospect 35:64-69.

Grimm, R.A., R.J. Bremer, and S.P. Stonestreet. 1980. GM Micro-Computer Engine Control System. SAE 800053. Society of Automotive Engineers, Warrendale, PA.

Gruenspecht, H. 2002. Regulatory Tailoring, Reliability, and Price Volatility with Stochastic Breakdowns. RFF Discussion Paper 02-37. Resources for the Future, Washington, DC. September [online]. Available: http://www.rff.org/Documents/RFF-DP-02-37.pdf [accessed July 28, 2005].

Guerry, W.M. 2004. State Practices in Setting Mobile Source Emission Standards. Presentation by OPEI (Outdoor Power Equipment Research Institute) at the First Meeting on State Practices in Setting Mobile Source Emissions Standards, June 28th, 2004, Washington DC.

Gumbleton, J.J., and L.L. Bowler. 1982. General Motors' Computer Command Control-System Development. SAE 820901. Society of Automotive Engineers, Warrendale, PA.

Guzy, G.S., and R. Perciasepe. 1999. Opinion on Issues Raised by AAMA v. Massachusetts Dep. Letter to Thomas F. Reilly, Attorney General, Commonwealth of Massachusetts, Boston, MA, from Gary S. Guzy, General Counsel, and Robert Perciasepe, Assistant Administrator, Office of Air and Radiation, U.S. Environmental Protection Agency, Washington, DC. Sept. 15, 1999.

Haagen-Smith, A.J., and M.M. Fox. 1954. Photochemical ozone formation with hydrocarbons and automobile exhaust. J. Air Pollut. Control Assoc. 4:105-109.

Hanz, P., and P. Hotz. 2005. Manufacturer's Perspective on State Standards in a Global Marketplace. Presentation by Briggs and Stratton at the Fourth Meeting on State Practices in Setting Mobile Source Emissions Standards, April 13-15, 2005, Boston, MA.

Harrington, W., V. McConnell, and M. Walls. 1996. Who's in the Driver's Seat? Mobile Source Policy in the Federal System. Discussion Paper 96-34. Resources for the Future, Washington, DC. September 1996 [online]. Available: http://www.rff.org/Documents/RFF-DP-96-34.pdf [accessed July 28, 2005].

Harrington, W., R.D. Morgenstern, and P. Nelson. 2000. On the accuracy of regulatory cost estimates. J. Policy Anal. Manage. 19(2):297-322.

Heck, R.M., and R.J. Farrauto. 2001. Automobile exhaust catalyst. Appl. Catal. A 221(ER 1-2):443-457.

HEI (Health Effects Institute). 2002. Research Directions to Improve Estimates of Human Exposure and Risk from Diesel Exhaust: A Special Report of the Institute's Diesel Epidemiology Working Group. Health Effects Institute. April 2002 [online]. Available: http://www.healtheffects.org/Pubs/DieselSpecialReport02.pdf [accessed May 18, 2005].

Heilprin, J. 2004. Trucking Industry Won't Fight Diesel. AP, November 17, 2004.

Heuss, J.M., D.F. Kahlbaum, and G.T. Wolff. 2003. Weekday/weekend ozone differences: What can we learn from them? J. Air Waste Manage. Assoc. 53(7):772-788.

Heywood, J.B. 1988. Pp. 567-667 in Internal Combustion Engine Fundamentals. New York: McGraw-Hill.

Hira, R., and H. Goldstein. 2005. Automakers top the list, drug companies trim back, and telecoms continue to gut their research budgets. R& D 100. IEEE Spectrum December 2005 [online]. Available: http://www.spectrum.ieee.org/dec05/2395 [accessed Jan. 25, 2006].

Holmes, K.J., and A.G. Russell. 2004. Why carbon monoxide still matters. Environ. Sci. Technol. 38(15):288A-294A.

IARC (International Association of Cancer Research). 1973. Benzo(a)Pyrene. Pp. 91-137 in Certain Polycyclic Aromatic Hydrocarbons and Heterocyclic Compounds. IARC Monographs on the Evaluation of Carcinogenic Risks of the Chemical to Man Vol. 3. Lyon, France: World Health Organization.

IARC (International Agency for Research on Cancer). 1989. Diesel and gasoline engine exhausts. Pp. 41-185 in Diesel and Gasoline Engine Exhausts and Some Nitroarenes. IARC Monographs on the Evaluation of Carcinogenic Risks to Humans Vol. 46. Lyon, France: World Health Organization [online]. Available: http://monographs.iarc.fr/htdocs/indexes/vol46index.html [accessed May 19, 2005].

Ingalls, M.N. 1989. On-road Vehicle Emission Factors from Measurements in a Los Angeles Area Tunnel. Paper 89-1373. Presented at the 82nd Annual Meeting of Air and Waste Management Association, Anaheim, CA.

Johnson, J.H, ed. 2002a. Diesel Particulate Emissions: Landmark Research, 1994-2001. PT-86. Warrendale, PA: Society of Automotive Engineers.

Johnson, J.H, ed. 2002b. Diesel Nitrogen Oxide Emissions: Landmark Research 1995-2001. PT-89. Warrendale, PA: Society of Automotive Engineers.

Johnson, J.H. 2005a. Emission Control and Fuel Economy for Port and Direct Injected SI Engines. PT-91. Warrendale, PA: Society of Automotive Engineers.

Johnson, J.H. 2005b. Combustion and Emission Control for SI Engines: Modeling and Experimental Studies. PT-121. Warrendale, PA: Society of Automotive Engineers.

Jolish, T.L. 1999. Negotiating the smog away. VA Environ. Law J. 18(3):305-373.

Jorgensen, R. 2005. Perspective of Cummins on California Emission Standards. Presentation by Cummins Engine Company at the Third Meeting on State Practices in Setting Mobile Source Emissions Standards, January 19, 2005, Dearborn, MI.

Kalhammer, F., A. Kozawa, C.B. Moyer, and B.B. Owen. 1995. Performance and Availability of Batteries for Electric Vehicles: A Report of the Battery

Technical Advisory Panel. Prepared for the California Air Resources Board, El Monte, CA. December 11, 1995.

Kim, T., W. J. Lee, J.S. Yoon, and C. K. Park. 2001. An exploratory study of THC emission reduction technologies compliant with SULEV regulations. Int. J. Automot. Technol. 2(2):63-75 [online]. Available: http://society.kisti.re.kr/~Eksae/_notes/data/pdf/v2n2_4.pdf [accessed Jan. 25, 2006].

Krier, J.E., and E. Ursin. 1977. Pollution and Policy: A Case Essay on California and Federal Experience with Motor Vehicle Air Pollution, 1940-1975. Berkeley, CA: University of California Press.

Kyocera. 2006. Oxygen Sensor Heaters. Environment-Friendly Technology [online]. Available: http://global.kyocera.com/application/automotive/product/compo/oxygen.html [accessed July 11, 2005].

Lawson, D.R. 1993. Passing the test – Human behavior and California's Smog Check Program. J. Air Waste Manage. Assoc. 43(12):1567-1575.

Lawson, D.R. 2003. The weekend ozone effect – The Weekly Ambient Emissions Control Experiment. EM (July):17-26.

Lents, J., K. Barnes, and N. Nikkila. 2000. The regulation of automobile emission: A case study. Chapter II in Environmental Regulation and Technology Innovation: Controlling Mercury Emissions from Coal-Fired Boilers, M. Tatsutani, and P. Amar, eds. Boston: Northeast States for Coordinated Air Use Management [online]. Available: http://www.nescaum.org/airtopics/mercury.html [accessed July 12, 2005].

Levin, R.L. 1995. Direct final rulemaking. Geo. Wash. L. Rev. 64: 1-34.

Lloyd, A.C., and T.A. Cackette. 2001. Diesel engines: Environmental impact and control. J. Air Waste Manage. Assoc. 51(6):809-847.

Los Angeles Times. 1947. 'Times' Expert Offers Smog Plan. Los Angeles Times, Final Ed., Jan. 19, 1947.

MACTEC. 2005. Midwest Regional Planning Organization (RPO), Identification and Evaluation of Candidate Control Measures." Prepared for The Lake Michigan Air Directors Consortium, by MASTEC Federal Programs/MASTEC Engineering and Consulting, Inc. April 14, 2005 [online]. Available: http://www.ladco.org/reports/rpo/MWRPOprojects/Strategies/Final%20Control%20Measures.pdf [accessed Jan. 21, 2006].

MA DEP (Massachusetts Department of Environmental Protection). 1999. Background Document and Technical Support for: Public Hearings on the Proposed Amendments to the State Implementation Plan for Ozone; and Public Hearing and Finding under the Massachusetts Low Emission Vehicle Statute. 310 CMR 7.40: The Massachusetts Low Emission Vehicle Program. Bureau of Waste Prevention, Division of Consumer and Transportation Programs, Department of Environmental Protection, Executive Office of Environmental Affairs, State of Massachusetts. October 1999.

MA DEP (Massachusetts Department of Environmental Protection). 2002. Technical Support Document for Public Hearings on Proposed Revisions to the Massachusetts State Implementation Plan for Ozone, Amendments to the: Reasonable Further Progress Plan for Eastern Massachusetts and Statewide Projected Inventory for Nitrogen Oxides in 2007. Bureau of Waste Prevention, Division of Planning and Evaluation, Department of Environmental Protection, Executive Office of Environmental Affairs, Commonwealth of Massachusetts. January 2002 [online]. Available: http://www.mass.gov/dep/bwp/daqc/daqcpubs.htm [accessed June 3, 2005].

Manson, C.J., M.B. Nelson, and J.E. Neumann. 2002. Assessing the Impact of Progress and Learning Curves on Clean Air Act Compliance Costs. Industrial Economics, Incorporated, Cambridge, MA. July 12, 2002.

Mark, J. 1999. Greener SUVs: A Blueprint for Cleaner, More Efficient Light Trucks. Union of Concerned Scientists. July 1999 [online]. Available: http://www.ucsusa.org/publications/ [accessed Jan. 25, 2006].

Marr, L.C., G.C. Morrison, W.W. Nazaroff, and R.A. Harley. 1998. Reducing the risk of accidental death due to vehicle-related carbon monoxide poisoning. J. Air Waste Manage. Assoc. 48(10):899-906.

Maxcy, G., and A. Silberston. 1959. The Motor Industry. London: Allen & Unwin.

McCarthy, G. 2005. States Perspectives on Adaptation of California Emissions Standards. Testimony of Gina McCarthy, Commissioner, Connecticut Department of Environmental Protection at the Fourth Meeting on State Practices in Setting Mobile Source Emissions Standards, April 15th, 2005, Boston, MA.

McKinnon, D.L., C.C. Webb, P.A. Weber, and B.B. Bykowski. 1999. Using Advanced Emission Control Systems to Demonstrate LEV II ULEV on Light-Duty Gasoline Engines. SAE Technical Papers 1999-01-0774. Warrendale, PA: Society of Automotive Engineers.

MECA (Manufacturers of Emissions Controls Association). 1997. Emissions Control of Diesel-Fueled Vehicles. Manufacturers of Emissions Controls Association, Washington, DC. March 1997 [online]. Available: http://meca.org/jahia/Jahia/cache/offonce/pid/229;jsessionid=507563C2B58931D5F172918265E828D8 [accessed July 11, 2005].

MECA (Manufacturers of Emissions Control Equipment). 1998. The Impact of Gasoline Sulfur on Catalytic Emission Control Systems. Manufacturers of Emissions Control Equipment, Washington, DC. September 1998 [online]. Available: http://meca.org/jahia/Jahia/cache/offonce/pid/229;jsessionid=507563C2B58931D5F172918265E828D8 [accessed July 11, 2005].

MECA (Manufacturers of Emissions Control Equipment). 2003. Tier 2/LEV II Emission Control Technologies For Light-Duty Gasoline Vehicles. Manufacturers of Emissions Control Equipment, Washington, DC [online]. Available: http://www.meca.org/galleries/default-file/Tier2813final_a.pdf. [accessed July 11, 2005].

ME DEP (Maine Department of Environmental Protection). 2004. Provisional Adaptation of Proposed Amendments to Chapter 127, New Motor Vehicle

Emission Standards Regulation (Zero Emission Vehicle Program). Memorandum to Board of Environmental Protection from Ronald Severance, Bureau of Air Quality, Maine Department of Environmental Protection, Augusta, ME. December 2, 2004 [online]. Available: http://www.state.me.us/dep/air/regulations/127zevadartmem.doc [accessed July 21, 2005].

Metzger, K.B., P.E. Tolbert, M. Klein, J.L. Peel, W.D. Flanders, K. Todd, J.A. Mulholland, P.B. Ryan, and H. Frumkin. 2004. Ambient air pollution and cardiovascular emergency department visits. Epidemiology 15(1):46-56.

Mondt, J.R. 2000. Clean Cars: The History and Technology of Emission Control Since the 1960s. Warrendale, PA: Society of Automotive Engineers.

Morris, R.D., E.N. Naumova, and R.L. Munasinghe. 1995. Ambient air pollution and hospitalization for congestive heart failure among elderly people in seven large US cities. Am. J. Public Health 85(10):1361-1365.

Mott, J.A., M.I. Wolfe, C.J. Alverson, S.C. MacDonald, C.R. Bailey, L.B. Ball, J.E. Moorman, J.H. Somers, D.M. Mannimo, and S.C. Redd. 2002. National vehicle emissions policies and practices and declining US carbon monoxide-related mortality. JAMA 288(8):988-995.

Muskie, E.S. 1990. NEPA to CERCLA The Clean Air Act: A commitment to public health. Environ. Forum 7(1):13-17 [online]. Available: http://www.cleanairtrust.org/nepa2cercla.html [accessed July 11, 2005].

MWCG (Metropolitan Washington Council of Governments). 2003. Plan to Improve Air Quality in the Washington, DC-MD-VA Region. State Implementation Plan (SIP) " Severe Area SIP". Prepared by Metropolitan Washington Council of Governments for the District of Columbia Department of Health, Maryland Department of the Environment and the Virginia Department of Environmental Quality. December 17, 2003 [online]. Available: http://textonly.mde.state.md.us/assets/document/CompleteSIP_Dec17.pdf [accessed June 3, 2005].

NARSTO. 2000. An Assessment of Tropospheric Ozone Pollution—A North American Perspective. Palo Alto, CA: EPRI.

NARSTO. 2004. Particulate Matter Assessment for Policy Makers: A NARSTO Assessment, P. McMurry, M. Shepherd, and J. Vickery, eds. Cambridge, England: Cambridge University Press.

NCAPC (U.S. National Center for Air Pollution Control). 1967. A Digest of State Air Pollution Laws. Public Health Service Pub. 101. Washington, DC: Division of Air Pollution, U.S. Public Health Service.

NERA (National Economic Research Associates). 2004. Re: Methodology for Estimating Potential Job Losses from Briggs & Stratton Relocation Off-shore. Memo to The National Academies of Sciences, Washington, DC, from National Economic Research Associates, Cambridge, MA. October 22, 2004.

NESCAUM (Northeast States for Coordinated Air Use Management). 2003. White Paper Comparing the Emissions Reductions of the LEV II Program to the Tier 2 Program. Northeast States for Coordinated Air Use Management, Boston, MA, and Cambridge Systematics, Inc., Cambridge, MA. October 2003 [online]. Available: http://www.nescaum.org/committees/mobile/ [accessed July 19, 2005].

NESCAUM (Northeast States for Coordinated Air Use Management). 2004. Re: LEV II v. Tier 2. Letter to Chester France, Director, Assessment and Standards Division, National Vehicle and Fuel Emission Laboratory, U.S. Environmental Protection Agency, Ann Arbor, MI, from Kenneth A. Colburn, Executive Director, NESCAUM, Boston, MA. July 29, 2004.

NESCAUM (Northeast States for Coordinated Air Use Management). 2005. Final Results of Comparing the Emissions Reductions of the LEV II Program to the Tier 2 Program. Letter to John Holmes, National Research Council, DC, from Arthur N. Martin, Executive Director, NESCAUM, Boston, MA. May 24, 2005.

NESCCAF (Northeast States Center for a Clean Air Future). 2004. Reducing Greenhouse Gas Emissions from Light-Duty Motor Vehicles. September 2004 [online]. Available: http://bronze.nescaum.org/committees/mobile/rpt040923ghglightduty.pdf [accessed Jan. 25, 2006].

Norton, T., S. Tucker, R.E. Smith, and D.R. Lawson. 1998. The Northern Front Range Air Quality Study. EM(January):13-19.

NRC (National Research Council). 1991. Rethinking the Ozone Problem in Urban and Regional Air Pollution. Washington DC: National Academy Press.

NRC (National Research Council). 1999. Ozone-Forming Potential of reformulated Gasoline. Washington DC: National Academy Press.

NRC (National Research Council). 2000. Modeling Mobile Source Emissions. Washington DC: National Academy Press.

NRC (National Research Council). 2001. Evaluating Vehicle Emissions Inspection and Maintenance Programs. Washington DC: National Academy Press.

NRC (National Research Council). 2002a. The Ongoing Challenge of Managing Carbon Monoxide Pollution in Fairbanks, Alaska. Interim Report. Washington, DC: National Academy Press.

NRC (National Research Council). 2002b. Estimating the Public Health Benefits of Proposed Air Pollution Regulations. Washington DC: National Academies Press.

NRC (National Research Council). 2003. Managing Carbon Monoxide Pollution in Meteorological and Topographical Problem Areas. Washington DC: National Academies Press.

NRC (National Research Council). 2004. Air Quality Management in the United States. Washington DC: National Academies Press.

NYDEC (New York Department of Environmental Conservation). 1992 Amendments to 6 NYCRR Part 218 Emission Standards for Motor Vehicles and Motor Vehicle Engines: Assessment of Public Comments. February 1992.

NYDEC (New York Department of Environmental Conservation). 2000. 6 NYCRR Part 218 Regulatory Impact Statement. December 2000.

NYDEC (New York Department of Environmental Conservation). 2003. 6 NYCRR Part 210 Regulatory Impact Statement. Adapted August 8, 2003.

Office of the Governor Washington. 2004. Governor Gary Locke Announces Strategy to Combat Global Warming. Press Release, December 8, 2004. Governor Locke's Communications Office, Olympia, WA [online]. Available: http://www.ecy.wa.gov/programs/air/globalwarming/Govs_press release_cars04.pdf [accessed July 20, 2005].

OMB (Office of Management and Budget). 2003. Regulatory Analysis. OMB Circular A-4. September 17, 2003.

OMB (Office of Management and Budget). 2004. 2004 Draft Report to Congress on the Costs and Benefits of Federal Regulations and Unfunded Mandates on State, Local, and Tribal Entities. Office of Information and Regulatory Affairs, Office of Management and Budget, Washington DC [online]. Available: http://www.whitehouse.gov/omb/inforeg/draft_2004_cbreport.pdf [accessed March 14, 2005].

Omnitek (Omnitek Engineering Corporation). 2006. Selective Catalytic Reduction (SCR) System. Emissions Control Technologies and Alternative Fuel Engines: Heavy Duty Trucks, Buses, Off-Road Equipment and Industrial Engines. Omnitek Engineering Corporation, San Marcos, CA [online]. Available: http://www.omnitekcorp.com/scr.htm [accessed April 11, 2006].

ORNL (Oak Ridge National Laboratory). 2004. Table 4.9 Light Vehicle Market Shares by Size Class, Sales Periods 1976-2003. Light Vehicle MPG and Market Shares System, Oak Ridge National Laboratory, Oak Ridge, TN [online]. Available: http://cta.ornl.gov/data/tedb24/Spreadsheets/Table4_09.xls [accessed May 19, 2005].

OTA (Office of Technology Assessment). 1995. Gauging Control Technology and Regulatory Impacts in Occupational Safety and Health: An Appraisal of OSHA's Analytic Approach. OTA-ENV-635. Washington, DC: U.S. Government Printing Office. September 1995 [online]. Available: http://www.dau.mil/educdept/mm_dept_resources/reports/OTA-Gauging-control-tech-and-impact-on-OSHA.pdf [accessed Jan. 24, 2006].

Park, K., C. Feng, D.B. Kittelson, and P.H. McMurry. 2003. Relationship between particle mass and mobility of diesel exhaust particles. Environ. Sci. Technol. 37(3):577-583.

Pechan/EEA, Inc. (E.H. Pechan & Associates, Inc./ Energy & Environmental Analysis, Inc.). 1991. Adapting the California Low Emission Vehicle Program in the Northeast States-An Evaluation. Prepared for Northeast States for Coordinated Air Use Management, Boston, MA, by E.H.

Pechan & Associates, Inc., Springfield, VA, and Energy & Environmental Analysis, Inc., Arlington, VA. September 1991.

Pierson, W.R., A.W. Gertler, and R.L. Bradow. 1990. Comparison of the SCAQS Tunnel Study with other on-road vehicle emissions data. J. Air Waste Manage. Assoc. 40(11):1495-1504.

Pierson, W.R., D.E. Schorran, E.M. Fujita, J.C. Sagebiel, D.R. Lawson, and R.L. Tanner. 1999. Assessment of nontailpipe hydrocarbon emissions from motor vehicles. J. Air Waste Manage. Assoc. 49(5):498-519.

Pokharel, S.S., G.A. Bishop, and D.H. Stedman. 2000. On-Road Remote Sensing of Automobile Emissions in the Chicago Area: Year 3. Prepared for Coordinating Research Council, Inc., Alpharetta, GA, by Department of Chemistry and Biochemistry, University of Denver, Denver, CO. April 2000 [online]. Available: http://www.feat.biochem.du.edu/assets/databases/Illinois/Arlhghts/Chicago_Year_3_CRC99.pdf [accessed Jan. 26, 2006].

Pokharel, S.S., G.A. Bishop, and D.H. Stedman. 2002. On-Road Remote Sensing of Automobile Emissions in the Chicago Area: Year 3. Prepared for Coordinating Research Council, Inc., Alpharetta, GA, by Department of Chemistry and Biochemistry, University of Denver, Denver, CO. January 2002 [online]. Available: http://www.feat.biochem.du.edu/assets/databases/Colorado/6th_I25/Denver_Year_3_CRC01.pdf [accessed Jan. 26, 2006].

Pokharel, S.S., G.A. Bishop, D.H. Stedman, and R. Slott. 2003. Emissions reductions as a result of automobile improvement. Environ. Sci. Technol. 37(22):5097-5101.

Polakovic, G. 2002. State Takes Sharp Turn on Emissions; Cars: With Electric Vehicles Still Impractical, Hybrids and Gasoline Engines Are Showing Unexpected Progress. L.A. Times, Sept. 15, 2002, Part 2, p. B1.

Pope, C.A., III, D.W. Dockery, J.D. Spengler, and M.E. Raizenne. 1991. Respiratory health and PM10 pollution. A daily time series analysis. Am. Rev. Respir. Dis. 144(3 Pt.1):668-674.

Pope, C.A., III, R.T. Burnett, M.J. Thun, E.E. Calle, D. Krewski, K. Ito, and G.D. Thurston. 2002. Lung cancer, cardiopulmonary mortality, and long-term exposure to fine particulate air pollution. JAMA 287(9):1132-1141.

Popp, P.J., S.S. Pokharel, G.A. Bishop, and D.H. Stedman. 1999. On-Road Remote Sensing of Automobile Emissions in the Denver Area: Year 1. CRC Project No. E-23-4-99. Prepared for Coordinating Research Council, Inc., Atlanta, GA, by Department of Chemistry and Biochemistry, University of Denver, Denver, CO [online]. Available: http://www.crcao.com [accessed July 12, 2005].

Pun, B.K., C. Seigneur, and W. White. 2003. Day-of-week behavior of atmospheric ozone in three U.S. cities. J. Air Waste Manage. Assoc. 53(7):789-801.

Reilly, W.K. 1991. Re: Questions from April 9, 1991 Letter from Chairman Dingell to Administrator Reilly, and EPA Responses. Letter to John D. Dingell, Chairman, Subcommittee on Oversight and Investigation Committee on Energy and Commerce, U.S. House of Representative, Washington, DC, from William K. Reilly, EPA Administrator, U.S. Environmental Protection Agency, Washington, DC. July 29, 1991.

Reilly, W.K. 1993. Letter to Thomas C. Jorling, Commissioner, New York State Department for Environmental Conservation, Albany, NY, from William K. Reilly, EPA Administrator, U.S. Environmental Protection Agency, Washington, DC. Jan. 7, 1993.

RI DEM (Rhode Island Department of Environmental Management). 2004. "Re: Proposed Amendments to Air Pollution Control Regulation No. 37, entitled "Rhode Island's Low Emission Vehicle Program", Decision." Office of Air Resources, Department of Environmental Management, State of Rhode Island [online]. Available: http://www.dem.ri.gov/pubs/regs/regs/air/levresp.pdf [accessed July 20, 2005].

Ritz, B., F. Yu, S. Fruin, G. Chapa, G.M. Shaw, and J.A. Harris. 2002. Ambient air pollution and risk of birth defects in Southern California. Am. J. Epidemiol. 155(1):17-25.

Rubin, E.S. 2005. The Government Role in Technology Innovation: Lessons for the Climate Change Policy Agenda. Presentation at the 10th Biennial Conference on Transportation Energy and Environmental Policy: Toward a Policy Agenda for Climate Change, August 24, 2005, Pacific Grove, CA [online]. Available: http://www.its.ucdavis.edu/events/outreachevents/asilomar2005/presentations/Rubin.pdf [accessed Jan. 25, 2006].

Sakurai, H., K. Park, P.H. McMurray, D.D. Zarling, D.D. Kittelson, and P.J. Ziemann. 2003. Size-dependent mixing characteristics of volatile and non-volatile components in diesel exhaust aerosols. Environ. Sci. Technol. 37 (24):5487-5495.

Samet, J.M., F. Dominici, F.C. Curriero, I. Coursac, and S.L. Zeger. 2000. Fine particulate air pollution and mortality in 20 U.S. cities, 1987-1994. N. Engl. J. Med. 343(24):1742-1749.

SCAQMP (South Coast Air Quality Management Plan). 2003. Final 2003 AQMP Appendix III, Base and Future Year Emission Inventories in 2003 Air Quality Management Plan. August 2003 [online]. Available: http://www.aqmd.gov/aqmp/docs/2003AQMP_AppIII.pdf [accessed May 31, 2005].

SCAQMD (South Coast Air Quality Management District). 2003. SCAQMD 2003 Air Quality Management Plan [online]. Available: http://www.aqmd.gov/aqmp/AQMD03AQMP.htm. [accessed July 20, 2005].

SCAQMD (South Coast Air Quality Management District). 2004. FY 2004-2005 AQMD Budget and Work Program. Governing Board Meeting June 4, 2004. Agenda No. 39 [online]. Available: http://www.aqmd.gov/hb/2004/040639a.html [accessed Oct. 14, 2005].

Schauer, J.J., W.F. Rogge, M.A. Mazurek, L.M. Hildemann, G.R. Cass, and B.R. Simoneit. 1996. Source apportionment of airborne particulate matter using organic compounds as tracers. Atmos. Environ. 30(22):3837-3855.

Schauer, J.J., M.J. Kleeman, G.R. Cass, and B.R.T. Simoneit. 2002. Measurement of emissions from air pollution sources. 5. C_1-C_{32} organic compounds from gasoline-powered motor vehicles. Environ. Sci. Technol. 36(6):1169-1180.

Schwartz, J., and R. Morris. 1995. Air pollution and hospital admissions for cardiovascular disease in Detroit, Michigan. Am. J. Epidemiol. 142(1):23-35.

Seagrave, J., J.D. McDonald, A.P. Gigliotti, K.J. Nikula, S.K. Seilkop, M. Gurevich, and J.L. Mauderly. 2002. Mutagenicity and in vivo toxicity of combined particulate and semivolatile organic fractions of gasoline and diesel engine emissions. Toxicol. Sci. 70(2):212-226.

Seinfeld, J.H., and S.N. Pandis. 1998. Atmospheric Chemistry and Physics From Air Pollution to Climate Change. New York: Wiley.

Shah, S.D., D.R. Cocker, III, J.W. Miller, and J.M. Norbeck. 2004. Emission rates of particulate matter and elemental and organic carbon from in-use diesel engines. Environ. Sci. Technol. 38(9):2544-2550.

Shelef, M. 1994. Unanticipated benefits of automotive emissions control: Reduction in fatalities by motor vehicle exhaust gas. Sci. Total Environ. 146/147:93-101.

Sierra Research. 1989. An Evaluation of Adapting the California Mobile Source Control Program in the Eight Northeast States. Prepared for Northeast States for Coordinated Air Use Management, Boston, MA by Sierra Research, Inc., Sacramento, CA. April 1989.

Sierra Research. 1994. The Cost-Effectiveness of Further Regulating Mobile Source Emissions. Prepared for the American Automobile Manufacturers Association by Sierra Research, Inc., Sacramento. February 1994.

Sierra Research. 1999. Cold Start, Plug-in and Mid-Trip Idling CO Emissions from Light-Duty Gasoline-Powered Vehicles in Alaska. Report No. SR99-06-01. Prepared for the Alaska Department of Environmental Conservation, by Sierra Research, Inc., Sacramento, CA.

Sierra Research. 2000. Analysis of Alaska Vehicle CO Emission Study Data. Report Number SR00-02-01. Prepared for the Municipality of Anchorage, by Sierra Research, Inc., Sacramento, CA.

Sierra Research. 2004. Review of the August 2004 Proposed CARB Regulations to Control Greenhouse Gas Emissions from Motor Vehicles: Cost Effectiveness for the Vehicle Owner or Operator; Appendix C to the Comments of The Alliance of Automobile Manufacturers. Report No. SR2004-09-04. September 22, 2004. Sierra Research, Inc., Sacramento, CA [online]. Available: http://www.deq.state.or.us/aq/aqplanning/orlev/docs/SR20040904AppendixC_AAM.pdf [accessed Feb. 22, 2006].

SMAQMD (Sacramento Metropolitan Air Quality Management District). 2004. Clean Air Plan Update Potential Control Measure Overview [online].

Available: http://www.airquality.org/cleanairplan/ws0306/ws0306Oview.shtml [accessed Jan. 26, 2005].

Smaling, R.M., S. Sung, and R. Bartlett. 1996. Washcoat Technology and Precious Metal Loading Study Targeting the California LEV MDV2 Standard. SAE 961904. Warrendale, PA: Society of Automotive Engineers.

Snyder, T.C. 2005. States Perspectives on Adoption of California Emissions Standards. Presentation at the Fourth Meeting of on State Practices in Setting Mobile Source Emissions Standards, April 15th, 2005, Boston, MA.

Sperling, D., D. Bunch, A. Burke, E. Abeles, B. Chen, K. Kurani, and T. Turrentine. 2004a. Analysis of Auto Industry and Consumer Response to Regulations and Technological Change, and Customization of Consumer Response Models in Support of AB 1493 Rulemaking. Prepared for the California Air Resources Board and the California Environmental Protection Agency, by Institute of Transportation Studies, University of California, Davis, CA. June 2004 [online]. Available: http://www.its.ucdavis.edu/publications/ 2004/UCD-ITS-RR-04-17.pdf [accessed July 11, 2005].

Sperling, D., E. Abeles, D. Bunch, A. Burke, B. Chen, K. Kurani, and T. Turrentine. 2004b. The price of regulation. Access 25(Fall):9-18.

Squitieri, R. 1998. Do We Always Overestimate Environmental Control Costs? Council of Economic Advisors, Washington, DC.

STAPPA/ALAPCO (State and Territorial Air Pollution Program Administrators and Association of Local Air Pollution Control Officials). 2001. Regulating Air Pollution from Diesel Trucks: A Model Rule for States. State and Territorial Air Pollution Program Administrators and Association of Local Air Pollution Control Officials, Washington, DC. July 2001[online]. Available: http://www.4cleanair.org/DieselTrucks.pdf [accessed August 29, 2005].

STAPPA/ALAPCO (State and Territorial Air Pollution Program Administrators and Association of Local Air Pollution Control Officials). 2004. Cleaning Up Diesel Trucks: A Model Rule for States. State and Territorial Air Pollution Program Administrators and Association of Local Air Pollution Control Officials, Washington, DC. September 2004 [online]. Available: http://www.4cleanair.org/CleaningUpTrucks.pdf [accessed August 30, 2005].

STAPPA/ALAPCO (State and Territorial Air Pollution Program Administrators and Association of Local Air Pollution Control Officials). 2005. HDDE Model Rule. State and Territorial Air Pollution Program Administrators and Association of Local Air Pollution Control Officials, Washington, DC [online]. Available: http://www.4cleanair.org/InnovationDetails.asp?innoid=15 [accessed July 13, 2005].

State of Oregon. 2004. Oregon Strategy for GHG Reductions, Governor's Advisory Group on Global Warming. Oregon Department of Energy, Salem, OR [online]. Available: http://egov.oregon.gov/ENERGY/GBLWRM/Strategy.shtml [accessed July 20, 2005].

Stedman, D.H. 1989. Automobile carbon monoxide emission. Environ. Sci. Technol. 23(2):147-148.

Stedman, D.H., G.A. Bishop, S.P. Beaton, J.E. Peterson, P.L. Guenther, I.F. McVey, and Y. Zhang. 1994. On-Road Remote Sensing of CO and HC Emissions in California. Final report. NTIS PB94-166345. Department of Chemistry, University of Denver, Denver, CO. Prepared for California Air Resources Board, Sacramento, CA.

Stedman, D.H., G.A. Bishop, and R.S. Slott. 1998. Repair avoidance and evaluating inspection and maintenance program. Environ. Sci. Technol. 32(10):1544-1545.

Summers, J.C., J.F. Skowron, and M.J. Miller. 1993. Use of Light-Off Catalysts to Meet the California LEV/ULEV Standards. SAE 930386. Warrendale, PA: Society of Automotive Engineers.

TCEQ (Texas Commission on Environmental Quality). 2004. Required control strategy elements. Chapter 5 in Revision to the State Implementation Plan (SIP) for the Control of Ozone Air Pollution: Houston/Galveston/Brazoria Ozone Nonattainment Area. Project No. 2004-042-SIP-NR. Texas Commission on Environmental Quality, Austin, TX [online]. Available: http://www.tnrcc.state.tx.us/oprd/sips/june2004hgb_EDrec.html#docs [accessed June 8, 2005].

Thurston, G.D., and K. Ito. 1999. Epidemiological studies of ozone exposure effects. Chapter 22 in Air Pollution and Health, S.T. Holgate, J.M. Samet, H.S. Koren, and R.L. Maynard, eds. San Diego: Academic Press.

TNRCC (Texas Natural Resource Conservation Commission). 2002. Houston/Galveston Nonattainment Area. All Major Emission Sources Categories-Table Summaries [online]. Available: http://www.tnrcc.state.tx.us/air/aqp/ei/rsumhg.htm#voc [accessed May 31, 2005].

Townsend, C.L., and R.L. Maynard. 2002. Effects on health of prolonged exposure to low concentrations of carbon monoxide. Occup. Environ. Med. 59(10):708-711.

Truett, L.J., and D.B. Truett. 2003. The Italian automotive industry and economies of scale. Contemp. Econ. Policy 21(3):329-337.

Truex, T.J. 1998. Development and Evaluation of Advanced Catalyst Technology for ULEV Emission Levels with Gasoline-Fueled Vehicles. Final report. Prepared for Mobile Source Air Pollution Reduction Review Committee, by Center for Environmental Research and Technology, University of California, Riverside, CA. February [online]. Available: http://www.cert.ucr.edu/research/pubs/98-ve-rt8k-001-fr.pdf [accessed Jan. 25, 2006].

TTM (Tightrope Technologies Motors, Inc.). 2004. Carbon Canister Solenoid Valve. Carbon Canister Solenoid Valve Diagnostics [online]. Available: http://www.ttmtechnical.com/carbon_canister_solenoid_valve.html [accessed July 11, 2005].

Turrentine, T., and K. Kurani. 2005. Automotive Fuel Economy in the Purchase and Use Decisions of Households. Presentation at the 2005 Annual Meeting of the Transportation Research Board.

University of Tennessee. 2003. Emission Inventories and Potential Emission Control Strategies for Ozone Early Action Compact Areas in Tennessee. Draft Report. Prepared for Tennessee Department of Transportation Division of Transportation Planning, and Tennessee Department of Environmental and Conservation, Division of Air Pollution Control, by University of Tennessee, Department of Civic and Environmental Engineering. April 13, 2003 [online]. Available: http://www.state.tn.us/environment/apc/eac/pdf/DrRpt041303.pdf [accessed Jan. 24, 2006].

U.S. Census Bureau. 1999. 1997 Economic Census, Vehicle Inventory and Use Survey. Geographic Area Series. EC97TV-US. U.S. Department of Commerce, Economics and Statistics Administration, U.S. Census Bureau [online]. Available: http://www.census.gov/svsd/www/97vehinv.html [accessed May 19, 2005].

Utell, M.J., M.W. Frampton, W. Zareba, R.B. Devlin, and W.E. Cascio. 2002. Cardiovascular effects associated with air pollution: Potential mechanisms and methods of testing. Inhal. Toxicol. 14(12):1231-1247.

UTI (Universal Trucks Isreal) 2005. EGR System. TI, Isuzu [online]. Available: http://uti.co.il/_uploads/imagesgallery/35EGR6.jpg [accessed Jan. 26, 2006. Isuzu Motors, Japan

Walsh, M. 2005. National and Global Impacts of California Mobile Source Emissions Standards. Presentation at the Fourth Meeting on State Practices in Setting Mobile Source Emissions Standards, April 15, 2005, Boston, MA.

Wang, Q., C. Kling, and D. Sperling. 1993. Light-duty vehicle exhaust emission control cost estimates using a part-pricing approach. J. Air Waste Manage. Assoc. 43(11):1461-1471.

Wayne, L.G., and Y. Horie. 1983. Evaluation of CARB's In-Use Vehicle Surveillance Program. CARB Contract No. A2-043-32. Prepared by Pacific Environmental Services, Inc., for California Air Resources Board, Sacramento, CA. October.

Wenzel, T. 2001. Evaluating the long-term effectiveness of the Phoenix IM240 program Environ. Sci. Policy 4(ER6):377-389.

Wenzel, T., M. Ross, and R. Sawyer. 1997. Analysis of Emissions Deterioration of in-use Vehicles, Using Arizona IM240 Data. Presented at the Society of Automotive Engineering Government/Industry Meeting, May 5-7, 1997, Washington, DC [online]. Available: http://enduse.lbl.gov/projects/vehicles/appendix_N.pdf [accessed Jan. 25, 2006].

White, L.J. 1982. The Regulation of Air Pollutant Emissions from Motor Vehicles. Washington, DC: American Enterprise Institute.

Wilson, R.D. 1992. Letter to Charles E. Walton, President of New Jersey Automobile Dealers Association, Trenton, NJ, from Richard D. Wilson,

Director ,Office of Mobile Sources, U.S. Environmental Protection Agency. July 2, 1992.

Witherspoon, C. 2004. California Perspective on Mobile Source Emission Standards. Presentation by California Air Resources Board at the First Meeting on State Practices in Setting Mobile Source Emissions Standards, June 28, 2004, Washington DC.

Wooley, D.R., and E. Morrs. 2004. CAA Handbook, 13Ed. St Paul, MN: Thomson West.

Zhu, Y., W.C. Hinds, S. Kim, and C. Sioutas. 2002. Concentration and size distribution of ultrafine particles near a major highway. J. Air Waste Manage. Assoc. 52(9):1032-1042.

Zielinska, B., J. Sagebiel, J.D. McDonald, K. Whitney, and D.R. Lawson. 2004. Emission rates and comparative composition from selected of in-use diesel and gasoline-fueled vehicles. J. Air Waste Manage. Assoc. 54(9):1138-1150.

Glossary

Acid Deposition – A comprehensive term for the various ways acidic compounds precipitate from the atmosphere and deposit onto surfaces. It can include: (1) wet deposition by means of acid rain, fog, and snow; and (2) dry deposition of acidic particles (aerosols).

Acute Exposure – One or a series of short-term exposures generally lasting less than 24 hours.

Acute Health Effect – A health effect that occurs over a relatively short period of time (e.g., minutes or hours). The term is used to describe brief exposures and effects that appear promptly after exposure.

Advanced Technology Partial Zero Emission Vehicle (AT-PZEV) – A vehicle that meets the PZEV emissions and warranty standards and, additionally, makes use of ZEV-enabling clean technology such as alternative fuels, electric drive, or other advanced technology systems

Adverse Health Effect – A health effect from exposure to air contaminants that may range from relatively mild temporary conditions, such as eye or throat irritation, shortness of breath, or headaches to permanent and serious conditions, such as birth defects, cancer or damage to lungs, nerves, liver, heart, or other organs.

After-Treatment Devices – Devices which remove pollutants from exhaust gases after the gas leaves combustion chamber (e.g., catalytic converters or diesel particulate filters). The term "exhaust gas aftertreatment" is considered derogatory by some in the emission control industry, but there is no consensus on the use of such alternatives as "post-combustion treatment" or "exhaust emission control."

Air-to-Fuel Ratio – The ratio, by weight, of air to gasoline entering the intake in a gasoline engine. The ideal (stoichiometric) ratio for complete combustion is approximately 14.7 parts of air to 1 part of fuel, depending on the composition of the specific fuel.

Air Quality Management Districts (AQMDs) – Administrative districts organized in California responsible for managing air quality on a regional or county basis. California is currently divided into 35 air districts.

Air-Quality Model – A computer-based mathematical model used to predict air quality based upon emissions and the effects of the transport, dispersion, and transformation of compounds emitted into the air.

Air Toxics – Air toxics refers to a host of carcinogens, respiratory toxicants, neurotoxicants, and other harmful atmospheric pollutants not included as criteria air pollutants. The Clean Air Act Amendments of 1990 listed 189 of these air toxics as hazardous air pollutants (HAPs) for future regulation. Also known as hazardous air pollutants

Alternative Fuel Vehicle (AFV) – Any dedicated, flexible-fuel, or dual-fuel vehicle designed to operate on at least one alternative fuel such as compressed natural gas.

Ambient Air – The air outside of structures. Often used interchangeably with "outdoor air."

California Air Resources Board (CARB) – The lead air quality management agency in California consisting of an eleven-member board appointed by the governor. CARB is responsible for attainment and maintenance of the state and federal air quality standards, and is fully responsible for motor vehicle pollution control in the state. It oversees county and regional air pollution management programs.

Carbon Monoxide (CO) – A colorless, odorless, tasteless, and toxic gas that results from the incomplete combustion of fuels containing carbon.

Carboxyhemoglobin (COHb) – A molecule formed when CO reacts with hemoglobin, the intracellular protein that transports oxygen in the blood. The presence of carboxyhemoglobin increases hemoglobin's affinity for oxygen, thereby reducing the transport of oxygen from the blood to the body's tissues.

Carcinogen – A cancer-causing substance.

Carl Moyer Fund – A multimillion dollar incentive grant program in California designed to encourage reduction of emissions from heavy-duty engines. The grants cover the additional cost of cleaner technologies for on-road, off-road, marine, locomotive and agricultural pump engines, as well as forklifts and airport ground support equipment.

Catalytic Converter – A mobile source emissions control device designed to reduce emissions of nitrogen oxides, hydrocarbons, and carbon monoxide. (See also "Two-Way Catalytic Converter" and "Three-Way Catalytic Converter".)

Chronic Exposure – Long-term exposure, usually lasting one year to a lifetime.

Chronic Health Effect – A health effect that occurs over a relatively long period of time (e.g., months or years). (See also "Acute Health Effect".)

Clean Air Act (CAA) – The original Clean Air Act in the US was passed in 1963, but the national air pollution control program is based on the 1970 version of the law. The 1990 Clean Air Act Amendments (CAAA90) are the most recent revisions of the law.

Closed Loop Fuel Control – A fuel metering system that uses real time feedback on combustion conditions for more effective emissions control. The closed loop fuel metering system of a contemporary vehicle uses sensors in the exhaust to evaluate the mixture exiting the engine and the catalyst to make adjustments to the air/fuel ratio through the use of an on-board computer that optimizes emissions performance.

Cold Start Emissions – Tailpipe emissions that occur before a vehicle is fully warmed-up. Vehicle emissions are higher during initial operations because the engine and catalytic converter must come to operating temperature before the emissions control system become effective. The time to catalyst "light off," when the catalyst is fully operational, is dependent on ambient temperature and vehicle technologies.

Compressed Natural Gas (CNG) – Natural gas compressed to a volume and density that is practical as a portable fuel supply.

Compression Ignition (CI) – A form of ignition that initiates fuel combustion in a diesel engine. The rapid compression of air within the cylinders generates the heat required to ignite the fuel as it is injected.

Criteria Air Pollutants – An air pollutant for which acceptable levels of exposure can be determined and for which National Ambient Air Quality Standards have been set. There are six common air pollutants (carbon monoxide, lead, nitrogen dioxide, ozone, particulate matter, and sulfur dioxide) that have been designated as criteria pollutants. The Clean Air Act states that the presence of criteria pollutants "in the ambient air results from numerous or diverse mobile or stationary sources. The term "criteria air pollutants" derives from the requirement that EPA must describe the characteristics and potential health and welfare effects of these pollutants. EPA periodically reviews new scientific data and may propose revisions to the standards as a result.

Diesel Oxidation Catalyst (DOC) – Catalyst promoting oxidation processes in diesel exhaust. Usually designed to reduce emissions of the organic fraction of diesel particulates, gas-phase hydrocarbons, and carbon monoxide.

Diesel Particulate Filter (DPF) – A device which physically captures diesel particulates preventing their discharge from the tailpipe. Collected particulates need to be removed from the filter, usually by continuous or periodic oxidation in a process called "regeneration."

Diesel Particulate Matter (DPM) – Sub-micron size particles found in diesel exhaust. Most emission regulations specify DPM measurement methods in which particulates are sampled on filters from cooled exhaust gas. The cooling causes condensation of vapors in the gas sampling train. Thus, the DPM is composed of both solid and liquid particles and is generally classified into three fractions: (1) inorganic carbon (soot), (2) organic fraction (often referred to as SOF or VOF), and (3) sulfate fraction (hydrated sulfuric acid).

Direct Injection – With direct injection in diesel engines, the combustion chamber is not divided and fuel is injected directly to the cylinder.

Design Standard – A technology-based standard that requires emitters to use a specifies technology to control emissions of a pollutant. These can also be called engineering standards.

Design Value – The monitored reading used by EPA to determine an area's air quality status; e.g., for the 1-hr ozone standard, the fourth highest reading measured over the most recent three years is the design value.

Dose – The amount of a contaminant that is absorbed or deposited in the body of an exposed person for an interval of time--usually from a single medium. Total dose is the sum of doses received by interactions with all environmental media that contain the contaminant. Units (mass) of dose and total dose are often converted to units of mass or contaminant per volume of physiological fluid or mass of tissue.

Dynamometer – A treadmill-like machine that allows cars to be tested under the loads typical of on-road driving.

Electronic Control Module (ECM) – A microprocessor that determines the beginning and end of each injection cycle on every cylinder. The ECM determines both fuel metering and injection timing in response to such parameters as engine crankshaft position and rpm, engine coolant and intake air temperature, and absolute intake air boost pressure.

Elemental Carbon (EC) – Inorganic carbon, as opposed to carbon in organic compounds, sometimes used as a surrogate measure for diesel particulate matter, especially in occupational health environments. Elemental carbon usually accounts for 40-60% of the total DPM mass.

Emissions Budget – Allowable emissions levels identified as part of a state implementation plan for pollutants emitted from mobile, industrial, stationary, and area sources. These emissions levels are used for meeting emission reduction milestones, attainment, or maintenance demonstrations.

Emissions Factor – For mobile sources, the emission factor is the relationship between the amount of pollution produced and the number of vehicle miles traveled. For stationary sources, the relationship between the amount of pollution produced and the amount of raw material processed or burned. By using the emission factor of a pollutant and specific data regarding activities (quantities of materials used by a given source or number of miles traveled), it is possible to compute emissions for the source.

Emissions Inventory – An estimate of the amount of a pollutant emitted into the atmosphere from major mobile, stationary, area-wide, and natural sources over a specific period of time such as a day or a year.

Emission Rate – The weight of a pollutant emitted per unit of time (e.g., tons / year).

Environmental Protection Agency (EPA) – The federal governmental agency that establishes regulations and oversees the enforcement of laws related to the environment.

Ethanol – Ethyl-alcohol, a volatile alcohol containing two carbon atoms (CH_3CH_2OH). For fuel use, ethanol is produced by fermentation of corn or other plant products.

Evaporative emissions – Hydrocarbon emissions that do not come from the tailpipe of a car, but come from evaporation, permeation, seepage, and leaks in a car's fueling system. The term is sometimes used interchangeably with *non-tailpipe emissions*.

Ex Ante – Analysis of the effects of a policy based only on information available before the policy is undertaken. Also termed prospective analysis.

Ex Post – Analysis of the effects of a policy based on information available after the policy has been implemented and its performance observed. Also termed retrospective analysis.

Exceedance – An air pollution event in which the ambient concentration of a pollutant exceeds a National Ambient Air Quality Standard (NAAQS).

Exposure – An event that occurs when there is contact at a boundary between a human and an environmental contaminant of a specific concentration for an interval of time; the units of exposure are concentration multiplied by time.

Four-Stroke Engines – A type of internal combustion piston engines where the engine cycle is completed after four strokes (up or down) of the piston, which distinguishes it from the two-stroke engine. In a typical four stroke engine, the air/mixture intake occurs on the downward stroke of piston, followed by compression on upward stroke. The compressed mixture is ignited, which drives the piston down on its third stroke. Finally exhaust gases are exited on the upward fourth stroke.

Federal Implementation Plan (FIP) – A plan prepared by the EPA in the absence of an approved state implementation plan that provides measures a nonattainment area must take to meet the requirements of the Federal Clean Air Act.

Federal Test Procedure (FTP) – A certification test for measuring the tailpipe and evaporative emissions from new vehicles over the Urban Dynamometer Driving Schedule, which attempts to simulate an urban driving cycle.

Greenhouse Gas – Atmospheric gases such as carbon dioxide, methane, chlorofluorocarbons, nitrous oxide, ozone, and water vapor that slow the passage of re-radiated heat through the Earth's atmosphere.

Gross Vehicle Weight Rating (GVWR) – The value specified by the manufacturer as the maximum design loaded weight of a single vehicle (i.e., vehicle weight plus rated cargo capacity).

Hazardous Air Pollutants (HAPs) – Air toxics listed under section 112(b) of the CAAA90.

Heavy-Duty Diesel Vehicle (HDDV) – An HDV using diesel fuel.

Heavy-Duty Vehicle (HDV) – Any motor vehicle rated at more than 8,500 pounds GVWR or that has a vehicle curb weight of more than 6,000 pounds or a frontal area in excess of 45 square feet. This excludes vehicles that will be classified as medium-duty passenger vehicles for the purposes of the Tier 2 emissions standards.

Hybrid Electric Vehicle (HEV) –Various types of electric vehicles that use another power source to propel the vehicle or generate power for an electric drive train, or a combination of the two types.

Hydrocarbons (HC) – Organic compounds containing hydrogen and carbon. (See also Appendix B.)

Inspection and Maintenance (I/M) – State emissions testing programs that attempt to identify vehicles with higher than allowable emissions or that have malfunctioning emissions control equipment and ensure that such vehicles are repaired or removed from the fleet

Light-Duty Vehicle (LDV) – A passenger car or passenger car derivative capable of seating 12 or fewer passengers. All vehicles and trucks under 8,500 GVWR are included (this limit previously was 6,000 pounds). Small pick-up trucks, vans, and sport utility vehicles may also be included.

Low-Emission Vehicle (LEV) – A vehicle that meets CARB's LEV I or LEV II standards or EPA's Clean Fuel Vehicle standards.

Low-Emission-Vehicle (LEV) Standards – Vehicles that meet the CARB low-emission-vehicle standards adopted in 1990 and covering 1994 through 2003 model year vehicles. The LEV standards for light duty vehicles include four categories: transitional low emissions vehicles (TLEV), LEV I vehicles, ultra low emissions (ULEV) vehicles, and zero emissions (ZEV) vehicles.

Low-Emission-Vehicle II (LEV II) Standards – Vehicles that meet the amended CARB low-emission-vehicle standards, known as LEV II standards, adopted in 1998 and covering 2004 and subsequent model years. The LEV II standards for light-duty vehicles include six categories: LEV II vehicles, ultra-low-emission (ULEV) vehicles, super ultra-low-emission (SULEV) vehicles, partial zero-emission (PZEV) vehicles, advanced-technology partial zero-emission (AT-PZEV) vehicles, and zero-emission (ZEV) vehicles.

Malfunction Indicator Light (MIL) – The instrument panel light used by the onboard diagnostic (OBD) system to notify the vehicle operator of an emissions related fault. The MIL is also known as the "service engine soon" or "check engine" lamp.

Marginal Benefit – The additional benefit gained from one more unit of output. In terms of reducing emissions, it represents the benefits from reducing emissions by one more unit.

Marginal Cost – The additional cost associated with producing one more unit of output. In terms of reducing emissions, it represents the cost of reducing emissions by one more unit.

Medium-Duty Passenger Vehicle (MDPV) – A new class of vehicles introduced with the federal Tier 2 emissions standards that includes sport utility vehicles and passenger vans rated at between 8,5000 and 10,000 GVWR.

Model Year – Vehicles are certified for sale, marketed, and later registered as a certain "model year" indicating the year a vehicle was produced and offered for sale. Model years typically begin in September or October of the prior year, and run for roughly 12 months. In the last decade, certain vehicles have been introduced as a 'pull-ahead' vehicle, appearing as early as January of the preceding year.

National Ambient Air Quality Standards (NAAQS) – Standards set by EPA for the maximum levels of criteria air pollutants that can exist in the outdoor air without unacceptable effects on human health or the public welfare. There are four elements of a NAAQS: (1) the pollutant indicator (such as $PM_{2.5}$), (2) the concentration of the indicator in the air, (3) the time over which measurements are made or averaged, and (4) the statistical form of the standard used to determine the allowable number of exceedances (such as the fourth highest value over a 3-year period).

National Low Emission Vehicle (NLEV) – A vehicle that meets voluntary low emissions tailpipe standards that are more stringent than can be mandated by EPA prior to model-year 2004. The NLEV program introduced California low emissions cars and light-duty trucks into the Northeast beginning in model year 1999 vehicles and the rest of the country in the model-year 2001 vehicles.

Nonattainment Area – A geographic area designated by the EPA to have concentrations of a criteria pollutant in excess of a NAAQS at some recent time. A single geographic area may have acceptable levels of some criteria air pollutants but unacceptable levels of others; thus, an area can be both an attainment area for one pollutant and a nonattainment area for another.

Nitrogen Oxides (Oxides of Nitrogen, NO_x) – A general term referring to nitric oxide (NO) and nitrogen dioxide (NO_2). Nitrogen oxides are formed when air is raised to high temperatures, such as during combustion or lightning, and are major contributors to smog formation and acid deposition.

Onboard Diagnostic (OBD) Systems – A system incorporated into new motor vehicles to monitor the performance of emissions control devices. When the system detects a problem, an onboard computer triggers a dashboard indicator light, referred to as a malfunction indicator light (MIL), alerting the driver to seek maintenance for the vehicle. The current OBDII system also communicates its findings to repair technicians by means of diagnostics trouble codes, which can be downloaded from the computer. OBD systems do not directly measure emissions.

Onboard Diagnostics Generation I (OBDI) – An onboard automotive diagnostic system required by the California Air Resources Board since 1988, which uses a microprocessor and sensors to monitor and control various engine system functions. A MIL illuminates when a malfunction is noted, but engine technicians cannot connect to the system and download trouble codes (MIL flash patterns communicate the problem).

Onboard Diagnostics Generation II (OBDII) – OBDII expands upon OBDI to include monitoring of both the emissions system and sensor deterioration and to provide diagnostic information to repair technician.

Open Loop Fuel Control – A system in which the air/fuel mixture is preset by design and contains no feedback correction signal to optimize fuel metering for emissions control. (See also "Closed Loop Fuel Control.")

Original Equipment Manufacturer (OEM) – Manufacturers of equipment (such as engines and vehicles) that provide the original product design and materials for its assembly and manufacture. OEMs are directly responsible for manufacturing and modifying the products, making them commercially available, and providing the warranty.

Oxygen Sensor – A sensor placed in the exhaust manifold to measure oxygen content. On some vehicles, oxygen sensors are located both before and after the catalytic converter.

Oxygenated Gasoline – Gasoline containing an oxygenate, typically methyl tertiary-butyl ether (MTBE) or ethanol, intended to reduce production of carbon monoxide, a criteria air pollutant. In some parts of the country, carbon monoxide emissions from cars make a major contribution to pollution. In some of these areas, gasoline refiners must market oxygenated fuels, which typically contain 2-3% oxygen by weight.

Oxygenates – Compounds containing oxygen (alcohols and ethers) that are added to gasoline to increase its oxygen content. Methyl tertiary-butyl ether (MTBE) and ethanol are the most common oxygenates currently used, although there are a number of others.

Ozone (O_3) – A reactive gas whose molecules contain three oxygen atoms. It is a product of photochemical processes involving sunlight and ozone precursors, such as hydrocarbons and oxides of nitrogen. Ozone exists in the upper atmosphere (stratospheric ozone), where it helps shield the earth from excessive ultraviolet rays, as well as in the lower atmosphere (tropospheric ozone) near the earth's surface. Tropospheric ozone causes plant damage and adverse health effects, and is a criteria air pollutant; it is a major component of smog.

Partial Zero Emission Vehicle (PZEV) – Vehicles that have achieved the CARB's cleanest tailpipe emission standard under the LEV II program, the super ultra-low-emission vehicle (SULEV) standard, and have nearly zero evaporative emissions and their emission control equipment warranted for 15 years/150,000 miles.

Particulate Matter (PM) – Any material, except uncombined water, that exists in the solid or liquid droplet states in the atmosphere. Particulate matter includes wind-blown dust particles, particles directly emitted as combustion products, and particles formed through secondary reactions in the atmosphere.

Photochemical Reaction – A term referring to a chemical reaction brought about by sunlight, such as the formation of ozone from the interaction of nitrogen oxides and hydrocarbons in the presence of sunlight.

Plug-In – An electrical device used to heat the engine under extreme cold conditions in order to facilitate engine starting and reduce the time for emissions control devices to be fully operational.

$PM_{2.5}$ – $PM_{2.5}$ refers to a subset of particulate matter collected by a sampling device with a size-selective inlet that has a 50% collection efficiency for particles with an aerodynamic diameter of 2.5 micrometers (μm). This fraction of PM penetrates most deeply into the lungs, and causes the majority of visibility reduction.

PM_{10} – PM_{10} refers to a subset of particulate matter collected by a sampling device with a size-selective inlet that has a 50% collection efficiency for particles with an aerodynamic diameter of 10 μm.

Primary Standard – A NAAQS for criteria air pollutants based on health effects.

Purge Test – A test used to determine if fuel vapors are properly drawn from the evaporative canister and the fuel tank into the engine for combustion. If the purge system is not working properly, the evaporative canister can become saturated and vent hydrocarbons into the atmosphere.

Regulatory Impact Analysis (RIA) – An analysis document produced by EPA for each major rulemaking listing the expected impacts of the rule including environmental impacts, health impacts, cost-benefit analyses, economic impacts, small business impacts.

Reformulated Gasoline (RFG) – Specifically formulated fuels blended such that, on average, the exhaust and evaporative emissions of HCs and related hazardous air pollutants (chiefly benzene, 1,3-butadiene, polycyclic aromatic hydrocarbons, formaldehyde, and acetaldehyde) are significantly and consistently lower than such emissions resulting from use of conventional gasolines. The 1990 Clean Air Act amendments required sale of reformulated gasoline in the nine areas with the most severe ozone pollution problems.

Regional Haze – The haze produced by a multitude of sources and activities which emit fine particles and their precursors across a broad geographic area. Federal regulations require states to develop plans to reduce the regional haze that impairs visibility in national parks and wilderness areas.

Remote Sensing – A method for measuring pollutant concentrations from a vehicle's exhaust with the use of a roadside monitoring device (known as remote sensing devices or RSDs). Infrared (IR) and Ultraviolet (UV) light is directed across the road and passively reflected back to detectors that monitor light intensity at characteristic wavelengths. The amount of characteristic infrared or ultraviolet light absorbed is translated into the exhaust concentration of the three regulated pollutants of interest, CO, HCs and NO_x.

Secondary Particulate Matter – Particulate matter that is formed in the atmosphere, and is generally composed of species such as ammonia or the products of atmospheric chemical reactions, such as nitrates, sulfates and organic material, in addition to some water. Secondary particles are distinguished from primary particles, which are emitted directly into the atmosphere.

Secondary Standard – A NAAQS for criteria air pollutants based on environmental effects such as damage to property, plants, visibility, etc.

Selective Catalytic Reduction (SCR) – Term frequently used as a synonym for catalytic reduction of NO_x in diesel exhaust or flue gases by nitrogen containing compounds, such as ammonia or urea. Such SCR systems are commercially available for stationary applications and are being developed for mobile diesel engines. Since "selective catalytic reduction" is a generic term which also applies to other reactions, its use may lead to confusion in some situations.

Spark Ignition (SI) – The form of ignition that initiates combustion in a gasoline, natural gas, or liquefied petroleum gas fueled engine or any other type of engine with a spark plug (or other sparking device).

State Implementation Plan (SIP) – A detailed description of the scientific methods and emissions reduction programs a state will use to carry out its responsibilities under the Clean Air Act for complying with the

NAAQS. The Clean Air Act requires that EPA approve each SIP after the public has had an opportunity to participate in its review and approval.

Super Ultra-Low-Emission Vehicle – Vehicles meeting CARB's super ultra-low-emission vehicle (SULEV) standard, the cleanest emission standard that a gasoline vehicle can meet under the LEV II program.

Supplemental Federal Test Procedure (SFTP) – The SFTP is a certification test for measuring the tailpipe and evaporative emissions from new vehicles that includes two driving cycles not represented in the FTP. The SFTP includes a test cycle simulating high speed and high acceleration driving (US06 cycle) and a test cycle that evaluates the effects of simulating air conditioner operation (SC03 cycle).

Technology-Based Standards – A type of standard that dictates polluters use specific techniques (e.g., a particular type of pollution abatement equipment) or follow a specific set of operating procedures and practices.

Technology Forcing – The establishment by a regulatory agency of a requirement to achieve an emissions limit, within a specified time frame, that can be reached through use of unspecified technology or technologies that have not yet been developed for widespread commercial applications and have been shown to be feasible on an experimental or pilot-demonstration basis.

Temperature Inversion – An atmospheric condition in which temperature in the lower part of the atmosphere increases with altitude, rather than decreasing with altitude, as is more typical. Inversion conditions can trap pollution near the surface because warmer, less dense air is resting above colder, more dense air.

Three-way Catalytic Converter – A catalytic converter designed to both oxidize CO and HCs and reduce NO_x emitted from gasoline-fueled vehicles.

Tier 0 Vehicles – Vehicles that meet federal Tier 0 tailpipe standards. For light-duty vehicles, these standards began with model-year 1981 vehicles and were phased out in model-year 1995 for passenger cars and most light-duty trucks.

Tier 1 Vehicles – Vehicles that meet federal Tier 1 tailpipe standards. For light-duty vehicles, these standards began with model-year 1994 vehicles.

Tier 2 Vehicles – Vehicles that will meet Tier 2 tailpipe standards. For light-duty vehicles, these standards begin with model-year 2004 vehicles.

Total Carbon – The sum of the elemental carbon and organic carbon. For diesel particulates, this is typically 80-85% of the total DPM mass.

Transitional-Low-Emission Vehicle (TLEV) – A vehicle meeting either EPA's clean fuel vehicle TLEV standards or CARB's California low-emission vehicle (LEV I) TLEV standards. TLEVs produce lower emissions than federal Tier 1 vehicles.

Transportation Control Measure (TCM) – Any control measure to reduce vehicle trips, vehicle use, vehicle miles traveled, vehicle idling, or traffic congestion for the purpose of reducing motor vehicle emissions. TCMs can include encouraging the use of carpools and mass transit.

Transportation Demand Management (TDM) Strategies – Strategies which use regulatory mandates, economic incentives, or educational campaigns to change driver behavior. TDM strategies attempt to reduce the frequency or length of automobile trips or to shift the timing of automobile trips.

Travel-demand Model – An analysis procedure using heuristics or formal systems of equations to estimate the number, distribution, mode choice, and/or route choice of trips made by a household or individual that can be aggregated to estimate the number of trips starting and/or ending in a specific geographical area. The model determines the amount of transportation activity occurring in a region based on an understanding of the daily activities of individuals and employers, as well as the resources and transportation infrastructure available to households and individuals when making their daily activity and travel decisions.

Two-Stroke Engines – A type of internal combustion piston engines where the engine cycle is completed after just two strokes (up or down) of the piston, which distinguishes it from the four-stroke engine. In a typical two-stroke engine, the air/fuel mixture is drawn into the crankcase as the piston moves up on its first stroke to compress the mixture above it. Then the compressed mixture is ignited, and hot gases are produced, which drive the piston down on its second stroke. As it moves down, it uncovers an opening (port) that allows the fresh fuel mixture in the crankcase to flow into the combustion space above the piston. At the same time, the exhaust gases leave through another port.

Two-way Catalytic Converter – A first generation catalytic converter designed to oxidize CO and HC emissions from gasoline-fueled vehicles.

Ultra-Low Emission Vehicle (ULEV) – A vehicle meeting CARB's California low-emission vehicle ULEV standards or EPA's clean fuel vehicle ULEV standards. ULEVs produce fewer emissions than LEV I or LEV II vehicles.

Vehicle Miles Traveled (VMT) – The number of miles driven by a fleet of vehicles over a set period of time, such as a day, month, or year. One vehicle traveling one mile is one vehicle mile.

Volatile Organic Compounds (VOCs) – Organic compounds that can include oxygen-, nitrogen-, and sulfur-containing compounds. Alkanes, alkenes and aromatic hydrocarbons are all VOCs (as well as being HCs). The simple carbon-containing compounds carbon monoxide and carbon dioxide are usually classified as inorganic compounds. A volatile organic compound is one that can exist as a gas at ambient temperatures. Many volatile organic chemicals are hazardous air pollutants; for example, benzene causes cancer. (See also Appendix B.)

Zero Emission Vehicle (ZEV) – A vehicle that emits no tailpipe exhaust emissions.

Sources: CARB at www.arb.ca.gov/html/gloss.htm; EPA at www.epa.gov/oar/oaqps/peg_caa/pegcaa10.html, and www.epa.gov/oms/stds-ld.htm.

Abbreviations

A/F	air-to-fuel ratio
AAM	Association of Automobile Manufacturers
ACP	alternative compliance plan
AQMD	Air Quality Management District
ARB	Air Resources Board in California (same as CARB)
ASTM	American Society for Testing Materials
AT-PZEV	advanced-technology partial zero-emission vehicle
CAA	Clean Air Act
CAFE	corporate average fuel economy
CARB	California Air Resources Board
CIDI	compression ignition direct injection
CNG	compressed natural gas
CO	carbon monoxide
CO_2	carbon dioxide
COHb	carboxyhemoglobin
DOE	U.S. Department of Energy
EGR	exhaust gas recirculation
EMFAC	emissions model used by CARB for mobile sources
EPA	U.S. Environmental Protection Agency
EV	electric vehicle
GVW	gross vehicle weight
HAPs	hazardous air pollutants
HC	hydrocarbon (see Appendix B for definition of various terms used for HCs)
HCCI	homogeneous charge compression ignition
HDV	heavy-duty vehicle

HEW	U.S. Department of Health, Education and Welfare
IC	internal combustion
LACPCD	Los Angeles County Air Pollution Control District
LDT	light-duty truck
LEV	low-emission vehicle
LPG	liquified petroleum gas
LVW	loaded vehicle weight
MDPV	medium-duty passenger vehicle
MDT	medium-duty truck
MDV	medium-duty vehicle
MECA	Manufacturers of Emissions Control Association
MOA	memorandum of agreement
MOBILE6	emissions model used by EPA for mobile sources
MVPCB	Motor Vehicle Pollution Control Board
$\mathbf{N_2}$	diatomic nitrogen
NAAQS	National Ambient Air Quality Standards
NAPCA	National Air Pollution Control Administration
NLEV	National Low Emission Vehicle Program
NMHC	nonmethane hydrocarbon (see Appendix B for definition of various terms used for HCs)
NMOG	nonmethane organic gas (see Appendix B for definition of various terms used for HCs)
$\mathbf{NO_2}$	nitrogen dioxide
$\mathbf{NO_x}$	oxides of nitrogen gases
NRC	National Research Council
$\mathbf{O_2}$	diatomic oxygen
$\mathbf{O_3}$	ozone
OBD	onboard diagnostics
OTC	Ozone Transport Commission
PAH	polycyclic aromatic hydrocarbon
PC	passenger car
PCV	positive crankcase ventilation
PM	particulate matter
PZEV	partial zero-emission vehicle
R&D	research and development
RIA	regulatory impact assessment
RHC	reactive hydrocarbon (see Appendix B for definition of various terms used for HCs)
ROG	reactive organic gas (see Appendix B for definition of various terms used for HCs)

SCAQMD South Coast Air Quality Management District
SCR selective catalytic reduction
SI spark ignition
SIP state implementation plan
SORE small off-road engine
SULEV super ultra-low-emission vehicle
TLEV transitional low-emission vehicle
UAM urban airshed model
UEGO universal exhaust gas oxygen (sensor)
ULEV ultra-low-emission vehicle
VOC volatile organic compound
ZEV zero-emission vehicle

Appendix A

Biographical Information on the Committee on State Practices in Setting Mobile Source Emissions Standards

David Allen is the Melvin Gertz Regents Professor in Chemical Engineering and the director of the Center for Energy and Environmental Resources at the University of Texas at Austin. Dr Allen is a member of the EPA Advisory Council on Clean Air Compliance Analysis. He has held regular faculty appointments at the University of California Los Angeles and the University of Texas, and he has held visiting faculty appointments at the California Institute of Technology, the U.S. Department of Energy, and the University of California Santa Barbara. Dr. Allen's research interests lie in environmental reaction engineering, particularly issues related to air quality and pollution prevention. Dr. Allen served on the NRC Board on Environmental Studies and Toxicology and has served on several NRC committees. He served as chair of the Texas Council on Environmental Technology, which provides advice to the state of Texas on the use of innovative air quality improvement technologies. He received his B.S. in chemical engineering, with distinction, from Cornell University and his M.S. and Ph.D. in chemical engineering from the California Institute of Technology.

John Bailar, III, is professor emeritus in the Department of Health Studies at the University of Chicago. He is a retired commissioned officer of the U.S. Public Health Service and worked for the National Cancer Institute for 22 years. Dr. Bailar previously held an appointment as a senior scientist in the EPAHealth and Environmental Review Division. He also held academic appointments at Harvard University and McGill University. Dr. Bailar's research interests include assessing health risks from chemical hazards and air pollutants and interpreting statistical evidence in medicine, with a special emphasis on cancer. He was editor-in-chief of the *Journal of the National Cancer Institute* for 6 years and a member of the editorial board of the *New England Journal of Medicine* for 7 years. Dr. Bailar was elected to the Institute of Medicine in 1993 and is a member of the International Statistical Institute. He received his M.D. from Yale University and his Ph.D. in statistics from American University.

Hugh Ellis is the chair of the Department of Civil Engineering at the Johns Hopkins University in Baltimore, and he holds a joint appointment in the Department of Geography and Environmental Engineering. His research interests focus on the development of uncertainty and risk-based approaches for environmental management, including the use of such techniques for assessing air quality and emissions control policies in Maryland. He served on the NRC Committee on the Effectiveness of Vehicle Emission Inspection and Maintenance Programs. He received his B.S., M.S., and Ph.D. in civil engineering from the University of Waterloo in Ontario, Canada.

Alison Geyh is an assistant professor at the Johns Hopkins Bloomberg School of Public Health. Her research interests focus on quantifying airborne contaminants, identifying their sources, investigating their chemical composition, and assessing the routes of human exposure. Her current projects include an assessment of personal exposure to mobile-source related pollutants, an examination of the metal content of ambient fine particulate matter, and an exploration of the potential relationship between exposure to fine particles and exacerbations of symptoms related to congestive heart failure. She is the principal investigator of a large ongoing project focused on assessing exposure and health outcome of workers who were involved in the cleanup and recovery effort at the World Trade Center disaster site. Dr. Geyh previously held an appointment as a staff scientist for the Health Effects Institute, which is funded jointly by EPA and the automotive industry. Dr. Geyh received her Ph.D. from Brandeis University in physical organic chemistry.

David Greene is a corporate fellow of Oak Ridge National Laboratory (ORNL). His research interests are in energy and environmental policy analysis in the transportation sector, including analyses of policies to mitigate greenhouse gas emissions from the transportation sector, modeling of energy and transportation demand, and assessment of the economics of petroleum dependence and market responses to advanced transportation technologies and alternative fuels. After joining ORNL in 1977, he founded the Transportation Energy Group in 1980 and later established the Transportation Research Section in 1987. Also, Dr. Greene was a senior research analyst in the Office of Domestic and International Energy Policy, U. S. Department of Energy. He is past chair and member emeritus of the Transportation Research Board's Energy Committee, past chair of the Section on Environmental and Energy Concerns, and a lifetime national associate of the National Academies. Dr. Greene received his B.A. from Columbia University, M.A. from the University of Oregon, and Ph.D. in geography and environmental engineering from the Johns Hopkins University.

James Lents is director of the Environmental Policy and Corporate Affiliates Program at the University of California Riverside. Dr. Lents joined the university after managing air quality improvement projects nationwide, including 11 years as executive officer of the South Coast Air Quality Management District in Diamond Bar, California. His experience includes work in defining the emissions inventory development, modeling, and emissions control process for Chattanooga, Tennessee, and Denver, Colorado. His work in Colorado included oversight of the efforts to evaluate air quality impacts of oil shale production during the oil shale boom of the late 1970s and early 1980s. He served on the NRC Committee on Air Quality Management in the United States. Dr. Lents received his Ph.D. from the University of Tennessee (Space Institute) in physics.

Gary Marchant is a professor of law and executive director and faculty fellow of the Center for the Study of Law, Science, and Technology in the College of Law at Arizona State University. Dr. Marchant teaches environmental law, science and technology, genetics and the law, and environmental justice. Before joining the ASU faculty, he was a partner at the Washington, DC, office of the law firm of Kirkland & Ellis, where his practice focused on environmental and administrative law. As part of this work, he represented the two major U.S. trade associations of motor vehicle manufacturers on a variety of regulatory and preemption litigation matters relating to federal, California, and Northeast States motor vehicle emission standards. He received his B.S. and Ph.D. in genetics from the University of British Columbia, his M.P.P. from the Kennedy School of Government of Harvard University, and his J.D. from Harvard Law School.

Virginia McConnell is a senior fellow in the Quality of the Environment Division of Resources for the Future (RFF). She is also a professor of economics at the University of Maryland, Baltimore County. Her recent work has centered on the evaluation of policies to reduce motor vehicle pollution, including the analysis of inspection and maintenance programs, old-car scrap programs, and emissions taxes. Dr. McConnell is a member of the EPA Advisory Council on Clean Air Compliance Analysis and has served on other EPA advisory committees. She has analyzed market-based policies for improving land use and the impact of environmental regulations on industry productivity. Dr. McConnell served on the NRC Committee on the Effectiveness of Vehicle Emission Inspection and Maintenance Programs. She received her B.S. in economics from Smith College and her Ph.D. in economics from the University of Maryland.

Alison Pollack is a Principal at ENVIRON Corporation, an environmental consulting firm. Her work is primarily in the analysis of on-road and off-road mobile-source emissions data, the estimation of mobile-source emissions inventories, the development and evaluation of on-road and off-road mobile-source emissions models, and the evaluation of mobile-source control measures. Ms. Pollack's research and consulting work has mostly been funded by state and federal agencies. She served on the NRC Committee to Review EPA's Mobile Source Emissions Factor (MOBILE) Model and the Committee on the Effectiveness of Vehicle Emission Inspection and Maintenance Programs. Ms. Pollack received her B.S. and M.S. in statistics from Princeton University and the University of Wisconsin-Madison, respectively.

Harold Schock is a professor in the Department of Mechanical Engineering at Michigan State University. Dr. Schock is the owner of Mid Michigan Research, LLC, an engineering research and design company. His research interests include flow and combustion phenomena in internal combustion engines using laser-based experimental techniques and numerical simulations, wear studies of mechanical parts using implanted radioisotopes, and flow physics in an automotive torque converter. Dr. Schock's research has been funded by a number of sources, including industry and federal agencies. He served on the NRC Committee on Mobile Electric Power Plant Technologies. Dr. Schock received his Ph.D. in mechanical engineering at Michigan Technological University.

Karl Springer was the vice president for Automotive Products and Emissions Research at Southwest Research Institute (retired). His research interests have focused on the measurement and control of air pollution emissions from on-road and off-road vehicles and equipment powered by internal combustion engines. Mr. Springer is a member of the National Academy of Engineering. He served on the NRC Committee on Carbon Monoxide Episodes in Meteorological and Topographical Problem Areas. He received a B.S.M.E. from Texas A&M and an M.S. in physics from Trinity University.

Appendix B

Acronyms and Names Used for Classifying Organic Compounds

Common Abbreviation	Full Name	Definition
VOC[1]	Volatile organic compounds	Organic compounds that are found in the gas phase at ambient conditions. Might not include methane.
ROG	Reactive organic gases	Organic compounds that are assumed to be reactive at urban (and possibly regional) scales. By definition, taken to be those organic compounds that are regulated because they lead to ozone formation. Does not include methane. The term is used predominantly in California.
NMHC	Nonmethane hydrocarbons	All hydrocarbons except methane; sometimes used to denote ROG.
NMOC	Nonmethane organic compounds	Organic compounds other than methane.
RHC	Reactive hydrocarbons	All reactive hydrocarbons; also used to denote ROG.
THC	Total hydrocarbons	All hydrocarbons, sometimes used to denote VOC.
OMHCE	Organic material hydrocarbon equivalent	Organic compound mass minus oxygen mass.
TOG	Total organic gases	Used interchangeably with VOC.

[1]Unless noted otherwise, HC is the term used in this report to represent the general class of gaseous organic compounds.

Source: NRC 1999.

Appendix C

Summary of Milestones in CARB Mobile-Source Emissions Regulations and Comparison with EPA

TABLE C-1 Heavy-Duty On-Road Vehicles

Model Year	Requirement	Result	EPA Model Year	EPA Comparison with California
1999	Carl Moyer Program	Incentive program to reduce NO_x and PM emissions (funded 7,000 projects to date)	—	EPA has no comparable program
2000	School bus program	Incentive program to retrofit or purchase improved buses (3,475 buses improved)	2004	Federal enforcement settlement with Toyota provided funding for retrofit of school buses
2004	Vehicle computer software upgrade	Required 1993-1999 trucks to modify software to reduce excessive NO_x emissions	—	EPA has no comparable program
2004	Particulate filters on refuse trucks	Reduced PM by up to 85% in residential neighborhoods	—	EPA has no comparable program
2005	Additional test procedures for certification	Reduce potential of excessive emissions during real-world driving	2007	Adopted a similar program (2 years later)
2007	First diagnostic system	Identifies failed emission control parts	—	EPA has no comparable program

TABLE C-2 Small Off-Road and Utility Engines

Model Year	Requirement	Result	EPA Model Year	Comparison with California
1995	First small off-road engine emission standards	Improved engine design reduced emissions 30-70%	1997	EPA adopted similar program (2 years later)
2000	First small off-road engine durability requirement	Use of catalysts, advanced two-stroke design and increased use of four-stroke engines for handheld equipment. Improved engine technologies for non-handheld equipment	2001	EPA adopted similar program (1 year later)
2005	California aligned with federal standards	Improved engine design and requirements similar in all 50 states for engines smaller than 50 cc	2005	Initiated at the federal level first and then adopted by CARB
2006	First small off-road engine evaporative emission requirements	Low fuel permeation fuel tank and fuel lines and use of carbon canister	—	Federal standards in development
2007-2008	HC and NO_x standards reduced for engines greater than 80 cc in displacement	Emissions reduced by 33-38% using catalyst	—	Federal standards in development

TABLE C-3 Off-Road Diesel Engines

Model Year	Requirement	Result	EPA Model Year	EPA Comparison with California
1996	First emission standards for engines over 175 hp (130 kW)	Improved engine design technologies	1996	Adopted California requirements from 50-175 hp (37-130 kW)
1999	Carl Moyer Program	Incentive program to reduce NO_x and PM (as noted earlier 7,000 projects to date)	—	EPA has no comparable program
2000	Agreement with manufacturers to align California and federal standards	Similar requirements in all 50 states	—	EPA and CARB standards aligned

TABLE C-4 Recreational Marine Engines

Model Year	Requirement	Result	EPA Model Year	EPA Comparison with California
2001	Set Tier 1 outboard marine and personal watercraft engines	Improved engine technologies reducing HC and NO_x emissions by 75% from uncontrolled	2006	EPA adopted similar requirements but will not fully match California until 2006 (5 years later)
2003	Tier 1 standard for inboard and sterndrive marine engines	Improved engine technologies with minor emission reductions	—	EPA has no comparable requirement but notice of advanced rulemaking issued in 2002
2004	Tier 2 outboard marine and personal watercraft engines	Improved engine design resulting in 80% HC and NO_x emission reductions compared to uncontrolled	—	EPA has no comparable requirement

(Continued)

TABLE C-4 *Continued*

Model Year	Requirement	Result	EPA Model Year	Comparison with California
2007	Tier 2 inboard and stern drive marine engine requirements	Use of catalysts producing 67% emission reduction compared with uncontrolled	—	EPA has no comparable requirements but notice of advanced rulemaking issued in 2002
2008	Tier 3 outboard and personal watercraft engine requirements	Increases use of four-stroke engines and use of direct injection two-stroke engines	—	EPA has no comparable requirements

TABLE C-5 Fuels and Vapor Recovery

Model Year	Requirement	Result	EPA Model Year	Comparison with California
1971	Fuel volatility limit of 9 RVP during ozone season	Large evaporative emission reductions	1989	EPA adopted similar program (18 years later)
—	—	—	1974	Federal regulations required large service stations to sell unleaded gasoline nationwide
1976	Stage 1 and Stage 2 vapor recovery required (transfer from cargo to service station and transfer to vehicle respectively)	Evaporative emission reductions	1990	Clean Air Act amendments of 1990 set similar requirements in non-attainment areas outside of California
1992	Phase 1 of cleaner burning gasoline	Full phase-out of lead, deposit control additives, RVP of 7.8, and wintertime oxygenates	1992	EPA adopted a partial program at the same time

(Continued)

TABLE C-5 *Continued*

Model Year	Requirement	Result	EPA Model Year	EPA Comparison with California
1993	Cleaner burning diesel fuel	Sulfur content limit of 500 ppmw and aromatic hydrocarbon limit of 10% by volume (20% for smaller refiners)	1993	EPA adopted a partial program at the same time
1996	Phase 2 of clean burning gasoline	Control of 8 major fuel properties: RVP 7.0 cap in ozone season, oxygenates, benzene limits, aromatic limits, olefin limits, sulfur limits, and 50% and 90% distillation limits	2000	EPA adopted a partial program 4 years later
2001	Enhanced Stage 1 vapor recovery (98% capture efficiency)	VOC emissions reduced	—	EPA has no similar requirement but many states have adopted the California requirement
2003	Phase 2 vapor recovery must be compatible with ORVR systems on vehicles	Improves nozzle design or processor to minimize tank pressure	—	EPA has no similar requirement but many states are adopting California requirement
2006	Reduce sulfur in diesel fuel to 15 ppmw	Enables the use of advanced after-treatment devices to meet 2007 heavy-duty vehicle requirements	2006	EPA adopted 1 year before California but not applied to nonroad sources until 2010, not to stationary sources, and no aromatic hydrocarbon limit

All tables adapted from CARB, unpublished material, 2005.

Appendix D

Statutory Sections Relevant to the Regulation of New Mobile-Source Emissions

[CLEAN AIR ACT SECTION 209 (42 U.S.C. 7543)]

Sec. 209.

(a) No State or any political subdivision thereof shall adopt or attempt to enforce any standard relating to the control of emissions from new motor vehicles or new motor vehicle engines subject to this part. No State shall require certification, inspection, or any other approval relating to the control of emissions from any new motor vehicle or new motor vehicle engine as condition precedent to the initial retail sale, titling (if any), or registration of such motor vehicle, motor vehicle engine, or equipment.

(b)(1) The Administrator shall, after notice and opportunity for public hearing, waive application of this section to any State which has adopted standards (other than crankcase emission standards) for the control of emissions from new motor vehicles or new motor vehicle engines prior to March 30, 1966, if the State determines that the State standards will be, in the aggregate, at least as protective of public health and welfare as applicable Federal standards. No such waiver shall be granted if the Administrator finds that-

(A) the determination of the State is arbitrary and capricious,

(B) such State does not need such State standards to meet compelling and extraordinary conditions, or

(C) such State standards and accompanying enforcement procedures are not consistent with section 202(a) of this part.

(2) If each State standard is at least as stringent as the comparable applicable Federal standard, such State standard shall be deemed to be at

least as protective of health and welfare as such Federal standards for purposes of paragraph (1).

(3) in the case of any new motor vehicle or new motor vehicle engine to which State standards apply pursuant to a waiver granted under paragraph (1), compliance with such State standards shall be treated as compliance with applicable Federal standards for purposes of this title.

(c) Whenever a regulation with respect to any motor vehicle part or motor vehicle engine part is in effect under section 207(a)(2), no State or political subdivision thereof shall adopt or attempt to enforce any standard or any requirement of certification, inspection, or approval which relates to motor vehicle emissions and is applicable to the same aspect of such part. The preceding sentence shall not apply in the case of a State with respect to which a waiver is in effect under subsection (b).

(d) Nothing in this part shall preclude or deny to any State or political subdivision thereof the right otherwise to control, regulate, or restrict the use, operation, or movement of registered or licensed motor vehicles.

(e) Nonroad Engines or Vehicles.-

(1) Prohibition on certain state standards.- No State or any political subdivision thereof shall adopt or attempt to enforce any standard or other requirement relating to the control of emissions from either of the following new nonroad engines or nonroad vehicles subject to regulation under this Act-

(A) New engines which are used in construction equipment or vehicles or used in farm equipment or vehicles and which are smaller than 175 horsepower.

(B) New locomotives or new engines used in locomotives. Subsection (b) shall not apply for purposes of this paragraph.

(2) Other nonroad engines or vehicles.-

(A) In the case of any nonroad vehicles or engines other than those referred to in subparagraph (A) or (B) of paragraph (1), the Administrator shall, after notice and opportunity for public hearing, authorize California to adopt and enforce standards and other requirements relating to the control of emissions from such vehicles or engines if California determines that California standards will be, in the aggregate, at least as protective of public health and welfare as applicable Federal standards. No such authorization shall be granted if the Administrator finds that-

(i) the determination of California is arbitrary and capricious,

(ii) California does not need such California standards to meet compelling and extraordinary conditions, or

(iii) California standards and accompanying enforcement procedures are not consistent with this section.

(B) Any State other than California which has plan provisions approved under part D of title I may adopt and enforce, after notice to the Administrator, for any period, standards relating to control of emissions from nonroad vehicles or engines (other than those referred to in subparagraph (A) or (B) of paragraph (1)) and take such other actions as are referred to in subparagraph (A) of this paragraph respecting such vehicles or engines if-

(i) such standards and implementation and enforcement are identical, for the period concerned, to the California standards authorized by the Administrator under subparagraph (A), and

(ii) California and such State adopt such standards at least 2 years before commencement of the period for which the standards take effect. The Administrator shall issue regulations to implement this subsection.

[CLEAN AIR ACT SECTION 177 (42 U.S.C. 7507)]

Sec. 177. Notwithstanding section 209(a), any State which has plan provisions approved under this part may adopt and enforce for any model year standards relating to control of emissions from new motor vehicles or new motor vehicle engines and take such other actions as are referred to in section 209(a) respecting such vehicles if-

(1) such standards are identical to the California standards for which a waiver has been granted for such model year, and

(2) California and such State adopt such standards at least two years before commencement of such model year (as determined by regulations of the Administrator). Nothing in this section or in title II of this Act shall be construed as authorizing any such State to prohibit or limit, directly or indirectly, the manufacture or sale of a new motor vehicle or motor vehicle engine that is certified in California as meeting California standards, or to take any action of any kind to create, or have the effect of creating, a motor vehicle or motor vehicle engine different than a motor vehicle or engine certified in California under California standards (a "third vehicle") or otherwise create such a "third vehicle."

[P.L. 108-199 DIVISION G]

Sec. 428. Regulation of Small Engines.

(a) In considering any request from California to authorize the State to adopt or enforce standards of other requirements relating to the control of emissions from new non-road spark-ignition engines smaller than 50 horsepower, the Administrator shall give appropriate consideration to safety factors (including the potential increased risk of burn or fire) associated with compliance with the California standard.

(b) Not later than December 1, 2004, the Administrator of the Environmental Protection Agency shall propose regulations under the Clean Air Act that shall contain standards to reduce emissions from new non-road spark-ignition engines smaller than 50 horsepower. Not later than December 31, 2005, the Administrator shall publish in the Federal Register final regulations containing such standards.

(c) No State or any political subdivision thereof may adopt or attempt to enforce any standard or other requirement applicable to spark ignition engines smaller than 50 horsepower.

(d) Exception for California.--The prohibition in subsection (e) does not apply to or restrict in any way the authority granted to California under section 209(e) of the Clean Air Act (42 U.S.C. 7543(e)).

(e) Exception for Other States.--The prohibition in subsection (c) does not apply to or restrict the authority of any State under section 209(e)(2)(B) of the Clean Air Act (42 U.S.C. 7543(e)(2)(B)) to enforce standards or other requirements that were adopted by that State before September 1, 2003.

Appendix E

Public Workshop Presentations

June 28, 2004

California Perspective on Mobile-Source Emission Standards
Catherine Witherspoon, Executive Officer, California Air Resources Board

State and Local Mobile-Source and Fuel Standards: A Briefing for the National Research Council
William Becker, Executive Director
State and Territorial Air Pollution Program Administrators and the Association of Local Air Pollution Control Officials

EPA's Perspective on the NRC Study Identifying State Practices in Setting Mobile-Source Emission Standards
Merrylin Zaw-Mon, Environmental Protection Agency, Director, Certification and Compliance Division
Chet France, Environmental Protection Agency, Director, Assessment and Standards Division

Presentation by Outdoor Power Equipment Institute
William Guerry, Jr., Outdoor Power Equipment Institute Counsel, Collier Shannon Scott, PLLC

Presentation by American Council for an Energy-Efficiency Economy
Therese Langer, ACEE, Transportation Program Director, American Council for an Energy-Efficiency Economy

State Mobile-Source Regulations

Peter Lidiak, American Petroleum Institute

Statement by the International Consortium for Fire Safety, Health and the Environment

George Miller, Chairman, International Consortium for Fire Safety, Health and the Environment

Presentation by Alliance of Automobile Manufacturers

Greg Dana, Vice President, Environmental Affairs, Alliance of Automobile Manufacturers

October 4, 2004

Local Air District Perspective

Barry Wallerstein, Executive Officer, South Coast Air Quality Management District

California Programs Adopted by New York

Peter Iwanowicz, Director of Environmental Health, American Lung Association of New York State

Presentation of the Engine Manufacturers Association: The Critical Role of Federal Preemption

Timothy French, Engine Manufacturers Association

Vehicle Regulations in California Science, Process and Results

Tom Cackette, Chief Deputy Executive Officer and Steve Albu, Chief Engineering Studies Branch, Mobile-Source Control Division, California Air Resources Board

January 19, 2005

Honda's Perspective on Compliance with Compliance with CARB's - CARB's LEV and LEV and LEV II Emission Standards LEV II Emission Standards

John German, Manager, Environmental and Energy Analyses, American Honda Motor Corporation, Incorporated

EPA's Process for Developing Mobile-Source Emissions Standards Regulations

Chet France, Environmental Protection Agency, Office of Transportation and Air Quality

Cummins' Perspective on State Practices in Setting Mobile-Source Emissions Standards

Bob Jorgensen, Cummins Incorporated

Perspective of General Motors on California Emission Standards

Robert Babik, Director, Vehicle Emission Issues, General Motors Corporation

Review of Past Motor Vehicle Pollution Control Estimates

Roland Hwang, Vehicles Policy Director, Air and Energy Program, Natural Resources Defense Council

April 14, 2005

A Manufacturer's Perspective: State Standards in a Global Marketplace State Standards in a Global Marketplace and the Briggs & Stratton History

Patricia Hanz and Peter Hotz, Briggs & Stratton Corporation

EPA's Process of Waiver/Authorization Determinations and Requirements for Opt-in States Under Section 177

Karl Simon, David Dickinson, Environmental Protection Agency, Office of Transportation & Air Quality

California Low Emission Vehicle Standards

Thomas Snyder, Director, Air and Radiation Management Administration
Maryland Department of the Environment

National and Global Impacts of California Mobile-Source Emissions Standards

Michael Walsh, International Consultant

Incremental Costs and Cost-Effectiveness of California Emissions Standards

Thomas Austin, Sierra Research, Incorporation

States' Perspectives on Adoption of California Emissions Standards Statements, Presentation and Panel Discussion

Arthur Marin – Executive Director, Northeast States Coordinated Air Use Management

Richard Valentinetti – Director, Air Pollution Control Division, Vermont Department of Environmental Conservation

Thomas Snyder – Director, Air and Radiation Management Administration,
Maryland Department of Environmental Protection

Robert Golledge – Commissioner, Massachusetts Department of Environmental Protection

Carl Johnson – Deputy Commissioner, New York Department of Environmental Conservation

David Littell – Assistant Commissioner, Maine Department of Environmental Protection

Gina McCarthy – Commissioner, Connecticut Department of Environmental Protection

Assessing the Emissions Benefits of LEV II as Compared to Tier 2

Coralie Cooper, Northeast States Coordinated Air Use Management